D0874212

Middle East Oil
and the Energy Crisis

Major Oilfields and Pipelines in the Middle East, 1972

Producing oilfield
Major oil pipeline

Source: CIA Atlas: Issues in the Middle East (Washington, D.C., 1973).

Middle East Oil
and the
Energy Crisis

—

Joe Stork

RECEIVED

DEC 8 1976

MANKATO STATE UNIVERSITY
MEMORIAL LIBRARY
MANKATO, MINN.

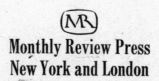

Monthly Review Press
New York and London

For my mother and father, Mary and Gerry Stork

HD
9576
N36
S73

Copyright © 1975 by Middle East Research
and Information Project, Inc.
All Rights Reserved

Library of Congress Cataloging in Publication Data
Stork, Joe.
    Middle East oil and the energy crisis.
    Bibliography: p.
    1. Petroleum industry and trade—Near East.
I. Title
HD9576.N36S73   338.2'7'2820956   74-7786
ISBN 0-85345-335-7

First Printing

Monthly Review Press
62 West 14th Street, New York, N.Y. 10011
21 Theobalds Road, London WC1X 8SL

Manufactured in the United States of America

Contents

394872

Introduction
by Richard J. Barnet

Joe Stork's account of seventy-five years of oil diplomacy is essential reading for anyone who wishes to understand the resource crisis that has shaken the foundations of the post-war world. The new strategy adopted by the OPEC countries at the time of the October War—the "oil weapon," as it has come to be known—cannot be understood except in its historical context. For three-quarters of a century the petroleum oligopolies have been conducting complex and highly successful diplomatic maneuvers in close conjunction with their home governments in order to secure continuing access to cheap oil. An understanding of this history is important not only for judging the equities involved in the contemporary struggle over the price of oil, but also for assessing what is happening and about to happen in this extraordinary drama. The foreign relations of Exxon and Shell are far more important in shaping the contemporary international system than the diplomatic efforts of most nations. Unfortunately, the companies are even more secretive than governments and internal documents of two generations ago still remain hidden from public view. But it is possible to reconstruct company strategies from a variety of open sources, and this is what Stork has done.

The petroleum companies were the pioneer multinational corporations. They were the first business enterprises to adopt a worldwide strategy of profit maximization, the first to locate a significant amount of their assets outside their home countries, and the first to depend on foreign-derived income for more than half their profits. The development of the oil companies into global enterprises with control over a variety of energy sources provides a good case study of the techniques and strategies now being used by other multinational enterprises to dominate many sectors of the world economy. There are no better examples of the use of oligopoly power to control markets, fix prices, and drive out competition. Nowhere else does the business-government interlock that guarantees that major governmental planning decisions are made by industry representatives "on loan" to regulatory agencies work better than in the energy sector.

The oil companies were the first to perfect the art of transfer-pricing on a grand scale. Having developed a corps of accounting virtuosos, they are able to arrange their affairs in such a way as to produce profits and losses where they can have the most favorable effect on the worldwide balance sheet. The space-age alchemists employed by the energy oligopolies are able to turn profits into losses, royalties into taxes, dividends into loans, and to perform other accounting miracles which keep the U.S. tax rate for the global energy giants well below 10 percent.

The oil companies were also the pioneers in the use of cross-subsidization, defined by John Blair, former counsel to the Senate Antitrust Committee, as the "use by a conglomerate of monopoly profits earned in another industry to subsidize sales at a loss or at an abnormally low profit." Because oil profits from the Middle East have been so extraordinarily high—Stork cites a Chase Manhattan Bank study of the early 1960s that computed the cost of producing a barrel of Middle East oil at 16 cents, as compared with $1.73 a barrel for U.S. oil—it has been possible for the "seven sisters" to use Middle East profits to finance their increasing hold on the distribution system in the United States and to gain control of alternative energy sources such as coal and uranium.

Finally, the history of the global energy companies illustrates the extraordinary planning power of the multinational corporation. Because vital information about oil reserves, real costs of drilling and distribution, and their own long-range strategies is in the exclusive hands of the companies, they become by default the only effective planners of energy policy. Even a federal government less beholden to the oil companies and more aggressive in developing alternative plans to serve the public interest would be crippled in its attempt unless it had access to crucial data about the supply of oil. (Richard Helms, former director of the CIA, testified before Congress that he could not get the companies to cooperate in providing such information.)

The worldwide energy crisis is not a problem of absolute shortages of resources. It is a political crisis over who shall control these resources; who shall decide where, when, and how they are to be distributed; and who shall share in the enormous revenues. The struggle for the control of the world's resources is just beginning. The emerging consciousness that the riches of the earth are not infinitely exploitable, that the cost of producing, distributing, and consuming them can no longer be set by the industrial nations, and that indeed such nations must compete with one another for natural resources to maintain their increasingly similar standards and styles of living marks a new stage in world politics. It is a time of danger because the capitalist countries may well be tempted to use military power to offset their declining economic power. But it is also a time of possibility. It is now clear that a death sentence hangs over millions unless there is a radical redistribution of the riches of the earth in this generation. It would be naive to underestimate the power and determination of the old cartelists—the energy companies, and the military establishments that protect them—or of the new cartelists—the shahs, sheikhs, and the bankers who serve them—to preserve their hold on world resources. But it would be more naive to underestimate the power and determination of those who struggle for survival.

Institute for Policy Studies
October 1974

Preface

The work that became this book began months before the October War of 1973 as a series of articles for *MERIP Reports* that would provide a concise but comprehensive political history of the Middle East oil industry, focusing on the developments of the last decade, in order to explicate the current confusion regarding the demands of the producing countries, through OPEC, for higher prices and greater national control. The project represented a fusion of my work with the Middle East Research and Information Project (MERIP) and my research on the domestic and international energy industry at the Institute for Policy Studies.

There was no reliable radical analysis of unfolding developments in the Middle East. Earlier work by Harvey O'Connor (*The Empire of Oil, World Crisis in Oil*) and Robert Engler (*The Politics of Oil*) only go as far as the early 1960s, and do not focus on the role of the Middle East. One useful book has appeared on the subject since that time: Michael Tanzer's *The Political Economy of Oil and the Underdeveloped Countries*. Although Tanzer offers a fine analysis of the predatory nature of the oil industry and provides the tools and insights necessary for understanding the contemporary scene, he does not focus on the Middle East and does not discuss events past 1968. The only other work on the subject worth mentioning is M.A. Adelman's *The World Petroleum Market*, which became available while I was working on this book.

Adelman's information and data, especially European market price data, are essential for understanding the contradictions at work in the world oil industry in the 1960s.

There has been a manifest need for a detailed analytical treatment of the events of the early 1970s which appreciates their dynamic historical context and clarifies the dimensions of the changes taking place. The oil industry dominates most sources of information and analysis and generally paints a picture of the companies as servants of the consumers, beleaguered by the wildly escalating and irresponsible demands of the producers. There is a reactive interpretation, too, which understands the phony nature of the "energy crisis" concocted in this country and seeks to apply this analysis to Middle East developments. In this view, OPEC and the producing countries are, or might as well be, puppet forces manipulated by the industry giants into establishing a new, higher plateau of energy prices and company profits.

Neither of these widely available interpretations grasps the specific dialectics of Middle East politics over the last several decades and the role of oil politics therein. Neither seems to understand the fundamental role of cheap energy resources in the political economies of the industrialized capitalist countries. As Fred Halliday points out in his new book, *Arabia Without Sultans*, the conventional views ignore the sharp class and national divisions within and among the oil producers in favor of a composite nation of "Arabs" or "sheikhs" "blackmailing" the oil-consuming nations. By the same token, this picture alternately exaggerates or ignores altogether the differences between the ruling classes of many of the oil-producing states and their counterparts in the Western industrialized countries. In contrast, the role of popular and revolutionary forces and movements is almost never noticed or understood.

These problems of reliable information and analysis have grown even more acute since the outbreak of the October

War, in which the war and the accompanying use of the oil weapon need to be understood fundamentally as class weapons, utilized by existing regimes to consolidate and expand their power and influence. At the same time, the war was a political tremor that unleashed forces and popular pressures building for decades and that disrupted political and economic hierarchies that have prevailed for almost two centuries. The need of the oil regimes to implement the embargo and production cutbacks at a time of intensifying crisis in the industrialized capitalist countries has long-term consequences not intended by the likes of Faisal and Sadat. I have tried to present an understanding of these events that shows them not as causes of the crisis but as milestones in the on-going struggle for a redistribution of the world's productive wealth and control of its resources, a struggle that is revolutionary in its trajectory and its implications.

This book began out of the collective work of the Middle East Research and Information Project and owes the most to the members of the collective, who helped and encouraged me through the writing of the original series of articles and then the book: Lynne Barbee, Karen Farsoun, Samih Farsoun, John Galvani, Peter Johnson, Chris Paine, Sharon Rose, and Rene Theberge. The Institute for Policy Studies deserves thanks for support of my early research into the energy industry and for providing the space and facilities for much of the actual writing. Tina Conner, Robb Burlage, Roger Lesser, and Jim Ridgeway proved to be good and helpful comrades through the work, and Dick Barnet contributed the introduction.

Of the many people who helped me complete this book, Lynne Barbee and Pat Huntington helped prepare the manuscript. Susan Lowes at Monthly Review has been a fine editor and has managed the book through all the pitfalls of production. Harry Braverman, Harry Magdoff, and Paul Sweezy at

Monthly Review, and Mike Tanzer, provided the initial and sustaining encouragement to extend my *MERIP Reports* articles into this book. The staff of the Middle East Institute Library in Washington facilitated my use of research materials there. Dr. Abbas Alnasrawi of the University of Vermont and Steven Duguid of Simon Fraser University have read parts of the manuscript and offered helpful criticisms at various stages. The final result, of course, is my responsibility alone. Praeger Publishers has given permission to use the industry financial table in Chapter 3.

Finally, but not least, I want to thank most warmly the people of my family who lived with me and who helped me most in ways I can't say—Suzie, Zafra, and Leila Ann; Joann, Blake, and Mariette.

—February 1975

1

Middle East Oil: The Beginnings

Oil, today the most important single commodity in the economic life of the industrialized and industrializing countries, had little or nothing to do with the first European incursions into the Middle East in the nineteenth century. At the same time, the relative ease with which the British, and later the Americans, took over these oil resources was directly related to the way the area had been softened up by centuries of Ottoman and later European oppression and exploitation. For economic viability, the area had historically been dependent on the commercial functions of the urban areas and on nomadic tribes, trading between Europe and Asia. Advances in European navigation and shipping during the fifteenth and sixteenth centuries circumvented the need to rely on the Middle East for transit rights and services and thus reduced one of the main sources of wealth; this coincided with Ottoman Turkish conquest of most of the Arab lands. By the eighteenth century, local and Ottoman rulers were reduced to a feudal-type exploitation of the peasantry in order to support their regimes.

European political and economic penetration in the nineteenth century led to the development of lopsided, export-oriented economies in Egypt, the Levant, and the Fertile Crescent. British and French capital expanded and intensified the production of agricultural raw materials—like cotton in Egypt, minerals and lumber in Iran, tobacco, wool, and silk in

Syria, Palestine, and Lebanon. The only serious effort at industrial development took place in Egypt under Mohammed Ali in the early part of the century, but had been crushed militarily by Britain and France by 1840. The introduction of capitalist land relationships brought about the formation of a parasitic class of absentee landlords concentrated in the urban areas, functioning in economies increasingly dominated by European capital. The Middle East, like the rest of the non-European world, was regarded as a source of raw materials and a market for Western manufactured goods.

The penetration of British and French finance capital accelerated after 1850 with such concessions as the Suez Canal, railway construction projects, and with other forms of government indebtedness. In the latter part of the century all the states went bankrupt, and European diplomatic residents and bank agents came virtually to rule the Arab countries, Iran, and, to a lesser extent, Turkey, redirecting their economies in order to pay the interest and principal on heavy loans. This had consequences not only for the local economies in general but for the peasant sectors in particular, as governments attempted to raise revenues to pay off these debts by intensifying the exploitation of the lower classes.[1]

Often superseding European interest in raw materials and markets was the rivalry to dominate politically and militarily the Middle East's strategic geographic position between Europe, Asia, and Africa. By the beginning of the twentieth century Britain had secured a dominant position, directly ruling Egypt, the Sudan, and the Persian Gulf as the price of maintaining a secure hold on India, Ceylon, and the other colonial territories east of Suez. French influence was guardedly tolerated in the area of Lebanon and Syria. Iran was divided between British and Russian spheres of influence in the south and north respectively. In these years before World War I Germany moved to secure a competing position of influence by obtaining a concession to build a railway from Europe to Baghdad. Except for Iran, most of the Middle East was still under the nominal sovereignty of the decrepit Ottoman Empire, ruled from Constantinople.

The history of the Middle East in this century is inseparably linked to the efforts of the Western powers to secure absolute control over its resources and trade. The primary resource is oil.[2] European (and later American) control of oil was achieved through a series of concessions to European companies or entrepreneurs. Local authorities conceded the sole right to exploit all mineral resources over a large area in return for an immediate cash payment and a stipulated low royalty per ton of petroleum once production was initiated. While there have been changes in the terms of the concessions from the initial negotiations with backward and impoverished local potentates, their basic character has remained unchanged until the present. It is only today, the 1970s, that the relationships between companies and local governments are being challenged and changed to any substantial degree.

The desire to control world oil resources did not become an important feature of Western politics in the Middle East until World War I. Up until then, the industrial economies of Europe were fueled mainly by indigenous coal resources. The oil industry was significant in the United States because large deposits of crude oil had been discovered there. Rockefeller's Standard Oil monopoly was based on control of pipelines, transportation facilities, and marketing rather than on direct control of the oil fields. Around the turn of the century Standard tried to extend its control of these facilities on a global basis, throughout Europe, Asia ("oil for the lamps of China"), and Latin America. Its only challenger in this role was the British-Dutch combine known as Royal Dutch-Shell.

By the outbreak of World War I, however, European powers, notably the British, were acutely aware of the growing importance of oil as a strategic military commodity essential for motorized ground transportation, newly emergent air forces, and more efficient and powerful navies. Many years later an American strategist wrote "that an adequate supply of oil was a vital element in military power was one of the most easily read lessons of [World War I] ."[3]

World War I was of momentous political consequence for the Middle East. The Ottoman Empire was destroyed and its Arab territories parceled out between Britain and France. Zionism became a factor of importance with the Balfour Declaration and the British Mandate over Palestine. Arab nationalism emerged as a potent political force. Control of petroleum resources became the overriding focus of Western rivalry in the area and the main impetus for the Western powers to establish and maintain political and economic control over the region.

The First Concession: Persia (Iran)

United States oil played a key role in supporting the Allied war effort, but just before the war the British government moved to gain control of access to the oil of Persia under the first significant oil concession in the Middle East. The concession originated in 1901 when a British engineer, William D'Arcy, was granted a sixty-year exclusive privilege to "search for, obtain, exploit, develop, render suitable for trade, carry away and sell natural gas, petroleum, asphalt, and ozocerite throughout the whole extent of the Persian Empire"[4] except for the five northern provinces that were considered to be under the Russian sphere of influence. The concession covered 500,000 square miles which, along with any machinery or other property that might be imported, were to be totally exempt from taxation during the long life of the concession. The price: £20,000 in cash, £20,000 in company stock, 16 percent of annual net profits, and a rent of $1,800 per year.

The concession was let by a regime that could most charitably be described as corrupt and irresponsible. The economy was based on a hierarchical social order of exploitation, the financial burden of which ultimately fell on the masses of peasants. The revenues thereby produced were frequently insufficient for the extravagant and pretentious needs of the court. Granting concessions was seen as a painless way of filling the royal treasury. A month after the concession was signed the grateful court issued a royal decree stating that

> it is accorded and guaranteed to the Engineer William D'Arcy, and to all his heirs and assigns and friends, full powers and unlimited liberty for a period of sixty years to probe, pierce and drill at their will the depths of Persian soil; in consequence of which all the subsoil products wrought by him without exception will remain the property of D'Arcy. We declare that all the officials of this blessed kingdom and our heirs and successors will do their best to help and assist the honorable D'Arcy, who enjoys the favor of our splendid court.[5]

From the very beginning the British government supported D'Arcy. The British ambassador in Teheran was advised by his government in London that shares in the proposed enterprise might be offered to influential ministers of the Persian Court. When initial failures to find commercially exploitable oil reserves led to the near bankruptcy of D'Arcy's company he turned to foreign sources of capital for support. The British Admiralty then approached the British-owned Burmah Oil Company and suggested that they form a syndicate with D'Arcy to provide the funds necessary to keep the Persian concession in British hands. Concessions Syndicate Ltd. was thus formed in May 1905. With the discovery of one of the world's biggest oil fields in April 1908, Concessions Syndicate formed the Anglo-Persian Oil Company.

The British government, indirectly responsible for the birth of Anglo-Persian, continued to play a protective role.

Through the Anglo-Indian government a force of Bengal Lancers was dispatched to protect drilling operations from attacks and harassment by local tribesmen. The Lancers were led by one Lieutenant Arnold Wilson, who would later play an important role in securing British control over Iraq after World War I and who, as Sir Arnold Wilson, would end up as general manager of the Anglo-Persian Oil Company "for the Persian and Mesopotamian area." In 1910 Wilson earned the commendation of the viceroy of India "for the excellent manner in which he championed the Oil Company's cause" in a tax dispute with the Persian government.[6]

Winston Churchill became First Lord of the British Admiralty in 1911 and embarked on a three-year naval expansion program unprecedented in size and cost that would, among other things, transform the navy to one based on oil rather than coal. The Admiralty turned to Anglo-Persian, which by now had fast-flowing wells, a pipeline, and a refinery in full operation. The Royal Commission on Oil Supply was created and Admiral Slade (later a vice-chairman of Anglo-Persian) and John Cadman (a mining expert, later chairman of Anglo-Persian) were sent to examine the company's operations. Based on their enthusiastic report, the Admiralty drew up an agreement that simultaneously set up a twenty-year supply contract between the navy and Anglo-Persian, and gave the government a controlling interest of 51 percent in return for £2.2 million.

The aim of this move, according to Churchill, was to free the navy from exclusive reliance on Standard Oil and Royal Dutch-Shell by assuring Britain control for fifty years of a petroleum-rich area "nearly as big as France and Germany put together." The commercial advantages of this move were not lost on Churchill either. As largest shareholder and principal customer of Anglo-Persian, he said, "what we [i.e., the British government] do not gain at one end of the process we recover at the other."[7] The welfare of Great Britain and the

Anglo-Persian Oil Company became cemented, as did their identity. With its new capital funds and assured markets, Anglo-Persian became a major petroleum producer, and Iran, after the war, became the fourth largest oil-producing country (after the United States, Russia, and Venezuela).

Following the war, in which over 1 million Iranian "noncombatants" were killed, the British sought to establish firm control over the Persian government and armed forces through the Anglo-Persian agreement of 1919. According to Lord Curzon, the chief architect of this plan,

> we cannot permit the existence between the frontiers of our Indian Empire . . . and those of our new protectorate [Iraq], of a hotbed of misrule, enemy intrigue, financial chaos and political disorder. Further . . . we possess in the south-western corner of Persia great assets in the shape of the oilfields, which are worked for the British Navy and which give us a commanding interest in that part of the world.[8]

The agreement was "negotiated in secrecy and eased along by substantial 'subsidies' to important Persian officials."[9] It provided for British advisors "endowed with adequate powers" in Persian government departments and the armed forces, as well as British equipment and munitions for the armed forces. All this was to be financed by "income at the disposal of the Persian Government," i.e., royalty payments from Anglo-Persian.

Nationalist opposition prevented Persian ratification of the agreement and it was formally repudiated after the coup by Reza Shah in February 1921.[10] The following years were marked by frequent disputes between the Persian government and the company over the terms and implementation of the oil concession, particularly its financial terms. Finally, in 1932, the Persian government announced cancellation of the concession. The British government immediately attempted to get the government to withdraw its cancellation, despite the Persian offer to negotiate a new, more equitable agreement. The dispute was eventually debated before the Council of the League of Nations, where Persia asserted that the original concession had been granted by a government with no constitutional basis and under the thumb of foreign powers. Although the concession area covered 500,000 square miles, company operations were restricted to an area just over 1 square mile. Petroleum products were sold in Persia at outrageously high world prices. As for the financial results of the concession, the Persian government contended that if they had permitted the company to work the area with no royalty payments at all, simply collecting regular taxes and customs fees, their receipts would have been 100 percent greater. The British government, of course, refuted all these charges. Anglo-Persian eventually negotiated a new concession with the government that provided a more certain and precise government income; in return the concession would be extended until 1993. The new concession brought on a truce that lasted until after the next world war.

Mesopotamia (Iraq)

The second important oil concession in the Middle East originated during the same period the British were taking control of Persian oil. In this case too the concession had to be secured by overt, armed British intervention to suppress a vocal, articulate, and determined nationalist movement. The final result was a concession to the Iraq Petroleum Company and its subsidiaries, amounting eventually to the whole territory of Iraq. The story of the British conquest of Iraq illustrates the fundamental British policy toward the Middle East before, during, and after World War I.

The country of Iraq was regarded, before its independence in 1925, as Mesopotamia and was nominally ruled by the Ottoman Sultan from Constantinople. A report by a flamboyant Armenian entrepreneur named Gulbenkian to Sultan Abdul Hamid II in 1904 on the oil potential of Mesopotamia led the Sultan to unilaterally transfer the oil-bearing lands in the north of the province from state ownership to the private ownership of the ruling family. German interests, through the Deutsche Bank, got the first sizeable concessions, but these were revoked shortly, allegedly for failure to exploit the concession.

D'Arcy, backed by the British government, applied for the lapsed German concession. Before these negotiations were concluded, D'Arcy had transferred all his claims in the area to the newly formed Anglo-Persian Oil Company. The Young Turk revolution in Constantinople in 1908 meant that negotiations had to begin anew. The British, alert to the commercial possibilities this change in regime might offer, organized the National Bank of Turkey to facilitate British economic interests in the Ottoman Empire. Gulbenkian, who was close to the Royal Dutch-Shell combine, was on the bank's board of directors.

Gulbenkian's approach was to unite the outside factions competing for Mesopotamian oil; he was the catalyst behind the formation of the Turkish Petroleum Company (TPC). The Deutsche Bank interests wanted to keep D'Arcy and Anglo-Persian out of Mesopotamia, while British interests wanted them in. The British Foreign Office sponsored an agreement in 1914 that fused German, British, and Dutch interests, with 50 percent of TPC in the hands of Anglo-Persian, 25 percent going to the Deutsche Bank, and 25 percent to Royal Dutch-Shell. Part of the agreement was that all parties would refrain from exploiting oil resources in the Ottoman Empire except through the united front of TPC. World War I interrupted final negotiations with the Ottoman authorities for the concession, and at the end of the war TPC could only assert a tenuous claim on a concession which had never been ratified by the Ottoman government.

The Turkish Petroleum Company would have been an empty and meaningless corporate shell without control of Mesopotamian oil. With the end of the war, however, Britain had secured a military hold over many of the Arab territories formerly under Ottoman rule, including Mesopotamia. British control over the company was consolidated when the British government expropriated the Deutsche Bank's 25 percent holding following Germany's defeat in the war. To top it off, the chief representative of the Royal Dutch-Shell interests, Henri Deterding, became a naturalized British subject and was promptly knighted for his prescience. Britain immediately moved to legitimize its hold over Mesopotamian oil by claiming that the prewar concession granted by Ottoman authorities (although never finalized) was valid. They were challenged by the French, who claimed a share of the Mesopotamian spoils as a result of the joint war effort, and by the United States, which claimed that the Open Door principle should apply to all territories acquired in the war, thus forbidding any restrictions on trade or investments by U.S. companies. Most importantly, the British were challenged in their designs by the people of Iraq, who struggled fiercely to achieve genuine and unrestricted national independence.

The division of the Arab territories between Britain and France following World War I, in the face of fraudulent British commitments to Arab independence, was certainly one of the more notorious exercises of imperialist power. Here we must confine ourselves to outlining the role of the oil interests in determining the final lines of the map of the Middle East that have survived, by and large, to the present day.

The British moved first to deal with French opposition. This was handled quite simply in a secret meeting in 1919 in which the French agreed that the province of Mosul, which contained most of the known oil reserves, would be included in Iraq and thus come under direct British control. In return the British guaranteed France a 25 percent share in the Turkish Petroleum Company (as it was still called), which would remain under British control. In addition, the British were informally committed to aid in the imposition of French rule on the territories of Lebanon and Syria.

The mandate system, concocted in the name of the League of Nations, was a modified colonial system established to facilitate the division of German and Ottoman territories following the war. In no case were the wishes of the local populations consulted, or, where they were known, respected. It was decided at a European conference table that France would rule Syria and Lebanon, while Britain would have the responsibility of bringing progress and civilization to Palestine and Iraq. In each case such rule was achieved only by large-scale suppression of local political forces insisting on the rights of the Arab peoples to self-determination.

The nationalist struggle in Syria was quickly suppressed in the early years. Feisal, the main (British-sponsored) spokesman for Arab nationalist demands, was run out of Damascus by the French army. Conveniently, Britain had need of a pliable king to help pacify the struggle that had erupted in Iraq in May 1920 with the announcement of the British Mandate. The insurrection posed the first serious challenge to British policy in the area, and it was met and crushed as such. Among other tactical innovations, the British used airplanes to bomb and terrorize peasant villages in the countryside.[11]

On the political front, the British arranged a "national referendum," after which Feisal was proclaimed king of Iraq by the British High Commissioner. This was all part of an elaborate scenario in which the Mandate was to be superseded by a treaty of alliance between Britain and an "independent" Iraqi government. The tactic was appropriately defined by the High Commissioner: "It was [Britain's] intention not that the proposed treaty should replace the Mandate, but rather that the Mandate should be defined and implemented in the form of a treaty."[12]

This charade may have fooled a lot of people, but it did not fool the Iraqis. Before King Feisal could ratify the treaty, protests became so vigorous that the cabinet was forced to resign and the British High Commissioner assumed dictatorial powers for several years. Nationalist leaders were deported from the country on a wide scale. In 1924 the commissioner arranged for the appointment of a new cabinet whose sole purpose was to ratify the treaty. It was so ratified, subject to the approval of a yet-to-be-elected National Assembly. The Assembly was elected in 1925 and met in March of that year, but widespread demonstrations in the capital delayed the treaty's approval. Only when the High Commissioner delivered a draft law to the king empowering him to dissolve the Assembly was that body persuaded to approve the treaty, and even then only thirty-seven voted in favor, twenty-eight were opposed, and eight abstained.

Under article four of the treaty, the king agreed to be guided by the advice of the High Commissioner "on all matters affecting the international and financial obligations and interests of His Britannic Majesty."[13] Even before the treaty was ratified, the British had moved to secure the Turkish Petroleum Company concession. Since the validity of the prewar concession was contested not only by the Iraqis but by interested American oil companies as well, the British decided to remove the basis for challenge by simply negotiating a new concession with the king they had installed. The concession's ratification by the cabinet preceded that of the treaty of alliance and even the promulgation of the country's constitution. The High Commissioner claimed that the concession was granted "with general public approval," apparently because only the ministers of justice and education resigned in protest.

Open Door: Americans Move into Middle East Oil

American interest in Middle East oil had been minimal prior to the end of World War I. Domestically the industry had been monopolized by Standard of New Jersey, and the Supreme Court antitrust action of 1911 did little to redistribute real economic power. The American industry, in contrast to the British, was based on extensive proven oil reserves within the country, and United States-owned foreign interests were confined to areas close at hand, notably Mexico and Venezuela. Since the United States was the source of most of the oil being traded internationally, Rockefeller circles were content to monopolize transportation, refining, and marketing facilities.

The extravagant and unrestrained production of domestic reserves, especially during the war, precipitated an "energy crisis" at the war's end. The United States had supplied 80 percent of the Allied war requirements for oil. The U.S. fleet had converted from coal to oil just before the war, and a program of naval expansion aimed at rivaling that of Britain was begun at the war's end. In 1920 President Wilson wrote:

> It is evident to me that we are on the eve of a commercial war of the severest sort and I am afraid that Great Britain will prove capable of as great commercial savagery as Germany has displayed for so many years in her commercial methods.[14]

The domestic economy, too, especially because of the newly emerging automobile industry, was becoming increasingly dependent on oil and petroleum products. American and British interests were bound to clash in the Middle East because of what one scholar has characterized as the "obsession" of the major oil companies to secure foreign sources.[15] It was clear that the British had the jump on the Americans in terms of concessions already secured or pending. Most disturbing to the Americans, though, was that the British were intending to use their role as a mandate power to exclude competing foreign interests from the areas under their control.

This was first demonstrated in Palestine, a British mandate area, where Standard Oil Company of New York (Socony) had obtained an exploration concession just prior to the war. The British military authorities, after the war, refused to allow the company to renew its activities on the grounds that the civil status of Palestine had not been determined. A predictable flurry of notes passed between Washington and London, but the British position was legally unassailable and nothing came of American pronouncements of an Open Door. Besides, Palestinian petroleum reserves were not expected to be significant enough to warrant a serious break in relations.

This was not the case in Mesopotamia (Iraq), where British favoritism toward its own companies and its exclusion of the Americans was much more flagrant and where the results were of much more consequence. The same flurry of notes concerning the Open Door and mandatory responsibilities shot back and forth. The Americans contended that the prewar TPC concession was invalid, and that subsequent agreements to share production with the French and exclude other interests were contrary to League provisions for unrestricted commercial activity in the mandated territories. The American stance was weakened by Britain's physical presence in Iraq, and by Washington's refusal to join the League of Nations. Given the fluid political situation in the Middle East, especially under conditions of local unrest, the dispute continued for several years while the British tried to consolidate their rule in Iraq. When Wilson left office in 1921 there had been no appreciable progress toward implementing the Open Door in the Middle East.

In the end it was the consolidated power of the U.S. oil companies, with the ready assistance of the government, that provided for American participation in exploiting Iraq's oil.

By the end of the war, the oil industry was a major part of the U.S. industrial sector, which had important consequences for national economic stability. The industry's close collaboration with the government had its roots in the war years. The National Petroleum War Service Committee had been set up under the War Industries Board in 1917. The head of the committee was Jersey Standard chairman A. C. Bedford, who voiced the enthusiasm of many business leaders for the new relationship with the federal government when he said: "We must keep our eyes on the goal of still more complete and wholehearted cooperation, of a more perfect coordinated unity of aims and methods."[16]

This period laid the foundation for the modern capitalist corporate state in America, and the oil industry was in the vanguard of this development.[17] No other industry, in fact, has ever created quite the same degree of symbiosis with the government, one reflected in the mutuality of policies and exchanges of personnel. Under these arrangements, and wrapped in the American flag, the companies were able to organize themselves as a cartel independent of government control in the interest of eliminating "inefficient" competition. The oil committee survived the war and became the American Petroleum Institute (API).

The major oil companies then became the chief and most visible advocates of a foreign policy that stressed American national interest in gaining access to foreign oil resources. They found enthusiastic supporters in the Republican administration of Warren Harding, including the new Secretary of State, Charles Evans Hughes (who later became chief counsel to the API), and Secretary of Commerce Herbert Hoover, himself a former mining engineer. Much publicity was generated about dwindling domestic oil reserves and the alleged needs of the new expanded American navy.[18]

Two proposals were put forward in Congress to deal with the alleged shortage (these were the years just prior to the fabulous discoveries in East Texas, Louisiana, and California). One was a bill to restrict American export of crude oil and petroleum products in the national interest and the other was to set up a federal oil corporation. But government cooperation with the industry was conceived as a one-way street in which any such restraints on private enterprise were out of place. Government support was not meant to include government control to any degree. The answer to the problem of a shortage, the industry contended, was to gain control of large untapped foreign sources, and the proper role of the government was to assist the American companies.

The U.S. industry's solution was the formation of a syndicate of companies interested in Iraqi oil, backed by the government. Commerce Secretary Hoover assumed a key role in implementing this policy. In April 1921 he wrote to the State Department: "I am in touch with the petroleum industry in the country in an effort to organize something specific that we can get behind."[19] An informal American Group, as it came to be known, was quickly formed under the leadership of Jersey Standard and included Socony, Sinclair, Texaco, Gulf, Atlantic, and Mexican.

In early 1922 leaders of the American Group met with the State Department's trade advisor, Arthur Millspaugh, to work out an effective use of their combined strength. These discussions were given impetus by hints from Anglo-Persian's chairman, Sir John Cadman, that an amalgamation of American and British interests in the Middle East might be desirable from a technological and financial, as well as a political, point of view. The British Foreign Office indicated its willingness to compromise by allowing Socony geologists to continue explorations in Palestine. At the same time, the Washington Naval Conference was working out agreements to stem Anglo-American naval rivalry.

The State Department was scrupulously impartial in public concerning the details of any American-British oil agreement, insisting only that the Open Door, "the broad principle of equality of commercial opportunity,"[20] be implemented and that the original TPC claim be renegotiated. When American oilmen responded that complete equality of opportunity would let in Japanese, Italian, French, and other interests, the government clarified its position by saying that it did not intend

> to make difficulties or to prolong needlessly a diplomatic dispute or so to disregard the practical aspects of the situation as to prevent American enterprise from availing itself of the very opportunities which our diplomatic representations have striven to obtain.[21]

By the end of that year, agreement on the principle of American participation had been reached between the American Group and Anglo-Persian/Royal Dutch-Shell, although the percentage share, means of compensation, and other important details remained unsettled for several years.

One must keep in mind the unsettled political situation in the Middle East in these years. In addition to the anti-imperialist struggles launched against the British and French in Iraq, Palestine, and Syria, a nationalist struggle had erupted in Turkey against the dismemberment of that country, the core of the old Ottoman Empire, by the Allies. What was known of Iraq's oil-bearing land was in the province of Mosul, claimed both by the Turkish nationalists and by the British for Iraq. During peace negotiations with the Turks in 1923, the British threatened to cede Mosul to the Turks unless the Iraqis agreed to the old TPC concession. This ability to drastically manipulate the very borders of a country was an important weapon in suppressing opposition in Iraq. The British nevertheless assured the American companies that no such move was seriously contemplated when the American companies began to wonder if they were negotiating with the wrong party, the British instead of the Turks. The Turkish nationalist government had in fact al-

ready granted the Mosul concession to an American entrepreneur named Colonel Chester. Chester's syndicate suffered
a financial collapse in late 1923, which Chester claims was
arranged by Rockefeller banking interests on Wall Street. In
any case, the American Group had no serious competition
from then on.

By the time the American Group signed on in 1928, some
of the original member companies had dropped out and
others had been bought out by the remaining two, Jersey
Standard and Socony. It incorporated as the Near East Development Company and shared a 23.75 percent interest in
Turkish Petroleum Company. Anglo-Persian, Royal Dutch-
Shell, and the French Compagnie Française des Pétroles
(CFP) each got 23.75 percent, and the remaining 5 percent
went to Gulbenkian. TPC was set up as a nonprofit corporation—that is, it would produce crude oil that would be sold
at cost to the parent companies. In 1929 it changed its name
to the Iraq Petroleum Company (IPC) and in 1931, it negotiated an entirely new concession with the Iraqi government.
The IPC-Iraqi government concession eliminated clauses
originally included at the insistence of the State Department
that theoretically would have allowed other firms to lease
blocks of the concession.

IPC control over the rest of Iraq was effected by the late
1930s by buying up a competing concessionaire during the
Depression (1936) and forming a wholly owned subsidiary,
Basra Petroleum Company (BPC) to control the southern
part of Iraq. Under the facade of the Open Door, American
oil monopolies gained a substantial share in all of Iraq's
actual and potential oil production under concessions that
were due to expire only at the end of the century.

By the time the American companies had formalized their
interest in Iraq, any real or imagined need for foreign oil in
the United States had vanished with the stupendous finds in

Texas, Louisiana, and California. The resulting glut of crude production forced prices down to the extent that some smaller companies and individual producers were severely hurt. Such surplus capacity and production, however, did nothing to dampen the major companies' moves to secure foreign reserves. On the contrary, their position was that global monopoly control of oil reserves was necessary to achieve "rational" production schedules that would eliminate competitive pressure to lower prices. The process of close cooperation between government and industry described earlier was further advanced with the creation in 1924 of the Federal Oil Conservation Board (made up of the secretaries of Interior, War, Navy, and Commerce). It saw its conservation role in the "avoidance of economic waste"—the restriction of supply to maintain profitable crude oil prices.[22] Represented by Charles Evans Hughes, the American Petroleum Institute pushed for the exemption of oil operators from antitrust restriction. The Oil Conservation Board supported this viewpoint in its first report, which called for legislation to allow producers to "coordinate" production (but *not* to legislate quotas) in order to avoid the "pressure of a competitive struggle."[23]

The third report of the Federal Oil Conservation Board appeared in February 1929 and is significant because it previews the long-range strategy of the oil industry and the government with regard to foreign oil reserves. The Board asserted that the United States was producing 68 percent of world reserves. It recommended both an increase in petroleum imports and the expansion of American investment abroad. In the name of conservation the Board noted that the domestic situation could only be understood in the context of international trends. The Board specified that it was not telling the oil companies where the foreign reserves were, since they already knew:

Rather, the intent is to acquaint the general public with the nature of these resources and to create that better understanding of the foreign fields which is essential to a sympathetic support of American oil companies now developing foreign sources of supply.[24]

Domestically, this price fixing under the name of conservation had to rely on voluntary of state-enforced production restrictions until state prorating in Texas and elsewhere was backed up by the federal government under the Connally "Hot Oil" Act of 1936. Internationally, however, possible antitrust repercussions did not stand in the way of immediate, if informal, cartelization of the international oil industry as a result of an agreement in which Jersey and Mobil (Socony's brand name) joined the Iraq Petroleum Company. The participants in the IPC pledged to refrain from any petroleum activity in the area of the former Ottoman Empire, including the whole Arabian Peninsula, except through the joint ownership of IPC. Kuwait and Bahrein were excepted because Britain intended to reserve them for British control and exploitation. The Red Line Agreement, as it came to be known, was designed to restrict production except at the express decision of the international petroleum cartel as it was then constituted in IPC.

To supplement this agreement, Jersey Standard, Shell, and Anglo-Persian entered an informal arrangement in the same year (1928) known as the As-Is Agreement. This was designed to achieve the cooperation in marketing that had already been achieved in production with the Red Line Agreement. World markets were pooled and divided on the basis of present shares. For example, if Standard and Shell each had 50 percent of the Far Eastern market for crude oil and petroleum products, they could expand output and marketing facilities only as much as was required to meet growing demand; they could not seek to augment their share of the market at the expense of another big company.

This proposal was drawn up by the Big Three and adopted as policy in 1929 by the API. The approval of the Federal Oil Conservation Board was assured by its members' presence at the API meeting. Only a warning by the Attorney General that the Board possessed no authority to grant antitrust immunity prevented this body from granting official government approval to the international oil cartel. This intervention did little to change history, however, and the American companies joined the cartel with no serious opposition from the government.

Bahrein, Kuwait, and Saudi Arabia

It was in the mid-1920s that the first oil explorations were conducted on the Arabian Peninsula. A key figure in the beginning was a Major Holmes, an eccentric entrepreneur and promoter from New Zealand who, under the corporate title of Eastern and General Syndicate, tried to negotiate concessions with local shiekhs and then sell them to the international companies. He sought his first from Abdul Aziz ibn Saud, who had led the Saudi tribe to victory in the Arabian Peninsula, uniting the area into a kingdom that still bears his tribal name.

The Saudi rulers were opposed to dealing with foreigners, especially non-Moslem foreigners, but the desire for an income besides the fees collected from Mecca pilgrims, along with scepticism about there being any oil, led Abdul Aziz to grant Holmes a concession in 1923 for £2,000 per year. Holmes organized a geological survey team and worked for two years but failed to interest established oil companies in the area. After finding no oil and no takers, the syndicate discontinued exploration and defaulted on its rental.

A decade later, with the decline during the Depression of his meager income from pilgrimages, Abdul Aziz displayed much more readiness to sell concessions. Meanwhile, Holmes had discovered oil in Bahrein, an island off the Persian Gulf coast of Saudi Arabia, which indicated that oil was likely to be found on the mainland.

The major companies involved in Iraq were not interested in concessions in Arabia as long as major competitors were unlikely to find sizeable reserves. In 1927 Holmes sold his Bahrein concession to a subsidiary of Gulf Oil. Since Gulf was still part of the American Group, which was then negotiating for Iraq, it could not exercise its newly acquired Bahrein option without IPC permission, which was not granted. Gulf sold its rights to Standard Oil of California (Socal). Socal got around British restrictions in Bahrein by organizing the Bahrein Petroleum Company (Bapco) as a Canadian subsidiary. In 1932 oil was found in Bahrein in commercial quantities.

Holmes had also interested Gulf in a possible Kuwait concession. Faced with the British policy of restricting the entry of non-British interests, Gulf enlisted the aid of the State Department to "open the door." The U.S. ambassador in London was instructed to secure "equal treatment for American firms."[25] Coincidentally, he was none other than Andrew Mellon, head of the Gulf empire, Secretary of the Treasury during the Teapot Dome scandal, and the man who had arranged for the introduction of the notorious oil depletion allowance as a favor to himself and his friends. Gulf was kept informed of the progress of the negotiations.

Gulf had a possible competitor in Anglo-Persian, but in the course of negotiations both companies decided they would be satisfied with a joint concession. Anglo-Persian was apparently moved to compromise by the threat to the stability of the world oil markets were Gulf to get an exclusive concession. Anglo-Persian therefore made an agreement with Gulf that Kuwait oil would not be used to "injure" either party and that Gulf would agree to be supplied by Anglo-

Persian (from Iraq or Iran production) at cost, in lieu of Kuwait production, if Anglo-Persian should determine it would "injure" the market.[26] Together, the companies also got better financial terms: royalties averaged ten cents a barrel less than in Persia, Iraq, or Saudi Arabia, and royalty payments were to be made in Indian rupees rather than guaranteed by gold.

Despite Holmes' initial failure to discover oil in eastern Saudi Arabia, Standard of California (Socal) became interested in a concession there after it struck commercial quantities in Bahrein. Anglo-Persian was also interested, but only to the degree that it could cheaply keep Saudi oil off the market through a preemptive concession. Standard seriously wanted the concession and was willing to pay accordingly. It was granted in May 1933 for a period of sixty years and covered most of eastern Saudi Arabia. The financial terms included £5,000 per year until oil was discovered, a £50,000 loan against future royalties, and a £100,000 loan when oil was discovered. A royalty of four shillings a ton was offset by the king's agreement to forego for all time the right to tax the company.

Concession

Oil was discovered in large quantities in 1938. Socal had previously dealt in domestic production and marketing. Now it possessed what would prove to be the world's richest oil concession. In order to avoid the expense and disruption building up its own world markets would entail, Socal had agreed in 1936 to acquire a half interest in Texaco's marketing facilities east of Suez. This corporate union was set up as a wholly owned subsidiary, California Texas Oil Company (Caltex). Texaco later solidified the union by taking a half interest in Socal's Arabian concession.

By the end of the 1930s, the U.S. oil giants had gotten a sizeable share (42 percent) of known Middle Eastern oil reserves, with no provision for relinquishing large concession areas, and had eliminated the threat of competing interests,

Those American and British companies with the capital to consider major foreign investments had, for the most part, already done so. Those that had not, like Texaco, were brought in for such stabilizing assets as their markets. At the beginning of World War II, then, all the big players were on the scene. The "seven sisters," as they were later dubbed, were locked into joint production and marketing arrangements that virtually guaranteed the absence of significant competition. The other major oil-producing center, the United States, had stabilized high prices through the joint efforts of large companies and state and federal governments to limit production. The oil world, such as it was then, belonged to the Big Seven: Jersey Standard, Mobil, Gulf, Socal, Texaco, Anglo-Persian, and Royal Dutch-Shell.

World War II and the Consolidation of American Oil Interests in the Middle East

During World War II, the Middle East, including Iran and North Africa, was an important military theater for several strategic reasons. One reason was that there were enormous oil reserves concentrated in the Persian Gulf region. This conflagration, like the one two decades before, increased American strategists' concern that those reserves remain under the direct control of the American oil companies. Because of the logistical and financial problems involved in establishing storage facilities for stockpiling, conventional wisdom prevailed—the only way oil could be safely and inexpensively stored was to leave it in the ground until it was needed. From this premise, according to the wartime economic advisor to the State Department: "It has been taken for granted . . . that American interests must have actual physical control of, or at the very least assured access to, adequate and properly located sources of supply."[1] In addition to Western Hemisphere sources like Venezuela, this meant primarily the Middle East.

The American oil companies were in a very solid position in the Middle East, controlling 42 percent of known and estimated oil reserves. The war prevented the expansion of production beyond meeting local needs and those of Allied forces in the area. Once the direct Axis threat to the Middle East was removed with the German defeat at al-Alamein in 1942, the oil companies turned to safeguarding their positions from potential changes in the policies of host govern-

29

MANKATO STATE UNIVERSITY
MEMORIAL LIBRARY
MANKATO, MINNESOTA

ments. In the words of the economic advisor quoted above, "They longed for assurance that the signed concessions should have a stronger basis than the will of a mortal and impetuous ruler."[2]

The aspired immortality of the oil companies seemed assured in Iraq, where the British installed a puppet government in 1941 in the face of a "pro-Nazi" coup by some nationalist Iraqi officers. In Iran, similarly, the "pro-Nazi" neutralism and nationalism of Reza Shah's regime was offset by the British and Russian occupation of the southern and northern halves of the country respectively.[3] In the later years of the war American officials became actively involved in Iran through the Lend Lease program, which supplied Russia through the Persian corridor. Primary American concern, however, was focused on the place where the American companies had sole control of the most important concession: Saudi Arabia.

Aramco and U.S. Interests in Saudi Arabia

During the early years of the war the production of Socal and Texaco in Saudi Arabia through their jointly owned producing company, later to be called Aramco (Arabian-American Oil Company), did not increase, and in fact dropped off, thereby sharply reducing royalty payments to the Saudi ruling family. A simultaneous interruption in the flow of pilgrims and their revenues to Mecca led the King in 1941 to demand $6 million annually in advance royalties. The company would only commit itself to $3 million for 1941, but knew that its lush future in Saudi Arabia was highly dependent on the political stability of the ruling family, which was in turn dependent on the funds with which Abdul Aziz could continue to buy the support of potentially hostile tribal chiefs. Genuinely interested in coming up with the necessary cash but not wishing to bear that burden alone, company officials turned to Uncle Sam.

The chosen intermediary was James Moffett, chairman of Caltex and a personal friend of President Roosevelt.[4] According to Moffett himself, he called the White House the day he was retained by Socal (so that he was actually working for both companies), and got an appointment with the President for the next day. At that meeting he raised the issue of the coincidence of corporate and national interests in Saudi Arabia and pressed for access to the U.S. Treasury through the Lend Lease program or the Export-Import Bank. Roosevelt responded by suggesting that Moffett draw up a proposal whereby the U.S. navy would contract to buy $6 million worth of oil products annually, thus indirectly providing the funds the King demanded. The navy declined, however, saying it had no need for that amount of oil. At the President's direction it was then decided that the funds should be put up by the British out of the $425-million loan they had just secured under Lend Lease. Reconstruction Finance Administrator Jesse Jones passed along the "suggestion" to the British that they "continue taking care of the King."[5] British payments amounted to $51 million during the war years.

This arrangement met Abdul Aziz's monetary needs for the next few years but created political complications. The company feared that channeling U.S. funds through Britain might provide the British with increased political influence over the Saudi Court, and continued to push for direct government aid. An Interdepartmental (Cabinet) Committee on Petroleum Policy, with representatives from State, War, Navy, and Interior, began meeting in early 1943. Its "most absorbing concern," according to one participant, was "whether and how to extend aid to the American oil companies established in the Middle East." The company proposed that in return for direct Lend Lease assistance to

Saudi Arabia *and* assumption of the King's debts by the British, up to 1 billion barrels would be reserved for U.S. military forces at lower than world prices. The goal was to get the stamp of "vital American interest" on the Saudi concession as a sort of political insurance, and the argument was accompanied by now familiar predictions that "maximum efficient production from all domestic wells will soon be insufficient to meet this country's expanding consumption and exports." Roosevelt's initial response was prompt, and in February 1943 he wrote to Secretary of State Stettinius that "in order to enable you to arrange lend-lease aid to the Government of Saudi Arabia, I hereby find that the defense of Saudi Arabia is vital to the defense of the United States."[6] This started a flow of funds that amounted to $17.5 million by 1946. Total aid to Saudi Arabia from both the United States and Britain amounted to $99 million by the end of 1948. This was without the company's having to set aside any reserves, and in fact after the war a congressional committee accused the company of overcharging the navy billions of dollars. The move to direct American government sponsorship of Saudi interests was given impetus by Saudi hints that its allegiance to the United States over Britain had a price. The American ambassador in Riyadh, Saudi Arabia, Colonel William Eddy, cabled this report of a conversation with one of the King's top advisors:

> The question is whether over and above the [Middle East Supply Center] program and limitations there is not some large area in which Saudi Arabia and America can collaborate *alone*, on a basis that leads far beyond the end of the war. Only thus can our combined effort assure continuity, stability and mutual benefit.[7]

All the talk about crude oil shortages in this country nearly backfired on Socal and Texaco. Taking the warnings seriously, Interior Secretary Harold Ickes wrote to Roosevelt in June 1943 proposing that the government buy a controlling interest in Aramco and otherwise "acquire and participate in the development of foreign oil reserves." Ickes asserted this was necessary to prevent dangerous shortages and to "counteract certain known activities of a foreign power which are presently jeopardizing American interests in Arabian oil reserves."[8] Roosevelt created the Petroleum Reserves Corporation, with Ickes as president and chairman of the board.

Ickes' original proposal to buy out Aramco lock, stock, and barrel scared company representatives so much that, according to Ickes, "they nearly fell off their chairs." Under company pressure Ickes whittled down the government's proposed share first to 70 percent, then 51 percent, and then 33 percent, but each was resisted by the company executives on the grounds that such a government role was "against the American tradition." The threat to American control of Saudi oil all of a sudden became manageable for the company. Ickes then proposed that the United States construct a pipeline from the Persian Gulf to the Mediterranean, at an estimated cost of $120 million. This would meet oil security needs without any ideological threat to "free enterprise." Caltex and Gulf, the main beneficiaries, were enthusiastic. Ickes also saw it as a way to "alert the British to the idea that we really mean business in the Middle East on oil."[9] Opposition from the rest of the oil industry, however, was intense and quickly did in Ickes' pipeline. After the war, the company proceeded to build the pipeline on its own.

Shortly after the company became Aramco in early 1944, negotiations began with Jersey Standard and Mobil to give them a cut in the lucrative Aramco concession, further enhancing American corporate control of Middle East oil. This increased interlocking control over the resources of the Middle East was fully supported in Washington, especially in the State Department. Many key petroleum-related positions were in the hands of oil company personnel. The Petroleum Administrator for War (PAW), Ralph Davies, received a government salary of $10,000 while continuing to draw

$47,500 from his old employer Socal. Max Thornburg, petroleum advisor to the State Department, got an official salary of $8,000 but continued to collect $29,000 from Socal. He would later turn up in Iran as the designer of that country's development plans. The director of the Army-Navy Petroleum Board, Admiral Carter, turned up after the war as head of Caltex's Overseas Tankship Corporation. Upper-level positions in the State Department, the Pentagon, and the Executive were held by corporate lawyers and Wall Street investment bankers whose outlook was virtually identical with that of their oil company colleagues uptown around Rockefeller Center. Certainly none would take exception to the State Department description of Saudi Arabia as a place "where the oil resources constitute a stupendous source of strategic power, and one of the greatest material prizes in world history."[10]

Middle East Oil and U.S. Postwar Strategy

Once the threat of Axis control of Middle East oil reserves was eliminated and sufficient supplies for Allied military and civilian use during the war were assured, American strategists and policy-makers concentrated on formulating and implementing policies. This was done in continued and close liaison with the heads of the oil industry for the purpose of translating this "stupendous source of strategic power" into concrete advantage in the postwar struggle to shape the global political economy in accordance with American interests. As early as 1943 Secretary of State Hull wrote to Ickes:

We strongly favor the full utilization of British oil resources and equipment in the Middle East to relieve the strain on American production. However, it should be kept in mind that the expansion of British facilities serves to build up their post-war position in the Middle East at the expense of American interests there. Accordingly, we believe that consideration should be given to any further increase of British oil facilities in the Middle East area only if such increase is clearly necessary from the military viewpoint and the need could not be met by providing for increased supplies of American Middle East oil.[11]

The postwar strategy received a broad but preliminary outline in a State Department memorandum of April 1944 that was simply entitled "Foreign Petroleum Policy of the United States." The "specific policy objectives" were noted as follows:

1. "To influence the flow of world trade in petroleum products in such manner as to substitute Middle Eastern oil for Western Hemisphere oil in Eastern Hemisphere markets," specifically Europe, Africa, and South Asia. This "may come about in consequence of natural economic forces once Middle Eastern production has been adequately stimulated."

2. To reach an intergovernmental understanding with Britain "on broad principles governing petroleum development and distribution with particular reference to . . . the Middle Eastern area," and "to eliminate the unilateral political intervention that has characterized Middle Eastern petroleum affairs heretofore." This was seen to involve specifically a scheduling of "aggregate exports from the Middle Eastern area to meet expanding demand" and "a flexible schedule of probable import requirements of Eastern Hemisphere markets. . . . The aggregate volume of export so determined should be allocated among the various producing countries on some equitable basis."

3. "To forestall those factors that might operate in the direction of alienating American-controlled concessions," including "failure to exploit such concessions" and "failure to foresee and guard against political complications that might develop."[12]

This strategy was further elaborated in a memorandum from Navy Secretary Forrestal to the Secretary of State:

> It is distinctly in the strategic interest of the United States to encourage industry to promote the orderly development of petroleum reserves in the more remote areas such as the Persian Gulf, thereby supplementing the Western Hemisphere sources and protecting against their early exhaustion. . . .
>
> . . .The prestige and hence the influence of the United States is in part related to the wealth of the Government and its nationals in terms of oil resources, foreign as well as domestic. . . . The bargaining power of the United States in international conferences involving vital materials like oil and such problems as aviation, shipping, island bases, and international security agreements relating to the disposition of armed forces and facilities, will depend in some degree upon the retention by the United States of such oil resources. . . . Under these circumstances it is patently in the Navy's interest that no part of the national wealth, as represented by the present holdings of foreign oil reserves by American nationals, be lost at this time. Indeed, the active expansion of such holdings is very much to be desired.[13]

While the Interior Department under Ickes futilely attempted to buy into the Saudi concession, the State Department moved to open negotiations with Britain in order to concretize the "foreign oil policy" outlined above. The result was an Anglo-American oil agreement, designed to secure American interests while avoiding a repetition of the post-World War I "disputes over the rights to develop the oil resources of the Middle East [which] had badly smeared the peace tables."[14] In the end, the agreement was never ratified by the U.S. Senate because of the objections of the politically powerful domestic oil industry, which saw the treaty as a wedge for opening the door to government control.[15]

The legislative fight over ratification dragged on for several years, by which time the principles of the "foreign oil policy" had been duly implemented, not by the governments of Britain and the United States but by the handful of men who ran the seven giant oil companies. The negotiations of Jersey Standard and Mobil for a piece of Aramco were finally concluded in 1948, resulting in 30 percent interest each for Socal, Texaco, and Jersey, and 10 percent for Mobil. The two newcomers, Jersey and Mobil, were able to match their extensive European and Asian markets with Aramco's prolific crude output. They brought the consortium the large chunk of capital ($100 million plus) necessary for the construction of the pipeline from the Saudi fields to the Mediterranean. The coincidence of interest between government policy planners and the oil companies is reflected in the State Department's record of a conversation with Jersey Standard Vice-President Orville Harden:

> Mr. Harden explained that his company's desire to acquire an interest in Aramco and build the pipe-line was part of a long range plan to obtain all the oil for their European and other Eastern Hemisphere markets from the Middle East, retaining Western Hemisphere oil exclusively for their Western Hemisphere markets.[16]

Another significant consequence of the new Aramco setup was the virtual abrogation of the Red Line Agreement of 1928 by which the Iraq Petroleum Company participants (Jersey, Mobil, Shell, Compagnie Française des Pétroles [CFP], and Anglo-Iranian) had pledged to refrain from any petroleum activity in Arabia except through the joint ownership of IPC. This move was most strenuously resisted by the French, whose CFP had the most to lose, but they were powerless to impede what was a reflection of the shift of political and economic power from British and European to American monopolies. The State Department, which saw no need to object to the Red Line restrictions in 1928, now obligingly "advised" the American partners, Jersey and Mobil, that "if the IPC Group Agreement needed to be reaffirmed, the restrictive clauses in the Red Line paragraph should be reconsidered."[17]

The interlocking dominance of Middle East oil production by the giant companies was further enhanced by a series of long-term supply contracts negotiated between the major firms, in which companies with surplus crude production capacity supplied oil at cost to those which were short in return for an even split of the monopoly profits. Potential competition between the giant concerns at the production and marketing ends of the industry was thus effectively precluded in order to maintain monopoly profits that would accrue to the companies at the expense of the producing and consuming countries. Gulf agreed to supply Shell with supplies ranging from 1 to 10 million tons of crude per year over a ten-year period, with all costs and profits pooled and divided equally. Similarly, Jersey Standard and Mobil agreed to purchase 134 million tons over twenty years from Anglo-Iranian (formerly Anglo-Persian) in Iran and/or Kuwait.

This further cartelization was carried out with the full knowledge and implicit support of the policy-makers in the administration and the State Department. Their awareness, moreover, of the consequences and possible alternatives to monopoly are well illustrated by a memorandum from the Assistant Chief of State's Petroleum Division to the special assistant to the Under-Secretary for Economic Affairs in February 1947:

> It is likely that with the dropping off of military demand there was no close relationship between the war-developed productive capacity of these companies and their peace-time market outlets. There can be no question but that this was true if the comparison were between their relatively small peace-time market position and their existing and potential production based upon a reasonable program of development. Thus, with these companies holding vast reserves pressing for market outlets, the prospects were favorable not only for increased competition for markets between

them and the principal established marketers—Shell, Jersey, and Socony—but also for the development of a competitive market for crude oil in the Middle East upon which other companies might rely for supplies in entering the international oil trade.

Under the three large Middle East oil deals, a major change in the conditions favorable to competition would seem to be in prospect. Instead of undertaking to develop their own market outlets for their present and prospective production, Gulf, Anglo-Iranian, and Arabian-American have aligned themselves through long-term crude oil contracts and partnership arrangements with the three large marketing companies, thereby at least contributing toward if not in fact precluding, the development of any bona fide competition for markets between the two groups of companies. . . .

. . . the question naturally arises as to whether some other feasible pattern of ownership of Middle East oil resources would not be in the public interest and more consistent with present United States foreign economic policy. It is believed that from these standpoints, the case is strong for doing what may be feasible toward making ownership of Middle East oil possible for a larger number of companies. It is believed that from the same standpoints the case is strong for discouraging, wherever possible, the further development of joint operations and joint interests between and among the large international oil companies.[18]

It need only be noted that the perceptive antimonopolist perspective of this low-level official found neither echo nor support among his superiors, leaving little doubt as to who determines the "public interest" and what is "consistent with present U.S. foreign economic policy."

The Truman Doctrine and Middle East Oil

Access to Middle East oil supplies and the protection and advancement of American interests there was a goal of U.S. foreign policy second to none. The end of the war left Britain and France in a tremendously weakened position and provided an opportunity for the extension of American interests at the expense of its erstwhile allies. This brought with it, however, the responsibility of protecting those interests against foreign and local threats—which were often conveniently treated as one. For instance, popular nationalist insurrections, in some cases led by leftist political forces, were seen as a threat to American interests in Greece and Iran and were assumed to be sponsored by the Soviet Union. The Truman Doctrine, with its provisions for American military aid to Greece and Turkey, was promoted as being essential to the security of Europe and the United States primarily because of Greece and Turkey's geographic position in relation to the Middle East and its oil. All unseemly references to "oil" and "natural resources" were expunged from the final text of Truman's message to Congress in favor of high-sounding abstractions like "democracy" and "freedom," but an early draft of the Truman Doctrine contained this contribution from Clark Clifford:

> If, by default, we permit free enterprise to disappear in the other nations of the world, the very existence of our own economy and our own democracy will be gravely threatened. . . . This is an area of great natural resources which must be accessible to all nations and must not be under the exclusive control or domination of any single nation.[19]

The Truman Doctrine, issued in March 1947, came after more than a year of futile attempts by Near East officers at the State Department to secure U.S. financial assistance in bulk to the Middle East. The basic situation after the war was concisely described by the British Foreign Minister Ernest Bevin:

> The present economic situation in many of the Middle Eastern countries is certainly not healthy. Countries like Egypt, Iraq and the Levant States are at present living to a large extent on the profits which they made during the war from the presence of Al-

lied forces. Huge fortunes have been made and the gap between rich and poor has been increased while inflation has made the lot of the poorer people more difficult.[20]

It was against this background that State Department's Middle East chief, Loy Henderson, pleaded:

The expenditure of a few millions of dollars to help secure the stability of backward countries by raising the standard of living would thus be a sound investment for the American taxpayer who would thus be buying in the Near East the same form of anti-war insurance which he is purchasing in Europe and the Far East.[21]

American policy during and after the war was to replace the British economically but to support a British military presence in Greece, Iraq, Jordan, and Libya. In Egypt and Iran, popular anti-British sentiment dictated Britain's assumption of a low profile position. The United States, for its part, would take over where necessary from the British—by maintaining the Sixth Fleet in the Mediterranean, for example. U.S. financial assistance, so desperately needed in Britain, was used as a lever to force the British to maintain troops in Greece in the face of popular opposition in England. U.S. strategists assumed that the British would maintain a substantial economic interest in Middle East oil that would serve as an "anchor" against British tendencies to withdraw from military and political commitments in the area. A background paper for the American negotiating team in the so-called Pentagon Talks of late 1947 put it this way:

It would be possible to describe the situation in terms of a simple bargain. If for political and strategic reasons we want them [the British] to hold a position of strength in the Middle East, then they must have from us economic concessions with respect to the area which will make it worth their while to stay there. . . . British economic interests are so important as to make it inconceivable that they would voluntarily pull out completely.[22]

On one level the Truman Doctrine was a rhetorical exercise designed to "scare hell out of the country"[23] in order to gain popular political consent for the administration's military and economic policies in the Middle East (and in Europe for the Marshall Plan). Although the Truman Doctrine provided for an aid program at first restricted to Greece and Turkey, the real thrust of American strategy concerned Iran, which was both a significant oil-producing country and one bordering on the Soviet Union. The military thinking behind the Truman Doctrine is well stated by the following assessment of the Iranian situation by the Joint Chiefs of Staff:

> The Joint Chiefs of Staff consider that as a source of supply (oil) Iran is an area of major strategic interest to the United States. From the standpoint of defensive purposes the area offers opportunities to conduct delaying operations and/or operations to protect United States-controlled oil resources in Saudi Arabia. In order to continue any military capability for preventing a Soviet attack overrunning the whole Middle East including the Suez-Cairo Area, in the first rush, it is essential that there be maintained the maximum cushion of distance and difficult terrain features in the path of possible Soviet advances launched from the Caucasus-Caspian area. Otherwise the entire Middle East might be overrun before sufficient defensive forces could be interposed. As to counteroffensive operations, the proximity of important Soviet industries, makes the importance of holding the Eastern Mediterranean-Middle Eastern area obvious. This is one of the few favorable areas for counteroffensive action. Quite aside from military counteroffensive action in the area, the oil resources of Iran and the Near and Middle East are very important and may be vital to decisive counteroffensive action from any area.[24]

Iran was the scene of the first postwar confrontation between the United States and the Soviet Union. The Soviet Union responded to the presence of American oil industry representatives in Iran (as the Shah's advisors for petroleum affairs!) by requesting a Soviet concession in the five northernmost provinces, those on or near the Soviet border. The

Soviet motive seems to have been strategic and political rather than economic, to offset the likelihood of another anti-Soviet regime in Teheran. The British were not opposed to such a move, probably because they felt it would justify their continued exclusive presence in the south. The United States, however, while adopting a formally neutral stance, supported anti-Soviet political forces in the Iranian parliament. They advised the Prime Minister, Qavam, that the Russians be kept out indirectly—by setting up impossible conditions rather than by rejecting their bid outright. The rhetorical commitment of American policy-makers to democracy did not prevent them from recommending to Qavam that elections be postponed as long as a substantial number of pro-Soviet deputies were likely to win office. The Soviet-American confrontation in Iran became the first "crisis" on the agenda of the newly formed United Nations. The issue was resolved with the eventual withdrawal of Soviet and British troops from Iran in a way which preserved, in Truman's words, "the raw material balance of the world."[25]

A more real threat to continued Western control of Iran's oil was the rising nationalist sentiment against continued exploitation of Iran's reserves by Anglo-Iranian, and the working-class militance that had grown up around the oil-producing centers in the southern part of the country. A mass strike against AIOC and the government, involving some 100,000 workers, broke out in July 1946. The British blamed Moscow radio and the local Communist organization (Tudeh) and referred to the strike as a "deliberate political act of hostility." The strikers accused the British of arming local ethnic Arab tribesmen to attack the strikers, killing 17 and injuring 150. The British dispatched 15,000 troops from India to the Iraqi post of Basra for purposes of intimidation and possible intervention "to protect British, Indian and Arab lives." The strike, which involved well over half of the total number of industrial workers in Iran at the time, ended when the Tudeh Party was brought into the cabinet. Strikes took place simultaneously in the Kirkuk fields in Iraq, where the British were disturbed by the appearance of "political elements similar to those in Iran."[26]

In Iran after the war, rising nationalist consciousness was revealed most strongly within the workers' movement. In the Arab world as well, workers and students participated in struggles against occupying powers, notably in non-oil-producing countries like Egypt. The most dangerous source of "instability" (from the American viewpoint) stemmed from the fierce conflict that had erupted in Palestine, but American interests were nowhere more cloaked than in Truman's policy in Palestine in 1947-1948. United States oil companies were opposed to any official support of Zionist aims, and this was expressed most strongly by such representatives of the American corporate elite as Defense Secretary Forrestal and Under-Secretary of State Acheson. U.S. support of the partition plan in the United Nations in 1947 lasted only as long as it seemed to hold the greatest possibility of avoiding armed conflict that might require outside armed intervention. It also had the virtue of satisfying to some degree the domestic Zionist lobby. When it became clear in early 1948 that war in Palestine would not be avoided through partition, United States support for that policy dissolved. The minimum goal of U.S. policy regarding Palestine was to facilitate the British departure and to prevent any Russian presence in the form of a UN peacekeeping force. As for the question of U.S. military intervention, the Joint Chiefs of Staff, in June 1946, urged that

> no U.S. armed forces be involved in carrying out the [Anglo-American Inquiry] Committee's recommendation . . . [that] the political shock attending the reappearance of U.S. armed forces in the Middle East would unnecessarily risk such serious disturbances throughout the area as to dwarf any local Palestine

difficulties . . . . the Middle East could well fall into anarchy and become a breeding ground for world war.[27]

What prompted Truman's instantaneous recognition of Israel on May 15, 1948, was his perception that it was the path of least resistance, both in terms of domestic political pressures and international complications. Strategic support of Israel for the advancement of specific American political and economic goals in the Middle East was a later development. An additional factor was the private assurance of the Saudi regime to U.S. officials and oil company executives that U.S. policy in Palestine would have no effect on the Aramco concession. In December 1947, Crown Prince Saud told the American ambassador in Jidda that if other Arab states (Iraq and Jordan) insisted on breaking relations with the United States, Saudi Arabia would break relations with them. In November 1948, the former Iraqi Prime Minister Saleh Jabr told the parliament that Saudi Arabia had blocked an Arab League plan to cut off oil production because of the question of the Palestine partition. Whether the United States policy on Palestine would have been the same in the face of a united front of Arab oil-producing states is a question that cannot be answered on the basis of the available evidence.[28]

As it was, the chief political repercussions of U.S. Palestine policy came in Palestine itself, where pipeline and refinery installations were a favorite target of Zionist commando operations, and in Syria, where the government declined to enter an agreement with the Trans-Arabian Pipeline (TAPline) Company for transit rights on Aramco oil headed for the Mediterranean. An agreement was reached in Beirut in September 1947, following a two-week full-dress conference of U.S. diplomatic personnel in the area, but the Israeli-Arab war of 1948 broke out before it was ratified by the Syrian parliament. Resistance or reluctance on the part of the Syrians to finalize an agreement with TAPline after the war was sidestepped when Colonel Husni Zaim seized control of the government at the end of March and dissolved the parliament a few days later. The United States quickly recognized the new regime and the agreement with TAPline was finally signed in Damascus in May 1949.[29]

The Political and Economic Strategy of Profit Sharing

The economic and political nationalism that characterized Middle Eastern and other Third World areas had the effect of winning incremental financial advantages from foreign oil interests. The United States, furthermore, in its campaign to win more oil and influence in the Middle East, followed a policy of buying the political support of local elites and regimes. It urged companies to improve their deals with host governments in the form of higher taxes and fees, in order to head off any fundamental challenges to the basic system of Western exploitation of Middle East resources. Another element in this policy, of course, was the substantial amount of direct government economic and military aid given to these regimes. The so-called profit-sharing agreements between the international companies and the producing countries represented an ingenious combination of these two approaches, whereby the taxes formerly paid to home countries (like the United States) were transferred by secret agreement between the companies and the U.S. government to the treasuries of the House of Saud, the Hashemite regime in Iraq, and other Middle Eastern potentates.

This arrangement was pioneered in Venezuela in 1948 and introduced into the Middle East by Aramco in Saudi Arabia in 1950. Under the old royalty payment system, funds paid to the ruling families and governments climbed rapidly with the postwar expansion of production. In Saudi Arabia,

Aramco produced 7.8 million barrels in 1944 (almost all for the U.S. military), earning the government $1.7 million. In 1949, it produced 174 million barrels and the House of Saud got about $50 million. Company profits climbed much more rapidly, though, from $2.8 million in 1944 to over $115 million in 1949. The regime was getting 21¢ per barrel royalty, while company income per barrel was over $1.10. The Saudis pushed for a more equitable division of these monopoly profits. The oil companies and their friends at the State and Treasury Departments came up with a foolproof scheme to satisfy Saudi court demands without costing the companies anything. The trick was to increase payments to the Saudis at a 50-50 rate, but to call the royalty an income tax. Under the foreign tax credit provisions of the U.S. tax code, these payments could then be deducted from income tax payments here. (Royalty payments, in contrast, can be deducted as a business expense, but the lower profits are still subject to U.S. taxes.) Thus, Aramco payments to the Saudi throne jumped from $39.2 million in 1949 to $111.7 million in 1950. Aramco taxes to the U.S. Treasury were reduced by a similar amount, to nearly zero, where they have remained ever since.

This creative accounting required the active and secret collusion of the U.S. government at the highest level. The reasons for agreeing to increase funds to Saudi Arabia through tax payments were recently laid out in congressional testimony by George McGhee, then Assistant Secretary of State for Near Eastern, South Asian, and African Affairs. The following selections from McGhee's testimony illustrate the highly political nature of the decision:

> At this time, the principal threat to the Middle East lay in the possibility of nationalist leaders moving to upset regimes which were relatively inept and corrupt, and not attuned to the modern world. . . .

Q: Was it the decision of the U.S. Government that it would be in the national interest to get more money into the hands of the conservative governments in the Arab world?

McGhee: Let's put it this way: I believe that Aramco and the U.S. Government reached this decision at the same time and independently by reasoning along parallel lines. We were, of course, basing our judgment on the same information. Internally, in Saudi Arabia, there were special problems, over and above the general problems of the region. There was, for example, a Finance Minister in Saudi Arabia, Abdullah Sulaiman, who was proving very difficult to deal with. He was always pressing to shut down the oil fields unless the Saudis got more money. . . .

. . . both the Aramco officials and the Department had, independently, reached the conclusion that something had to give. Some greater sharing of profits with Saudi Arabia must take place, otherwise there would be an increasing threat to the regime and to Aramco's ability to maintain its concessions.

. . . .

Aramco was, of course, quite happy with their existing profit split. They were receiving roughly $1.34 a barrel before taxes, which at that time were 25 cents a barrel, so they were netting around $1.10. In moving from this position to profit-sharing they stood to lose, of course, approximately half of this income. . . .

Q: But upon the recommendation of the National Security Council, the Treasury made the decision to permit Aramco to treat royalties paid to Saudi Arabia as though they were taxes . . . the impact on the national treasury was direct and dramatic. . . . the effect of the decision was to transfer $50 million out of the U.S. Treasury and into the Arabian treasury. That was the way it was decided to give Arabia more money and to do it by the tax route. Isn't that correct?

McGhee: Yes, that is one way of looking at it.

. . . .

The ownership of this oil concession was a valuable asset for our country. Here was the most prospective oil area in the world. Every expert who had ever looked at it had said that this was the "jackpot" of world oil. To have American companies owning the concession there was a great advantage for our country. . . .

We felt it exceedingly important from the standpoint of the stability of the regimes in the area and the security of the Middle East as a whole and the continued ownership of our oil concessions there and the ability to exploit them, that the Government of Saudi Arabia receive an increased oil income.[30]

Iran: Opening the Last Door

The American policy objective of supplanting British power where possible and suppressing popular demands for control of national resources in countries where those demands could not be bought off was most clearly revealed in Iran, the only Middle East country where the United States had no direct role in existing petroleum operations. This was not for lack of interest. American oilmen were among the retinue when the United States moved to increase its diplomatic presence in Iran after 1944. Lend Lease activities were transferred from British to American agencies to strengthen the U.S. military mission there "for the protection and advancement of our interests . . . ," according to Secretary of War Stimson.[31] In the course of the dispute over Soviet concessions in the north initiatives toward concessions for American companies were tactically suppressed by the administration because "we are anxious that impression not be obtained that we have been influenced in our recent actions before Security Council by selfish interest in Iranian petroleum."[32] Even after the diffusion of the Soviet-American crisis, popular opposition to the old concession arrangement with AIOC, which surfaced with the large-scale strikes of 1946, continued to be manifested in the actions and demands of various contending political forces in Iran. In February 1949, for example, thousands of students surrounded the Majlis (parliament) in Teheran and refused to disband until new legislation was enacted regarding the AIOC concession consistent with the oil law passed in 1947 to justify rejection of Russian concession demands. Between 1913 and 1951, Anglo-Iranian grossed $3 billion, only $624 million of which went to the Iranian government. The remaining $2.4 billion was transferred abroad as profit. Between 1944 and 1950, AIOC's profits increased tenfold and government revenue only fourfold.[33]

The demands of the majority of politically conscious Iranians could not be met simply by getting a larger chunk of Anglo-Iranian's fabulous profits. Fundamentally, what was at stake was the question of national sovereignty, exacerbated by four decades of abuse and exploitation at the hands of British imperialists and the Iranian ruling class. In December 1950, a special oil committee of the Iranian parliament, chaired by Dr. Mohammed Mossadeq, the nationalist leader who had led the struggle against the Russian concessions, began to study the option of nationalization. Anglo-Iranian responded by offering to accede to the 50-50 profit-sharing arrangements then in vogue, but this last minute offer failed to head off the Mossadeq committee's recommendation that the oil industry be nationalized. In April 1951, the Majlis passed a bill for nationalization, and in May Dr. Mossadeq become Prime Minister. The British government, the major shareholder in the expropriated company, found itself unable to play a traditional gunboat role largely because of the opposition of the United States, which surmised that British military action would meet with armed popular resistance and would be likely to invite a Russian military presence as well. Instead, the British insisted on compensation that would cover Anglo-Iranian not only for its investments but also for the profits on the estimated oil underground—terms so excessive they were certain to be rejected. They launched a campaign of economic warfare by instituting a boycott that effectively prevented Iran from transporting and marketing its own oil. Anglo-Iranian then shifted its production to its Kuwait and Iraq concessions.[34]

The United States' role in bringing on the crisis was indirect. Since the war, the United States had promoted itself as an alternative ally to the Iranians. Iran had become a prime recipient of Truman's Point Four aid program. The country's economic future was being designed by a comprehensive development plan drawn up by Overseas Consultants, Inc., in 1950. Overseas Consultants was headed by Max Thornburg, former State Department oil affairs advisor also on salary to Standard of California. This had given Mossadeq reason to think that the United States might provide some measure of support in the face of British intransigence. An Iranian "prominent public figure" (probably Mossadeq himself) presented the nationalization move as a step necessary to "halt the growth of Communism."[35] Assistant Secretary of State McGhee was quickly dispatched to Iran, while higher level American officials huddled with their British counterparts in Washington. Harriman and Acheson began playing roles as self-appointed mediators between Teheran and London, with both sides suspicious of their motives and eager for their favor.

Although Iranian nationalization removed Anglo-Iranian from its exclusive position and opened up the possibility of some kind of American presence, the U.S. government and the oil companies necessarily viewed the developments as a crisis as well as an opportunity. In the first place, there was a fundamental hostility toward such assertions of national sovereignty. This was well put by the State Department's petroleum division chief in 1945:

> While recognizing the sovereign right of any country to assume ownership . . . of the petroleum industry or any of its branches, this Government must nevertheless recognize and proclaim that international commerce, predicated upon free trade and private enterprise (which is the conceptual core of United States econom-

ic foreign policy), is, in the long run, incompatible with an extensive spread of state ownership and operation of commercial
properties.[36]

More directly, the Truman administration was acting
against the background of its own vigorous attempts to reverse
the exclusion of American oil interests from nationalized
fields in Mexico and Bolivia, attempts which failed in the first
and succeeded in the second. Truman confided to *New York
Times* columnist Arthur Krock that he privately felt the Mexican action was justified, but "if . . . the Iranians carry out
their plans as stated, Venezuela and other countries on whose
supplies we depend will follow suit. That is the great danger
in the Iranian controversy with the British."[37] A larger dimension of the significance of the crisis is provided by a
series of articles in the *New York Times* in May 1951 which
stressed the role of Middle East oil in the struggle between
the capitalist and socialist countries known as the Cold War.
"The oil of the Near East," the *Times* analyst wrote,

> could be used just as advantageously by Russia for the develop-
> ment of China and other parts of the Far East friendly to her, as
> by the Western world in Europe and elsewhere. It is for that
> reason that the U.S. and Europe could hardly afford to see any
> important part of the Near East oil resources pass to the control
> of Russia, *either directly or indirectly through nationalization.*
>
> Given an abundant supply of cheap oil for power, Russia in
> the course of a few years could industrialize a large section of
> Asia. With the enormous manpower already available there it
> would take but a relatively short period to bring a virtually un-
> beatable military force into being.
>
> . . . With Near East oil available to Europe in steadily increas-
> ing quantity its industrialization is furthered and its economy is
> strengthened. So long as such a condition exists the radical
> elements in Europe will be held in check, since communism
> makes slow progress in prosperous areas.[38]

The major American oil companies demonstrated virtually total solidarity with Anglo-Iranian's boycott of Iran. The results of this were seen in the precipitous drop in Iranian production, from 243 million barrels in 1950, to 8 or 9 million barrels in 1952 and 1953. Although Iran had been totally dependent on Anglo-Iranian, the situation was not reciprocal; the company's strength lay in the fact that it did not depend on Iran for all of *its* crude oil. In Iraq, where AIOC held a 23.75 percent share, production jumped by 189 million barrels.[39]

The boycott not only created intended economic hardships in Iran but also the likelihood that political power would continue to shift to more radical and leftist forces. There was overwhelming and unwavering popular support for Mossadeq and his policies. In July 1951, two months after nationalization, he received a unanimous vote of confidence in the Majlis. Subsequent votes of confidence were tempered by some reluctance to grant demands that he be allowed a temporary period of rule by decree. When his opposition to the Shah brought forth parliamentary opposition, he dissolved parliament and won a popular plebiscite with more than 99 percent of the vote.

While the masses supported Mossadeq and his National Front government unequivocally, other elements were less enthusiastic. Economic problems brought on by the boycott led Mossadeq to take measures against some of the vested landed interests. Beginning in August 1952, he established commissions to collect unpaid taxes from the rich, and to throw them in jail and confiscate their property if they did not pay. By decree, he cut off feudal dues received from the sharecroppers. Symbolic was his decision in early 1953 to cut into the Shah's multimillion-dollar income and government allotment. The Shah and "his" Prime Minister had been struggling for control over the armed forces, some of whose officers were unhappy at the increasing revenues being diverted from the military to more socially beneficial budget categories. The army and police had been under the tutelage of the American military ever since the war.[40]

The story of the CIA coup that restored the Shah to power in August 1953 is too well known to serve up the gory details here. It was staged after the Eisenhower administration took office, with the Dulles brothers in charge of the State Department and the CIA. John Foster came from the law firm of Sullivan and Cromwell, which represented the Rockefeller oil interests. Allen had an association with the Middle East going back to World War I.[41] Kermit Roosevelt, the CIA man in charge of the Iranian operation, later left "public service" to become vice-president of Gulf Oil. Herbert Hoover, Jr., of Union Oil Company, became the State Department intermediary between Anglo-Iranian and the American companies. (It will be remembered that Herbert Hoover, Sr., then Secretary of Commerce, played a role in the American entry into the Iraq Petroleum Company in the 1920s.) The Shah took as his advisor on petroleum affairs Thorkild (Tex) Rieber, who had been Texaco's board chairman until forced to resign in 1940 for alleged pro-German sympathies.[42]

After the coup, American strategists' first priority was to conclude the long-simmering negotiations between Anglo-Iranian (which changed its name to British Petroleum at about this time) and the American oil giants, in order to provide revenues for the new "friendly" regime of the Shah and his Prime Minister, General Zahedi. In January 1954, the National Security Council advised the U.S. Attorney General that

> the security interests of the United States require the United States petroleum companies to participate in an international consortium to contract with the Government of Iran, within the area of the former AIOC concession, for the production and refining of petroleum and its purchases by them, in order to per-

mit the reactivation of the said petroleum industry, and to provide therefrom to the friendly Government of Iran substantial revenues on terms which will protect the interests of the western world in the petroleum resources of the Middle East.[43]

Under the national security blanket the companies got 40 percent of the new consortium and exemption from a pending criminal antitrust suit launched by the Department of Justice.[44] Eight smaller American companies were later brought in at 1 percent each. For the new members, large or small, any stake in this immensely profitable but virtually riskless venture was "like getting a license to print money." The consortium agreement paid lip service to Iranian national sovereignty, but invested full management and commerical rights in the hands of the companies. The Iranian government got no additional revenues on industry assets it already owned but which were used freely by the companies, making the Iran agreement "one of the most attractive contracts to the oil industry in the Middle East, as far as terms of payment are concerned."[45] The consortium arrangement in Iran represented the height of the control of Middle East oil by the five American oil giants.

3

The Bonanza Years: 1948-1960

The reference to the postwar years as a bonanza time should not imply that previous oil operations were anything but very profitable for the seven giant companies. For the years 1913 to 1947, aggregate financial data for Middle East oil operations show that total receipts came to more than $3.7 billion, of which just over $1 billion met the costs of fixed assets and ongoing operations, and $510 million was distributed in payment to local governments, mainly as rents, royalties, and bonuses. This left the oil companies with a net income of $2.2 billion; only $425 million of this was reinvested in the area while $1.7 billion was transferred abroad as profit.

In the bonanza years, these figures were literally dwarfed. Of $28.4 billion in total receipts, operating costs accounted for $4.8 billion, net investment in fixed assets for $1.3 billion, payments to local governments for $9.4 billion, and income transferred abroad (i.e., profits) accounted for $12.8 billion.[1]

Crude Oil Prices: Europe After World War II

The oil industry is legendary for its high profits and allegedly high risks. In the Middle East, the profits are higher and the risks lower than any place else. The basis for fantastic

Gross Receipts, Expenditures, and Distribution of Net Income in the Middle East, 1948-1960
(millions of dollars)

Item	1948	1949	1950	1951	1952	1953	1954	1955	1956	1957	1958	1959	1960	1948-60
Middle East														
A. Gross receipts														
Local sales of crude and refined products	34.4	46.7	50.9	28.2	9.0	12.3	13.9	30.0	35.7	39.8	48.3	53.8	58.9	461.9
Exports of refined products	682.5	578.2	674.2	561.4	376.9	425.4	488.0	614.1	719.3	801.9	845.9	840.8	865.7	8,474.3
Exports of crude	313.9	392.9	566.6	852.5	1,097.2	1,379.4	1,570.4	1,798.9	1,883.3	1,946.7	2,463.6	2,450.5	2,740.8	19,456.7
Total	1,031	1,018	1,292	1,442	1,483	1,817	2,072	2,443	2,638	2,788	3,358	3,345	3,665	28,393
B. Expenditures														
Cost of petroleum operations	254	275	287	265	238	261	278	388	446	495	491	552	575	4,805
Payments to local governments	154	157	239	284	400	586	719	903	986	1,035	1,229	1,322	1,420	9,434
Net income of oil companies	623	586	766	893	845	970	1,075	1,152	1,206	1,258	1,638	1,471	1,670	14,153
Total	1,031	1,018	1,292	1,442	1,483	1,817	2,072	2,443	2,638	2,788	3,358	3,345	3,665	28,392
C. Distribution of net income														
Net investment in fixed assets	243	276	51	79	107	58	25	6	55	79	121	(120)	(100)	1,320
Transfers of investment income abroad	380	310	715	814	738	912	1,050	1,146	1,151	1,179	1,517	1,351	1,570	12,833
Total	623	586	766	893	845	970	1,075	1,152	1,206	1,258	1,638	1,471	1,670	14,153

Source: Charles Issawi and Mohammed Yeganeh, *The Economics of Middle Eastern Oil* (New York, 1962), pp. 188-89.

profits is the incredibly low cost of producing Middle East oil, combined with the artificially high price structure of the world oil industry. The latter is based on the high production costs (and not insubstantial profits) of oil production in the United States. "Prices at the Persian Gulf have only an arbitrary relationship to production costs," according to one oil economist. "They contain a substantial element of what may be called monopoly profits."[2] A Chase Manhattan Bank study in the early 1960s concluded that the average cost of maintaining and expanding production in the Middle East is 16¢ per barrel, while average Venezuelan costs per barrel are 51¢ and U.S. costs are $1.73. Although Middle East costs have moved down over the years and Venezuelan and U.S. costs have moved up, these figures convey the magnitude of the difference.[3] One way of illustrating the workings of the pricing system is to recount the controversy that developed around the price of Middle East oil in Europe during the postwar recovery and reconstruction period.

Profits, crudely speaking, represent the difference between costs and prices. The price of oil controlled by the big international companies has historically been based on the price of oil in the United States, which, until World War II, was the major supplier of crude oil and petroleum products on the world market. The U.S. price has been consistently maintained at a high level by means of prorating, or production restrictions. Under the basing-point formula, with Texas as the basing point, the price of crude oil around the world, even in Teheran or Baghdad, was set as the Texas price *plus* the cost of transportation from the Gulf of Mexico to the purchase point.

Under this formula, which remained in effect until the end of World War II, inexpensive crude oil purchases in the Persian Gulf was priced as if it had come from Texas. It was based on the expensive Texas price plus the utterly fictional cost of tanker transport from the Gulf of Mexico, rather than on the actual tanker cost from the Persian Gulf to the point of delivery. The prime beneficiaries of this system were the international oil companies. The integrated structure of the international companies—by which they controlled all aspects of the industry from well to market—made this possible, as did their joint plans to control rather than compete for various markets. The companies and not their consumers enjoyed all the benefits of lower production and transportation costs.

This system lost its contrived logic by the end of the war, when the Persian Gulf became a main source of oil for Europe and the Eastern Hemisphere. In 1944 the British navy, which had been paying the formula price (Texas Gulf plus transport) for oil purchased in the Persian Gulf, launched an investigation into oil prices. Rather than supply the British government with the relevant cost data, the companies discreetly decided to lower the price of Persian Gulf supplies to match Texas Gulf prices—to eliminate the "phantom freight" tanker charge. In effect, the Persian Gulf became a second basing point, but prices still bore little relation to actual costs and instead remained the same as the Texas price.[4]

The reason for this is clear, from the companies' point of view. With extensive oil operations in the United States, still by far the largest market, they had a direct stake in stabilized prices. Within a year after the war, Middle East production had tripled and its reserves had barely been tapped. Any pricing system based on actual costs would have forced prices downward. Competition has never been a goal of the oil industry in its constant battle against the law of supply and demand. The contradictions built into this carefully rigged system were felt acutely in Europe, then trying to rebuild itself after the war.

The postwar economic reconstruction of Europe and Japan required unprecedented amounts of petroleum as whole national economies were built anew. Economies based on domestic coal resources were transformed at every level so that petroleum would provide a larger share of basic fuel and energy needs. Transportation, with the rise of the automobile and the change in railroads from coal to diesel fuel, came to depend almost exclusively on petroleum products. In contrast to the United States, Europe and Japan were almost totally dependent on foreign sources of oil, the bulk of which was controlled by U.S. companies. The predominant source was going to have to be the Middle East since Western Hemisphere production was consumed by the United States. A report of the petroleum experts sent to Saudi Arabia in 1944 as part of Ickes' federally sponsored Petroleum Reserves Corporation noted that:

> The center of gravity of world oil production is shifting from the Gulf [of Mexico]—Caribbean area to the Middle East, to the Persian Gulf area, and is likely to continue to shift until it is fairly established in that area.[5]

 Two of the American oil giants, Jersey Standard (Esso) and Mobil, had extensive marketing facilities in Europe predating the war, which had been supplied by tanker from the Caribbean. In addition, the European recovery program was being financed by the United States which saw that American companies got the lion's share of the contracts. All in all, the American oil majors were in an excellent position to capitalize on the activities of the Marshall Plan and the Economic Cooperation Administration (ECA), both in terms of supplying immediate needs and asserting preeminent marketing positions for the future.

Some of the tactics used to implement America's overall strategy of predominance in Western political economy relate to the oil industry. In brief, there was a consistent attempt to increase the European economies' dependence on those industrial or resource sectors in which U.S. industry possessed comparative advantage or controlling strength. One of these sectors was oil. The European economies, for example, had traditionally relied on rail transportation systems, but European requests for freight cars under the Marshall Plan were reduced from 47,000 to 20,000. By contrast, the Americans insisted on allocating 65,000 trucks, although none were requested.[6]

By mid-1950, 11 percent of the value of all ECA shipments to Europe consisted of oil. According to Walter Levy, who resigned from Mobil to head the oil division of the ECA (and now as a private consultant in Rockefeller Center serves as one of the industry's chief consultants), "ECA has maintained outlets for American oil in Europe . . . which otherwise would have been lost."[7] The ECA, though, experienced its own contradictions, given that scarce dollar reserves of the European countries were being soaked up by the oil companies, leading to a severe monetary crisis in mid-1949.

Any disruption to the European recovery program would have had unfavorable repercussions on the American economy, which was then entering a recessionary period. Rising oil imports into the United States, coinciding with the economic downturn, were generating intense opposition on the part of independent domestic producers. At the same time Britain, faced with a severe dollar shortage, announced restrictions on oil purchases from American companies in favor of those from Anglo-Iranian and Shell. Although Britain's overall economic weakness forced a retreat on this demand, the possibility of a severe and long-standing crisis between Britain and the United States on this question was interpreted ominously. As it turned out, mobilization for the Korean War absorbed the oil surplus and temporarily ended the crisis over markets. An American economic strategist summed up the issue:

> The international oil companies cannot expect to maintain their
> position abroad over a period of systematic restriction of foreign
> dollar resources and dollar spending. As sellers of a dollar com-
> modity and ultimately even as holders of dollar investments

abroad, their position would prove vulnerable. . . . When progress [toward economic integration in the Atlantic community] seemed to slacken in 1949, the foreign operations of the American oil companies were immediately endangered. When dollar supply and world demand improved in 1950, the international surplus came to a rapid end. Fresh evidence of this country's intent to lead in the defense and the economic development of the community of free nations made it less likely that crises of that sort would recur in the near future.[8]

While ECA supported and promoted American oil interests in order to, in Levy's words, "maintain for American oil a . . . competitive position in world markets," it was forced to confront and deal with the absurdity of the industry's pricing system. This basing-point formula put the net Persian Gulf price for European oil (prior to transport) at $2.22 per barrel. Increased deliveries of Middle East oil to the United States, however, under the same formula, were priced at $1.75 a barrel at the Persian Gulf in order to compete with Texas and Venezuela deliveries on the U.S. East Coast. This discrepancy in the pre-transport price was too much even for the ECA, and under pressure the oil companies lowered the European f.o.b. prices at the Persian Gulf to $1.75. It should be noted that this price was still not even remotely based on the cost of producing the oil. Substantial monopoly profits were still going to the oil companies, although an increasing share, after 1950, accrued to the host governments. In the mid-1950s this price differential between Texas and Persian crude remained fairly steady, although the arbitrary nature of the price itself was revealed on several occasions when the Persian Gulf price went up (or, less often, down) just about the same amount as the Texas price.[9]

The Structure of the Industry: Vertical Integration

The basis for maintaining this monopolistic pricing system lies in the structure of the international (that is, American and British) oil industry. One significant feature of this structure has been the complete integration within each of the major companies of all phases of the industry: exploration, production of crude, refining, transport, and marketing. Competition between the major companies, at each point and for the industry as a whole, has been virtually eliminated by the further integration of the companies with each other at various points in the process. In 1949, according to a report of the Federal Trade Commission, the seven giants owned 65 percent of estimated crude reserves in the world and 92 percent of reserves outside the United States, Mexico, and the USSR. They controlled 88 percent of crude production, 77 percent of refining capacity, at least two-thirds of the tanker fleet, and all of the major pipelines outside the United States and the USSR.[10]

Let us first analyze the significance of the vertical integration of all industrial phases within each company. Concentration of monopoly power is high at virtually every stage: the companies, separately and jointly, have monopoly control over the sources of oil in the Middle East through concessions granting them exclusive rights for 60 to 100 years over large tracts, sometimes covering a whole country. This control has been the basis for dominance in other phases of the industry as well. Refining has been less concentrated than production and transport, but owing to the high capital cost of refinery operations, and, more importantly, to the severely restricted access of independent refiners to crude supplies, entry into the industry has been restricted at this level too. Concentration in the marketing phase basically stems from the monopoly position of the major companies at all the previous levels.

Another way to put it is that historically there has been no free market for crude oil in the world oil industry. Because of the concentrated and integrated nature of the major firms, a price can be set which has an arbitrary relationship to cost. For the integrated company, only the combined production costs and the final aggregate product selling price matters. How profits are distributed among the various phases of the operation is a bookkeeping exercise. An integrated company, for example, could place a low price on Middle East crude and a high price on refined products. Or it could have its tankers and pipelines charge rates higher than cost. If the difference between total cost and market price of a gallon of gasoline at the service station is forty cents, for example, it makes little difference to the overcharged consumer whether that profit margin is made by the producing affiliate, the refining affiliate, the pipeline or tanker affiliate, or the marketing affiliate of the giant firm.

Historically, the industry has chosen for several reasons to base its monopoly and its profit in the production stage. In the first place, an exclusive concession is a monopoly that is practically invulnerable to any intrusion or competition short of government expropriation. Even then, as we have seen in Iran, the resources of a single company are not concentrated in any one country. There is no such "natural" monopoly in the refining or marketing stages.

Second, a monopoly at this initial stage decreases the prospect for competition at a later stage. With a relative monopoly on crude, a company can charge an outrageously high price to its refining affiliate without in any way minimizing the ultimate financial return to the company as a whole. The refinery may show a normal and not excessive profit margin, but the owners and shareholders are only concerned with the outstanding profit performance of the entire company. By the same token, the company will charge the same high crude price to a potential competitor, but that price, along with a low rate of return on refinery operations in the industry as a whole, will tend to make any refinery venture by a nonintegrated enterprise not worth the capital risk.

The consequences of this vertical integration have been severe for virtually all the countries of Africa, Asia, and Latin America. For the few oil-producing countries, this integration forecloses the possibility of marketing crude oil (or its products) except through the giant firms. For the oil-consuming countries, such as India, attempts to develop locally or publicly owned refining and marketing operations have until recently been thwarted by the giant firms' control of most or all sources of crude; the companies sell these sources to their own local subsidiaries at prices which they control and which have always ranged much higher than those in the Western industrialized countries. According to one seasoned observer of the industry, "The poorer a country was, the higher were the prices it paid for oil."[11]

Third, the tax loopholes for which the industry is notorious are concentrated at the production stage. Most notable, and lucrative, is the infamous depletion allowance, according to which 22 percent (until recently 27½ percent) of the profits from the production stage are exempt from income taxation. There is no comparable tax advantage at any other stage of the industry.

Foreign subsidiary producers have an even greater incentive to show a profit at the production stage. With the advent of the 50-50 profit-sharing formula, worked out with the producing governments around 1950, the U.S. government conveniently ruled that if such payments to the producing countries were called income taxes they would be credited against any income tax owed to the U.S. treasury.

The Structure of the Industry: Horizontal Integration

Another dimension of the monopoly power of the major oil companies is their global scale and ability to act as a single entity rather than as separate units. Because of the high degree of integration between the different firms at the production, transport, and marketing levels, there is no significant competition among them and they can be legitimately analyzed as a single entity. This horizontal integration takes place in three principal ways: joint ventures, long-term supply contracts, and market-sharing arrangements.

Joint Ventures. With the entry of the major American firms into the Iranian consortium in 1954, every major oil concession in the Middle East, without exception, was operated by two or more of the seven vertically integrated firms. In other words, every company had a significant, if not a controlling, interest in two or more countries. According to the Federal Trade Commission:

> Each of the companies has pyramids of subsidiary and affiliated companies in which ownership is shared with one or more of the other large companies. Such a maze of joint ownership obviously provides opportunity, and even necessity, for joint action. With decision power thus concentrated in the hands of a small number of persons, a common policy may be easily enforced. . . . For example . . . the directors of Standard of N.J. and Socony Vacuum [Mobil], who determine the policies of Arabian-American Oil Co. [Aramco], are the same men who help to shape the behavior of the Iraq Petroleum Company. The directors of Anglo-Iranian Company who assist in making high oil policy for Iraq and Iran, participate, along with the directors of Gulf, in planning the price and production policies in Kuwait.

Through similar control over pipeline facilities, markets, and even technological patents, "control of the oil from the well to the ultimate consumer is retained in one corporate family or group of families." The boards of directors that manage the myriad jointly owned companies are "in effect . . . private planning boards where differences are resolved and where an oil policy for the world can be established."[12]

A more recent and cautious assessment of the joint venture strategy is that the combined effect "is to reduce the independence forced on each participant, who knows the investment and output plans of his rivals."[13]

Long-Term Supply Contracts. The second means of joint monopoly control of resources and markets is through contracts among the big companies to purchase and sell large quantities of crude oil. In practice this serves much the same function as the joint venture. As we have seen, one of the motivations for the joint venture is that individual giant companies seldom have perfectly balanced supplies and markets. Jersey and Mobil, for example, became part of Aramco because they had extensive marketing outlets for the immense amounts of Saudi crude, markets that Socal and Texaco did not have. By thus matching needs and assets, the profits of all the companies were magnified and market stability was assured.

The major long-term supply contracts in the Middle East were made in 1947 between Gulf and Shell and between Anglo-Iranian and both Jersey and Mobil. The characterization provided in the FTC cartel study is precise:

> Sales of oil covered by the contracts can often be utilized as an instrument to divide production, restrain competition in marketing, and protect the market positions of both the buyer and the seller. . . . They tend to keep surplus crude out of the hands of independent oil companies. The existence of these contracts in an atmosphere of joint ownership of production and marketing, the long periods for which they run, the manner in which prices are determined under them, and the marketing restrictions often written into them, indicate that they are something more than ordinary commercial purchase and sale contracts. . . . Joint ownership and crude oil supply contracts have, in effect, served to

complement each other in protecting the mutual interests of the international oil companies in the production and marketing of world oil. . . . Under the Gulf-Shell contract [Gulf and Shell, along with Jersey, also control most of Venezuela's oil] Shell acquired control over 1¼ billion barrels of Kuwait oil . . . to be delivered over an open-end contract period of at least 12 years. . . . No price is stated but elaborate provisions were written providing for the division of profits between the two parties. The profits are determined and shared for the entire integrated process of producing, transporting, refining, and marketing for a minimum period of 12 years. Thus to all intents and purposes, Gulf and Shell are joined together in a long term integrated oil enterprise. . . . Under its 1933 joint ownership agreement with Anglo-Iranian, Gulf was restricted as to the markets which it could enter. . . . This restriction . . . was carried forward in the Gulf-Shell agreement, and detailed instructions were added, specifying the market territories in which Shell could distribute the oil and marketing organizations through which the oil would be sold. . . . The effect of the restrictions . . . was to limit Gulf to those markets in which *it* held a historic marketing position but to allow it, through the profit-sharing arrangement, to participate in the marketing of oil in other territories in which *Shell* held a marketing position.[14]

Division of Markets. The international oil industry has always operated on the assumption that control of world oil production was not sufficient to maintain its desired profit level—all production arrangements must be supplemented by agreements to prevent "wasteful" market competition. We have already referred to the As-Is-Agreement of 1928 between Jersey, Shell, and Anglo-Iranian. Other major producers were brought into such agreements of principle between 1928 and 1934, and subsequent entry into Middle East production was included through the joint ownerships mentioned above. From these "principles" there followed, beginning around 1934, the organization of local marketing cartels.

effectuated only after a lengthy and continuous process of negotiation and consultation by local managements in each marketing
area. . . . These local arrangements were to be governed, administered and modified in the light of current conditions by the
local officers of the international oil companies, who were
directed to meet frequently for that purpose.[15]

A detailed study of the workings of one such local cartel is
provided in the case of Sweden, where a legislative committee
investigated the industry's marketing arrangements. The
major conclusions of the Swedish study, as summarized in
the FTC study, are:

1. The subsidiaries of the internationals operating in
Sweden—Jersey, Shell, Anglo-Iranian, Texaco, and Gulf—
effectively dominated all petroleum product markets.

2. The parent corporations supervised and directed the
cooperative efforts of their subsidiaries.

3. The administration of local cartel arrangements was
carried on at regular weekly meetings of the managing directors of the companies. Uniform rebates, commissions, bonuses, discounts, and other selling terms were fixed, and were
consulted in advance before making tenders to state institutions or other large purchasers, and were assigned distribution
quotas in each product market. Over- and under-trading was
compensated for by steering or transferring customers from
one firm to another.

Profits and Profit Sharing

The overall effect of the profit-sharing plan was to keep oil
company profits in the realm of the outrageous rather than
the absurd. Between 1948 and 1960, as we mentioned at the
beginning of this section, oil company profits in the Middle
East amounted to $12.8 billion. Payments to local govern-

ments, despite profit sharing, came to only $9.4 billion, including royalties, bonuses, and other payments. Virtually none of this wealth found its way to the people of these oil-producing countries. Even the oil workers themselves had to struggle against company policies like that of Bahrein Petroleum Company (Socal and Texaco), which "has based successive increases in the minimum daily wage largely on the cost of a diet containing the requisite number of calories to keep a man and his family properly nourished. . . ."[16]

We have already referred to the remarkably low cost of finding and producing oil in the Middle East when compared with other parts of the world. Even with the large-scale investment after the war, the pre barrel cost of finding and producing oil declined in the Middle East while it rose elsewhere. By 1960, Middle East oil represented 68 percent of the "free" world's proved crude reserves, 26 percent of its production, 7 percent of its refining capacity, but only 5 percent of the industry's total fixed assets.[17]

The oil companies' share in the gross income of the industry decreased from an average of 81 percent prior to 1948 to an average of 55 percent in the 1955-1960 period. The size of the gross income on which this share is based increased from 78 percent of net assets in 1948 to 130 percent in 1958-1960. As a result, despite the increasing share that went to the host governments, the net income (after taxes and profit sharing) of the oil industry as a percentage of net assets rose from 61 percent in 1948-1949 to 72 percent in 1958-1960. This rising rate of return on capital is in sharp contrast with the low and declining return rate in Venezuela and the United States. This was due to a fivefold increase in production and a simultaneous decrease in per barrel costs by about half. These figures average in the income from refining and pipeline transportation operations which, although considerable, are not nearly as profitable as the crude production stage alone.[18]

Oil industry propaganda has never talked about the profitability of foreign, and particularly Middle East, operations. The companies prefer to talk in terms of overall operations which, they say, bring only modest returns quite in line with the profit margins of manufacturing industries as a whole (10 to 20 percent). To the extent that this is true and not a product of the "creative accounting" that the oil industry has pioneered, it indicates how much this industry—and in many ways the economies of the United States and of other developed capitalist countries—is subsidized by cheap and easy access to the resources of the Third World. Rather than dispel suspicions and criticisms of the oil industry, this broader perspective should serve as an indictment of the larger system which is so integrally dependent on economic exploitation.

Oil Concessions
and Facilities
in the Middle East,
January 1951

LEGEND

OIL CONCESSIONS:

① Anglo-Iranian Oil Co., Ltd.

② Arabian-American Oil Co.

③ Bahrein Petroleum Co., Ltd.

④ Iraq-Petroleum Co., Ltd.,
and affiliated companies.

  Asterisk indicates areas where
  I.P.C. holds exploration permits only.

⑤ Kuwait Oil Co., Ltd.

----- Concession boundary lines (approx.).

Source: U.S. Senate, Hearings Before the Subcommittee on Multinational Corporations of the Committee on Foreign Relations, On Multinational Petroleum Companies and Foreign Policy, Part V, p. 292.

Ownership Links Between the Major International Oil Companies
(Including Compagnie Française des Pétroles)
and the Major Crude-Oil Producing Companies in the Middle East

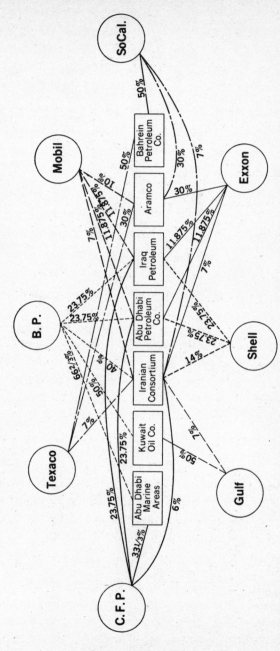

Source: U.S. Senate, Hearings Before the Subcommittee on Multinational Corporations of the Committee on Foreign Relations, On Multinational Petroleum Companies and Foreign Policy, Part V, p. 290.

4

Oil Politics and Economic Nationalism: The Background to OPEC

The 1950s were bonanza years for the oil industry in the Middle East, as is reflected in the nearly $15 billion in profit the companies took out of the area from 1948 to 1960. But these years also saw the emergence of a number of serious challenges to the economic and political power of the oil companies. In the beginning of the decade that power was enough to crush the challenge posed by Mossadeq, and it could be concluded that "underdeveloped countries with rich resources now have an object lesson in the heavy cost that must be paid by one of their number which goes berserk with fanatical nationalism."[1] It took the combined power and intrigue of the United States and Britain to stabilize the situation, as well as provide more than $1 billion over the next decade to strengthen the repressive apparatus of the Shah's military regime. Even so, the popular pressures that Mossedeq represented forced some modifications in the consortium's concession that would prove important in the next decade. The Mossadeq episode was, in a sense, a harbinger of things to come. For the oil industry, it was a great victory, but in some ways it also marked the beginning of the end.

Nationalism and Revolution in the 1950s:
From the Truman Doctrine to the Eisenhower Doctrine

In view of the critical importance oil holds for the Western capitalist world as a commodity, and for the oil industry as a source of income, the political conflicts between the Western countries and the people of the Middle East take on a clearer focus. Before the decade was out, Western governments twice resorted to large-scale armed intervention in the Middle East to secure their interests. The first was the British-French-Israeli aggression against Egypt in 1956, following the nationalization of the Suez Canal. The second was the American and British military intervention in Lebanon and Jordan, following the Iraqi revolution in July 1958. To be sure, oil was not the only interest, and in the first instance it was not the obvious issue. But there is little doubt that in both cases control of oil resources was the underlying and substantial interest.

The challenge to Western control of Middle East oil in the early 1950s was assumed in the prevailing Cold War atmosphere to emanate primarily from the Soviet Union, either directly as a military-political threat, or indirectly through subversion and nationalization. The response was appropriate to that perception. Beginning in 1950, American and British planners launched a drive, under British leadership, to form an anti-Soviet Middle East Defense Organization (MEDO). This concoction never got past the strong anti-British political sentiment in Egypt which demanded British evacuation from Suez, an obstacle which became more formidable with the July 1952 coup that brought Gamal Abdul Nasser to power. United States policy-makers saw Nasser as a nationalist but fundamentally pro-Western political force who could be supported in order to block potentially more radical elements from taking over. One of the first projects of the new Secretary of State, John Foster Dulles, was a tour of the

Middle East in early 1953. From this experience Dulles concluded that the Arab states of the Middle East were not sufficiently aware of the Soviet threat and were too close to recent Western domination to participate in a Western-sponsored military pact, and that this role should therefore fall to the northern tier states: Turkey, Iran, and Pakistan. This, of course, amounted to no more than an extension of the Truman Doctrine.

Dulles's initial perceptions notwithstanding, the United States could not refrain from trying to line up every possible ally in the struggle against the menace of "International Communism." The conflict with Britain over the military base at Suez kept Egypt out of the picture for a while, but Egypt's erstwhile rival for inter-Arab political hegemony, Iraq, then under a reactionary pro-British monarchy run by Prime Minister Nuri as-Said, proved to be a ready conscript. With the backing and encouragement of the United States, Turkey concluded bilateral mutual assistance pacts with Iraq and Pakistan, forming the nucleus of the Baghdad Pact. Iran's participation had to await the quieting of the controversy over the new consortium agreement. By August 1954, the United States had obtained rights to military bases in Spain and Libya, had secured French agreement to Tunisian and Moroccan independence, and had backed negotiations between Britain and Saudi Arabia over the potentially explosive Buraimi border dispute. The Shah's regime in Iran had survived more than a year since the CIA coup against Mossadeq. With the conclusion of an agreement between Britain and Egypt for British military withdrawal from Suez, the United States began to press Nasser to join the emerging anti-Soviet alliance, but with little success. When Egypt came to the United States for arms following the devastating Israeli attacks of February 1955, the United States insisted that the price was a full-scale military-political agreement. Nasser refused, and the Egyptian-Soviet arms deal resulted. The

United States retaliated by withdrawing offers to assist in the construction of the Aswan dam, and the confrontation reached a peak with the nationalization of the Suez Canal, and the military action that followed. By this time, the struggle between Nasser and the Iraqi regime for political hegemony in the Arab world had erupted openly, with the Baghdad Pact as the primary issue.

In the months between the Suez nationalization and the Western military attack on Egypt, the United States continued to try to balance the contradictory thrusts of its policies. Nasser's willingness to make deals with "International Communism"—even going so far as to recognize the People's Republic of China—was not only a public affront to U.S. policy, but was overwhelmingly popular among the masses of the Arab world. While Syria had long ago been written off as the heartland of Arabism, now pro-Nasser political forces were coming to the fore throughout the Arab world. In Jordan, King Hussein had been forced to dismiss the British army commander, Glubb Pasha, and elections in October 1956 saw a resounding victory for Arab nationalist forces.

The oil-rich countries were not immune either. As far away as Bahrein, the administrative center of British political rule in the Gulf, there were widespread strikes by oil and other workers in April 1956, demanding, among other things, the dismissal of the British political agent, Sir Charles Belgrave. The culprit in every case was, by Western lights, Radio Cairo and Egyptian agents. Oil worker militancy and union organizing, which had played an important role in Iran, had spread to the Arab countries. There were significant strikes in Iraq and Saudi Arabia in 1953, each crushed by the use of troops and martial law regimes, and in both cases followed by shipments of arms from the United States. In June 1956, a visit of King Saud to Dhahran sparked a spontaneous protest against Aramco. Shaken by the massive character of the demonstration and its nationalist and anti-imperialist slogans, the king ordered a crackdown on any and all trade union or political activity by workers. The possibility of another Mossadeq-type situation was compounded, in Western eyes, by the increasingly militant and popular character of Nasser's appeal throughout the Arab world. Against this background, the short-term threat to Europe's oil supplies represented by the Egyptian nationalization of the Suez Canal was overshadowed by a long-term threat: encouragement of similar moves in the oil-producing countries. For that reason it had to be dealt with resolutely.[2]

Strong United States opposition to the British-French-Israeli aggression against Egypt was primarily tactical, and should not obscure the shared objective: to isolate, and if necessary eliminate, Nasser as a political force. The United States correctly saw that direct military action would probably have the opposite effect and would enhance Nasser's prestige among the masses and increase the political difficulties of the puppet regimes in Jordan, Iraq, and elsewhere. The United States, though, did all it could to foment economic and political instability in Egypt. It froze Egyptian assets in the United States and placed an embargo on food and other nonmilitary assistance, even from private agencies like CARE. After the invasion, Dulles made a big show of pulling out of the joint committee that had been set up to deal with oil supply disruptions—formal reactivation of the committee did not take place until November 30, after British and French withdrawals had begun. In the meantime, both the French and British publicly accused Washington of "oil blackmail." This proved to be a masterful facade to avoid reprisals in the Middle East: two days after the British-French invasion, on November 8, U.S. domestic oil exports to Europe began to climb to nearly 1 million barrels per day (b/d), up from the normal 44,000 b/d. In Syria, the British-dominated IPC pipeline was sabotaged, but Aramco's TAPline was left unharmed and oil ouptut from Saudi Arabia and the Gulf was otherwise unimpeded.[3] An emerging right-wing alliance between Iraq and Saudi Arabia to counteract nationalist forces in their countries was facilitated. The stage was set for the next phase of American policy; it commenced with the Eisenhower Doctrine.

The Eisenhower Doctrine, announced in early January 1957, was primarily an exercise in psychological warfare, merely calling for the application of $200 million in economic aid that had, in fact, already been appropriated by the Congress. By defining the political struggle in the Middle East in terms of a "power vacuum" and "International Communism," the announcement invited the beleaguered pro-Western regimes in the area to ascribe their own domestic political difficulties to inimical outside forces controlled by the Soviet Union or its unwitting agent, Egypt. The political polarization of the Middle East was thus heightened. King Saud, formerly allied with Egypt against Iraq, was still recovering from the shock of the Saudi workers' uprising the previous June, particularly the general strikes in Bahrein and Kuwait, and the wave of popular support for Nasser in the wake of Suez. At the behest of Aramco executives, Saud was invited to Washington to try out for the role of a desperately needed popular political figure who could compete with Nasser for Arab hearts and minds. The difficulty of the public relations effort needed for this task could be surmised by the State Department's need to issue a statement on the eve of Saud's arrival, noting that their experts estimated the slave population in Saudi Arabia to be "fewer than 1,000."[4]

The first and only successful application of the Eisenhower Doctrine occurred in Jordan in April 1957, where the "plucky little king," invoking the shibboleth of "International Communism," arrested the popularly elected prime minister, dissolved the parliament, outlawed political parties, established martial law, and threw his "enemies" into specially erected concentration camps. The United States' contribution to this "happy ending" (in the words of the semiofficial Council on Foreign Relations version) included sizable arms deliveries, millions in economic aid, and appropriate naval and troop maneuvers on the part of the Sixth Fleet. Emboldened by the success of this putsch, American strategists moved to implement a similar strategy in Syria, first by recruiting plotters within the Syrian army. When that was exposed, they tried to provoke an incident that would permit military intervention by American-armed-and-trained Turkish troops massed on Syria's northern border. This game plan was foiled as well, in part because of strong Soviet diplomatic backing for the Syrian regime. In Lebanon, meanwhile, the right-wing Chamoun government invoked the mantle of anti-Communism to secure U.S. help in its struggle against local nationalists, and the United States responded with large amounts of aid, especially police and military equipment. After the Syrian fiasco, however, there was a clear reluctance to become more committed to Chamoun's political survival.[5]

With the evaporation of the Syrian crisis at the end of 1957, American strategists were unenthusiastic about the future usefulness of the Eisenhower Doctrine, and a new, conciliatory and cooptative approach to Nasser was being formulated in what Dulles liked to call an "agonizing reappraisal." The formation of the United Arab Republic in February 1958 was interpreted ominously by some industry sources, since on the surface it enhanced Nasser's ability to disrupt oil operations by controlling the IPC and Aramco pipelines through Syria, as well as the Suez Canal. This was offset, however, by the realization that the alternative to the UAR setup might well have been a more radical and perhaps Communist-dominated government in Syria. There followed much company and government talk about building a politically "safe" pipeline from Iraq and the Gulf north, through Turkey to the Mediterranean, bypassing Syria; but nothing came of it. More concretely, work on an Israeli pipeline (built with French financial help) from the Red Sea to the Mediterranean was speeded up and plans for a second one were drawn up, exploration was increased in Libya and Algeria, and the construction of large super-tankers for the long haul around Africa was also pushed ahead. The main political consequence of the formation of the UAR was an increased prestige and influence for Arab nationalist forces, notably in Lebanon.

When the Lebanese crisis continued into 1958 with no sign of resolution favorable to the United States, Dulles admitted on July 1 that military intervention would be "a sort of measure of last resort."[6] Scarcely two weeks later there was a military coup in Iraq, the first against a puppet regime in an oil-producing country. The "last resort" had come to pass. Obviously caught completely by surprise by the Iraqi coup, and unsure of how to minimize, contain, or even reverse its political and economic impact, United States and British strategists reacted impulsively to protect their oil interests by sending troops to Lebanon and Jordan, partly to guard against nationalist upheavals in those countries, and largely to prepare for a possible invasion of Iraq. On July 17, three days after the coup, after repeated declarations by the new regime that oil interests would not be touched and after it was clear that no counterrevolutionary political resistance was being mounted there, the United States and Britain decided against military intervention. They were then left with the problem of how to withdraw with least embarrassment from Lebanon and Jordan. Whether or not the Iraqi revolutionaries' decision not to interfere with Western oil interests was prompted by the potential military intervention, we can only speculate. If they had called the United States' bluff, an invasion might have triggered off an upheaval throughout the Arab world. The revolutionaries did stay in power, however, and in the

1960s Iraq took the lead in the Arab world in asserting popular sovereignty over its national resources. The U.S.-British military intervention demonstrated the lengths to which they were willing to go once their oil interests were directly threatened. On the other hand, their rather ignominious withdrawal served to emphasize again that the balance of political forces was changing in the Middle East and in the world, and that Western control over Third World resources could no longer be assured by military means. After July 1958, the protracted struggle for national political and economic sovereignty continued but on a less spectacular and visible plane.[7]

Independents, Joint Ventures, and Emerging Contradictions in the Oil Industry

The turbulent and often spectacular confrontations between the United States and Europe and the emerging nationalist forces in the Middle East tended to obscure related changes taking place within the oil industry during the bonanza years. The enormous oil company profits were based on a monopoly position in all segments of the industry, from production to marketing, all over the "free" world. As long as profits were based on the great discrepancy between production cost and selling price, there was a tremendous incentive for other companies to try to secure crude production assets themselves. The discrepancy between the cost and price of crude was greatest in the Middle East, where surplus production capacity was also the greatest. To move into heretofore monopolized Middle East production, smaller, integrated private and state-owned oil firms known as independents used two methods: (1) They tried to make deals that were better for the producing countries, offering them greater shares of production revenues through joint-venture arrangements instead of traditional concessions; and (2) they tried to cut prices in the crude market in order to win customers away from the majors. The gap between costs and prices, especially in the Middle East, was so great that these smaller companies could well afford to increase costs and cut prices; they would still end up with handsome profits. In addition, as relative newcomers they didn't share the conservative concern of the oil giants to preserve existing markets and profit margins.

Eight American independent companies were cut in on the Iran consortium deal, sharing a small but profitable 5 percent of that operation. The first significant industry challenge to the dominance of the American and British majors, however, came from one which was left out: the Italian state enterprise Ente Nazionale Idrocarburi (ENI), headed by Enrico Mattei, "who more than any other man of his generation left an imprint on the international oil industry."[8] Mattei wanted ENI to become at least partially independent of the oil majors as a source of Italy's crude oil. Prior to Iran's nationalization in 1951, ENI had depended on Anglo-Iranian for the bulk of its crude. Mattei remained loyal to Anglo-Iranian during the nationalization crisis and refused to provide a market for Iran's oil. Despite this, ENI was kept out of the consortium in 1954, and Mattei resolved to obtain an independent source of crude in a manner which would cause as much trouble for the international companies as possible. His opportunity came in the upsurge of Middle East nationalism following the British-French-Israeli aggression against Egypt in 1956. In Iran, the aggression against Egypt, combined with strong wellsprings of resentment against the West for imposing the 1954 consortium on them, led to the passage of the Iranian Oil Act in July 1957. This move was expediently supported by the Shah. The act transformed the National Iranian Oil Company (NIOC), set up by Mossadeq in 1951 but accorded strictly nominal ownership of Iranian oil by the consortium agreement, into a functioning arm of the state in planning and executing an independent oil policy. The act specifically authorized NIOC to enter into joint ventures with foreign concerns (to be approved by the cabinet and parliament), covering all phases of the industry from exploration and production to marketing. The first such venture, with ENI, came only a month later, in August 1957.

The Iranian Oil Act of 1957 aimed at national control of exploration for and exploitation of oil in areas outside of the consortium's concession. Under the joint-venture arrangement, Iran's national aspirations and Mattei's commercial ambitions resulted in an agreement in which the foreign company (ENI) undertook all exploration expenses, to be shared only if and when commercial reserves were discovered. Thereafter, each partner was to share in the capital expenses of producing and exploiting the oil. Since the joint-venture company was to pay half its profits (50 percent) as a tax to the Iranian government and half of the remaining profits (25 percent) to the National Iranian Oil Company, the actual government take was 75 percent of net profits—compared to the 50-50 profit-sharing arrangement under all traditional concessions. What for the Iranians was the "most progressive legislation in any Middle Eastern country," was for the oil majors "a kick in our corporate guts." When the majors had caught their breath, they proceeded to deride the agreement as an uneconomic and desperate move on the part of Mattei to sabotage the traditional concession arrangements. But when the Iranian government announced three months later that another area would be open for joint-venture bids, fifty-seven companies from nine countries applied (at $2,700 each), and the company selected was not a maverick of Mattei's stripe, but a solid old American company, Standard Oil of Indiana.

Some independents were already on the Middle East scene—notably Getty in the Saudi-Kuwait neutral zone—but the joint-venture technique hastened the process. American and Japanese private companies and European state enterprises vied for opportunities opening up in Saudi Arabia, Iran, Egypt, and elsewhere. This is significant for several reasons. First, the eagerness of the companies for deals under which they would receive no more than 25 percent of the profit indicates the size of the profits the majors were trying to protect. Second, the agreements lessened to a slight degree the dependence of the host countries on a single concessionaire and provided a small, almost symbolic, measure of national control over the exploitation of national resources. Third, these agreements broke up, again to a very slight extent, monopoly control at the production and marketing ends of the industry. Since production was no longer completely under the control of the majors, some oil was marketed in competition with the majors' crude at lower prices. As more and more companies entered this enormously profitable phase of Middle East crude production, price levels tended to erode and profit margins declined. The more competition, the greater was the threat to the monopoly profits of the industry.

The movement of independent companies into Middle East production did not cause, but only aggravated and accelerated, a process already at work. Since by 1950 postwar shortages had been made up by increased Middle East production, the chronic problem was that of surplus production capacity and, as a result, more crude oil on the market than was called for by the high monopoly prices. The monopoly structures described in the previous chapter operated imperfectly in the face of contradictory pressures on the major companies which led to surplus production and price competition. Throughout the 1950s this basic trend had been temporarily offset by a series of crises that either sharply raised demand (the Korean War) or curtailed production (Iranian nationalization, the Suez crisis). By the end of the decade, however, the fundamental nature of these contradictory forces, and the consequent weakening of the monopoly price and profit structure, were unmistakable. The imperfect means at the disposal of the companies to deal with this contradiction can be seen in the course of events that led to the imposition of mandatory import controls in the United States in 1959.

Import controls had been in effect on a "voluntary" basis in the United States ever since Middle East production had boomed after World War II. One important force behind the import quotas was the domestic independent oil producer who relied on the absence of low-cost foreign crude to maintain the artificially high prices of domestic supplies. Though there was a domestic industry whose aims and needs were not always the same as the international majors, it should not be forgotten that the international majors controlled a significant portion of domestic production and refining as well. The top eight controlled 44 percent of domestic crude production in 1960; the top twenty controlled 63 percent, and thus were highly motivated to refrain from eroding the domestic price structure. But all the giants were not in the same position. Gulf, for example, had enormous crude reserves in Kuwait and a crude deficit domestically. All of the international companies, to one degree or another, could have profited by importing low-cost Middle East crude into the high-priced U.S. market, but their own anticompetitive and protectionist impulses, combined with threats from congressmen (representing domestic producers) to impose quotas if necessary, kept imports at a tolerably low level.[9]

What we are trying to describe here is not a simple monopoly-conspiracy picture of the oil industry, but a complex set of forces working over a period of time (the 1950s) in which the monopoly character of the industry was partially eroded by contradictions within the industry as a whole, between international majors and domestic independents, and between the majors themselves. The profit per barrel of low-cost Middle East oil was a tremendous enticement to produce and sell as much as possible in order to maximize profits, a pressure only partially offset by the need to avoid "wasteful" competition for markets. As late as 1957, the monopoly position of the majors was strong enough to enact a price-hike of nearly 22 percent in the face of declining production costs. This only increased the pressure to produce more at this higher profit per barrel, led to surplus crude production, and increased imports into the U.S. market—directly by the international majors but also by smaller, independent companies who purchased the new supplies f.o.b. in the Middle East. This increase led to the mandatory import quotas of 1959.

The imposition of the quotas protected the high-priced U.S. market, but only increased the supply of crude forced to seek markets in Europe and elsewhere. Since the overall increase in crude production was more a response to higher prices than increased demand, it led to a situation of oversupply which further increased the pressure to cut prices. The pressure to lower European prices was increased again by the reemergence of the Soviet Union as a major supplier. Primarily and initially through deals with Mattei's ENI, Soviet crude was priced just low enough to compete successfully with that controlled by the majors. In the dominant Cold War atmosphere, this was portrayed in the U.S. media and in congressional hearings as a "Soviet offensive" using "cutthroat pricing" to undermine the oil industry, that bastion of "free enterprise."[10]

Soviet entry into the European market is incidental to the main forces eroding the monopoly position of the majors in maintaining market prices. It soon became clear to company executives that the 1957 price-hike was not going to hold in the face of increased production. In February 1959, and again in August 1960, official or posted prices at the Persian Gulf were cut to reflect the weakening market prices. The revenue of the producing states was directly tied to the posted price per barrel of crude multiplied by the number of barrels produced. This created no problem as long as the posted price was more or less equivalent to the selling price. The financial interest of the countries only conflicted with the companies once the actual market price was less than the posted price on which the 50-50 profit split was calculated. By lowering the posted prices, the majors were able to pay less in taxes to the producing countries, thus making them pay for the erosion of monopoly profits in the European crude and product markets. The 1959 and 1960 price cuts were undertaken unilaterally by the companies, without informing, much less consulting, the governments involved. Enraged at this arrogant action that severely disrupted their national budgets, representatives of the Arab producing states met in Baghdad with their colleagues from Iran and Venezuela after the second price cut to attempt collective action in defense of their countries' economic interests. The result was the formation of the Organization of Petroleum Exporting Countries (OPEC).

"Oil Consciousness": Prelude to OPEC

The creation of OPEC marks a turning point in the struggle of the Middle East countries to control their own resources. It marks the emergence in the 1950s of what might be called an "oil consciousness" among small but important sectors of the local populations: an awareness of the strategic and economic value of Middle East oil to the West; of the mechanisms of corporate power and control over the producing countries; and of the amount of waste and exploitation that characterized the industry in its squandering of irreplaceable natural wealth. In decades past, the governments had been totally dependent on the companies for the economic information and analyses used in formulating concession terms and revisions. As late as 1951, for example, none of the Iraqi negotiators knew anything about the function of the posted prices on which their country's revenues were to be calculated, nor did they have access to economic studies that would help them formulate a negotiating position not effectively predetermined by the company. Ironically, it was the publication of the 1952 Federal Trade Commission study on the international petroleum cartel that provided some factual basis and analytical insight for many Arab nationalists trying to understand the operations of the oil industry in the Middle East.[11]

The discovery took place in a climate of growing nationalism and had an electrifying effect on the political outlook of the Arab nationalist movement, as reflected in the pamphlets and books that began circulating in Cairo and Beirut in the mid-1950s. Up until the end of the decade, however, information and radical analysis had little effect on the policies of the oil-producing or transit states. The Arab League had set up an Oil Experts Committee in 1951, and then a Department of Petroleum Affairs in 1954, but the only concrete result was the application of an oil boycott against Israel. Attempts to formulate any sort of joint oil policy for particular economic or political objectives had to await the appearance of a small but growing number of Arabs (and Iranians, etc.) with Western technical training in petroleum engineering and economics. These people began to fill positions in the companies and in state oil ministries and organizations in the middle 1950s. The most important and influential was Abdullah Tariki, who, after training at the University of Texas, returned to Saudi Arabia in 1954, at age twenty-nine, to become the first Director General of Petroleum Affairs, initially part of the Ministry of Finance and later a separate ministry.

As the person responsible for government relations with Aramco, Tariki's dedication to work—to educating the public about oil affairs, to improving the terms of the concession and opening up positions for Saudis in the company management, to using the country's wealth for the benefit of all the people—made him stand out in the Saudi regime, where venality and corruption were the rule, where no administrative structure existed, and where power and influence depended on one's lineage in the Royal House of Saud. Tariki placed a great deal of emphasis on educating the public about oil matters, and published a weekly newsletter from his office to further that aim. He defined his goals with regard to Aramco as three: first, he pressed for greater Saudi control of Aramco affairs at all levels; second, he pressed for Aramco to become an integrated company, so that Saudis would eventually control not only oil-producing facilities, but transportation, refining, and marketing facilities as well; third, he moved immediately to get new and more favorable concessions on territories Aramco would be forced to relinquish. His first and third aims were partially, although minimally, met during his tenure. He oversaw the placement of Saudis at middle levels of management where they would acquire the experience necessary for eventual takeover. He negotiated a new concession with a Japanese company in late 1957 that provided improved profit terms and Saudi participation in an integrated venture, but the integration provisions were never implemented. As far as his second aim of integrating Aramco operations was concerned, it was impossible given Aramco's role as a producing subsidiary for four American giants. By raising this impossible demand, however, Tariki helped to educate others about the inherently contradictory interests of the oil companies and the producing states.

Tariki realized from the beginning that real changes in company-government relations could only come about through concerted action by the producing countries. As early as 1957, he was proposing that any new pipelines be constructed by a joint-stock company in which all the Arab states would own shares. Around the same time he began pressing for the convening of an Arab Oil Congress under the auspices of the Arab League. Support for this idea first foundered on Iraq's opposition to a session that would likely be politically dominated by Egypt, and then was postponed until after the turmoil occasioned by the Iraqi revolution and the American-British invasion. The First Arab Petroleum Congress was finally held in April 1959 in Cairo, despite the last minute decision of the new Iraqi regime not to participate. Observers were present from Venezuela, Iran, and Kuwait (then British-ruled and not a member of the Arab League). The only significant oil-producing country that was a full participant was Saudi Arabia, whose delegation was led by Tariki.

The congress was the first concrete attempt to formulate a joint oil policy that would transcend the political rivalries of the states involved. The presentations and arguments were models of moderation, technical precision, and apolitical astuteness. The only real controversy was prompted by the paper presented by the American advisor to the Saudi oil ministry, which argued that a producing state has an inherent right to modify concession terms unilaterally when it is manifestly in the public interest to do so. This paper fully represented Tariki's views, of course, but coming from the mouth of an American, it represented to the companies a more serious threat for which Tariki could not be held directly responsible. Beyond this, the real issues and controversies of the day, notably the first series of price cuts in 1959, were discussed privately and off-the-record by delegates and observers alike. This communication and contact unquestionably facilitated the convening of the special meeting of oil ministers in Baghdad in September 1960, after the second price cuts, the meeting from which OPEC emerged.

Many controversial and radical approaches to the oil industry that were noticeably absent from the First Arab Petroleum Congress were reflected in the second congress of the Federation of Arab Labor Unions held shortly afterward in Cairo. The resolutions of that congress included demands that: (1) Arab oil wealth be considered the property of the whole Arab nation; (2) Saudi Arabian oil revenues be allocated to meet the needs of all the Arab people; and (3) oil concessions be reconstituted to reflect current national aspirations. While the oil congress was a model of civility and moderation in dealing with the companies and its resolutions were much more general and much less threatening, the oil companies could ignore only at their peril the popular attitudes and political sentiments expressed by the demands of the labor unions congress and other openly political movements. Equally threatened by the prospect of a more equitable distribution of wealth and power were the regimes of the oil-producing states themselves. The political transformation of the Middle East during the 1950s was characterized by increasingly radical popular demands regarding the future of the oil industry that threatened companies and governments alike. There could be no question of whether things would change, but rather what forces would dominate and control those changes and how extensive they would be. This was the setting for the birth of OPEC.

5

Middle East Oil
in the 1960s

The story of Middle East oil in the 1960s is, in part, the story of OPEC and the struggle of the oil-producing states to achieve some measure of political and economic control over the industry that dominates the political economy of the area. Underlying this broad struggle, and fundamental to understanding the role of OPEC and its successes and failures, was the continuing conflict among states of the area (producers and nonproducers alike) for political hegemony. These conflicts resulted in the protracted military-political confrontation between Saudi Arabia and Egypt, growing out of the 1962 Yemen revolution, and in the Arab-Israeli war of June 1967. The United States involved itself directly in both of these conflicts. The role of "friendly" oil-producing states like Saudi Arabia in supporting U.S. objectives was manifest in what OPEC was able to achieve (or, more to the point, what it was not able to achieve) vis-à-vis the companies. Finally, the picture of Middle East oil in the 1960s is not complete without a discussion of the various forces at work in the world oil industry, and of the political and economic nationalism in the Middle East, both of which affected the continued profitability and expansion of the giant firms in the Middle East and the world.

OPEC's Early Years

The countries that joined OPEC first were Saudi Arabia, Kuwait, Iraq, Iran, and Venezuela. The organization was created with the express purpose of giving these chief oil-exporting countries the collective strength to restore posted prices on crude oil to the levels set before the 1959 and 1960 price cuts. This would provide some measure of stability and predictability to the revenues on which they were all highly dependent. Abdullah Tariki, then Oil Minister of Saudi Arabia, emerged as the leading spokesman for the Arab oil countries and was a prime mover of OPEC in its early days. Tariki was outspoken in his "extreme" views, which obviously included eventual nationalization in his own country. His able performance in a debate with oil industry representatives at the Second Arab Oil Congress in Beirut in October 1960, one month after the founding of OPEC, firmly established him as the hero of the hour and helped generate popular political pressure for the oil countries to support OPEC. The companies, meanwhile, refused to acknowledge OPEC's existence and contemptuously referred to it as "this so-called OPEC," a tactic with which they persisted for several years.[1]

OPEC's initial declaration on price restoration was not followed up with a concrete policy to achieve that objective until the Fourth OPEC Conference, two years later. Even then, by opting for a tactic of negotiation with the companies rather than unilateral price action, and by agreeing to company demands for individual country-company negotiations, the OPEC countries discarded their joint political and economic strength in favor of relying on the civility, rationality, and sense of fair play of the companies. Civility was cheap and available in quantity. Rationality and fair play meant convincing the countries involved that their revenues should depend on the mythical free market. As a result, 1958 posted prices were not restored until 1971, under a whole new set of political and economic forces.

Even though OPEC failed in its endeavor to restore posted prices, and in effect helped to diffuse the popular political momentum that had led to its creation, the organization, merely by its continued existence and potential power, did succeed in preventing any further cuts in the face of continually declining oil prices on the European market. This resulted in a slowly growing gap between the now-stable posted price on which the governments' 50 percent share was calculated, and the declining price for which the companies actually sold the crude. In other words, the official or posted price quickly came to function as a tax-reference price only and, in the face of declining market prices, increased the governments' share. By 1963, for example, Persian Gulf posted prices ranged as high as $1.80, which after calculating deductions for operating costs and assorted company allowances and discounts, left a pro-forma profit of $1.33. Half of this, or 67¢, went to the government. The actual selling price in 1963, though, was not $1.80, but more like $1.59. For the company, the cost is 67 ¢ in taxes and 10 ¢ in operating costs. The company's profit at the producing level, including the allowance and discounts noted above, comes to $1.59 minus 77¢ , or 82¢ per barrel. This is still more than 50 percent, but it would have been higher by 11¢ if posted prices had been reduced with realized prices. By 1969, the market price had fallen to a low of $1.20, while posted prices remained unchanged. This resulted in an actual division of profits more like 62-38 than 50-50, in the producing countries' favor.[2]

Given the history of oil companies' complete control of pricing and their propensity to manipulate producers and consumers in order to maximize profit, OPEC's prevention of further cuts was not negligible. But there was still no formal recognition of OPEC's existence on the part of the companies, or of the principle of collective bargaining. The industry's recognition of OPEC as a political force was indirect. They did not formally surrender their administrative discretion with regard to prices. They simply never again tried to lower them directly.

The second phase of OPEC's campaign to secure better arrangements for the producing countries involved, first, the elimination of various allowances and discounts by which the companies secured greater profits by calling them something else, and second, pressing for acceptance of the principle of royalty expensing. Prior to the so-called profit sharing arrangements begun in 1950, payments to host governments were in the form of royalty per barrel, usually 12½ percent of posted price. This was customarily treated by the companies as a necessary production cost in calculating profits. With the 50-50 deals, the companies calculated the royalties as part of the government's 50 percent, thereby making the government's actual share much less. The royalty, in effect, became a payment to the companies rather than the countries. In the words of the first Secretary-General of OPEC,

> either the . . . companies are paying income tax at the full rate prescribed by law, but no royalty; or they are effectively paying royalty but their income tax payments amounted to about 41 percent of income and not 50 percent.[3]

Eventually the companies agreed to the principle of expensing royalties: that is, treating them as a cost to be deducted before profits are calculated and divided, thus increasing the actual government take per barrel of crude. Along with the elimination of other company allowances, the government share of profits rose to 74¢ per barrel in the late 1960s, plus the royalty, bringing total government revenue per barrel to around 85¢, compared with company profits of 30-40¢ per barrel. Although these advantages won by OPEC were not dramatic or highly visible outside the industry, they were financially significant for the producing countries. They were, however, won at a political price that cannot be ignored when evaluating the overall achievement of OPEC in this period. We can begin this evaluation by briefly considering the negotiations on royalty expensing.[4]

In the first place, there turned out to be a significant gap between OPEC's original position, calling for a fully expensed royalty rate of 20 percent, which would have resulted in about 20¢ additional government revenue per barrel. The final settlement stipulated a royalty of 12.5 percent, the expensing of which would be diluted by discounts off posted prices, bringing the governments' additional revenue to less than 5¢ a barrel. Of more consequence was that in the course of the negotiations OPEC first set aside its demands for posted price restoration to concentrate on the royalty negotiations. Then it agreed to the posted price discounts in the royalty agreement and effectively subverted those price demands by implicitly accepting company claims that market conditions would not allow for any price increases.

The second part of our evaluation concerns the negotiating practice of OPEC. Negotiations were allowed to drag on interminably, in a way which left the companies with the initiative and the countries only with the option of accepting or rejecting company offers. There was no real negotiating strategy used by OPEC beyond repeated declarations of its "sense of responsibility and good faith" instead of "being a menace and a threat to the security of the international oil industry."[5] OPEC repeatedly went along with company ultimatums and deadlines, and just as readily waived its own deadlines whenever the companies made some last-minute token gestures of accomodation. This policy of capitulation was capped by the decision of the Seventh OPEC Conference, in November 1964, to leave final settlement to individual country-company negotiations, thus forsaking the very principle of collective bargaining in a way that left Iraq, in particular, alone in its attempt to assert national sovereignty over its oil. One can only conclude that "the principle of negotiation" and a rather small increase in government revenues were secured at the cost of severely weakening the "principle of collective bargaining" and reducing OPEC's political credibility to near zero among the masses of the oil-producing countries and the companies as well. The question remains, after OPEC's promising beginning in the fall of 1960: why?

The Politics of OPEC

"Oil has aroused certain emotions in the Middle East which must be satisfied, and this is OPEC's task," wrote the *Middle East Economic Survey* in January 1964. This candid appraisal of OPEC's role by a source with close ties to both oil companies and producing-country governments contrasts with the repeated declarations of the companies and most of the governments that oil is not a "political commodity," and that "politics" should not be allowed to influence purely rational and scientific decisions about the industry structure, the role of the producing countries, and so on. By 1964, as the "apolitical" direction of OPEC became manifest in the negotiations over royalty expensing, a body of informed and politically conscious public opinion emerged that was sharply critical of OPEC's timid and capitulatory approach to the giant oil companies. These attacks were led, ironically, by the man who, along with Perez Alfonso of Venezuela, had been responsible for OPEC's creation, the former Saudi Oil Minister, Abdullah Tariki.

The political situation in Saudi Arabia in the late 1950s was anomalous. The political struggle for power between brothers Saud and Faisal had left a good deal of political space for a committed and capable man like Tariki to educate and mobilize public opinion in his own country and throughout the Arab world. The struggle had been resolved in 1962 when Crown Prince Faisal staged a palace coup with the support of the United States and regained his position as prime minister. One predictable result was the dismissal of Tariki, his expulsion from the country, and his loss of a direct role in the political evolution of OPEC.

The forces at work in the world oil industry supply some explanation for the contrast between OPEC's meek performance and the militant expectations that accompanied its birth. The surplus crude production and weak market prices that accompanied industry expansion in the late 1950s made the companies more indispensable to the oil producers in maintaining stability of markets and prices, and preventing the "wasteful competition" that would result from increased production. This interpretation was largely accepted by OPEC personnel and was supported by the reality of slowly declining prices in the European and other markets. This surplus productive capacity could be (and was) used by the companies to pressure any producing country to come to terms by simply cutting back production in that country, and thus tax revenues, in favor of a more compliant or "realistic" producer. OPEC continually accepted this understanding, and acted accordingly. As an explanation, however, it is rather circular: obviously most, if not all, of the surplus crude was being produced by OPEC members, and specifically the Middle East members. Collective action with regard to prices was seen from the beginning by people like Tariki to require collective action on production. This would entail an agreement among all or most of the producing countries to limit production expansion in the member countries according to an agreed-upon schedule that would eliminate surplus production, strengthen prices, and increase the collective political power of the oil-producing countries. The failure of OPEC to agree on this kind of international prorationing program, as advocated by Tariki and others, must be seen as part of the problem rather than as the answer.

Part of the explanation of OPEC's political character lies in the role of the emerging group of Middle Eastern oil technocrats—the professional engineers, technicians, and lawyers trained in the intricacies of the industry—who were dominant in creating OPEC and shaping its objectives and techniques. Tariki was the most prominent of these, but his strongly held Arab nationalist political beliefs made him somewhat unrepresentative of the group as a whole. These men moved into important administrative positions in their countries in the late 1950s and their expertise and judgment has affected oil policy since. For many of these technocrats, if not for Tariki himself, one of the motives for creating OPEC was to preempt the creation of an oil organization under the auspices of the Arab League which would then be under the political influence of Egypt in particular as well as the Arab nationalist ideology then in the ascendancy in the Arab world. Tariki perhaps naively saw no contradiction between the formation of an exclusive organization of oil-producing states and the possibility of achieving national sovereignty over oil resources as demanded by the popular political movements in the Arab world from which OPEC strove mightily to isolate itself. The moderation and "apolitical" approach of most of the oil technocrats was based partly in the divisive character of Arab politics, with conflict between progressive and reactionary states, and even among the progressive states (e.g., Egypt and Iraq), in the struggle for political influence and hegemony. These differences would have to be surmounted in order for them to deal with the companies from a position of strength. In addition, though, there is a conservatism characteristic of the technocrats that is rooted not so much in political realities as in their Western training which inculcated the sovereignty of a scientific, rational, and supposedly apolitical and nonideological approach in dealing with economic and political problems. In a political spectrum defined in terms of "traditional" versus "modern" the technocrats invariably saw themselves in the political vanguard. Only later did some of them realize that "removing oil matters from the realm of ordinary politics" was in itself a highly political act which best suited the reactionary and conservative oil-producing states by "depoliticizing" oil policy, the one area in which they were most politically vulnerable.[6]

It is possible, however, to overemphasize the political character of the newly emerging technocrats. Though the concepts of cultural and intellectual dependency correspond to a reality which should not be overlooked, in the end the technocrats could not function except as agents of national governments—as representatives of regimes who owed their power in most cases to the oil companies and the Western governments. They perceived any strong political movement directed at weakening the oil companies as a direct threat to themselves. For the conservative oil-producing regimes, the goal was the establishment of stronger political and economic links with the companies, along with a "more equal" role in this partnership. For some of these countries, like Saudi Arabia, OPEC served as a means of temporarily defusing Arab nationalist political opposition, particularly the outside opposition. For others, like Iran, participation in OPEC was necessary in order to prevent Arab nationalist forces like Iraq, or individuals like Tariki, from determining its political role and potential. The conservative character of OPEC is thus rooted in the material interests of the oil-producing states rather than in the special characteristics of the oil technocrats. The maximization of technocratic influence was limited by OPEC's inability to surmount the determining influence of its most conservative members. This led OPEC in the 1960s to pursue diversionary objectives with desultory tactics.

In the most significant and important struggle for control of Middle East oil in the 1960s, Iraq was isolated politically by the other oil-producing states both inside and outside of OPEC. The power of the oil companies to subvert and contain the attempts of a people to gain control of their resources was based in no small way on their alliance with the reactionary regimes in the area. The companies' ability to play off one producing country against another, and to significantly affect the economic and political stability of a country, is seen in contrasting the policies of the companies, in alliance with Iran and Saudi Arabia, toward Iraq.

The Struggle for Iraqi Oil

Even before the revolution in Iraq in July 1958, there were many areas of conflict between the government and the Iraq Petroleum Company consortium that controlled the country's entire actual and potential resource base. The initial conflicts in Iraq (as in the other producing countries) centered around the cost accounting practices of the company in calculating the profits to be shared with the government. Kassem's 1958 revolution temporarily interrupted contacts between the government and company representatives as Kassem hastened to assure the United States and Britain that the new regime would not nationalize the oil fields. In 1959 and 1960, the posted price reductions (referred to earlier) added another dimension to the conflict. The ability of the governments of Iran and even Saudi Arabia to enter into new and better deals with the companies through joint ventures on territory not under concession accentuated the fact that in Iraq the IPC group controlled virtually the whole country, thus leaving no areas on which new agreements might be made.[7]

With the new regime in power, negotiations between IPC and the government resumed in 1959. One of the prime targets for the regime was IPC's relinquishment of 60 percent of its total concession area so that new joint-venture deals might be worked out. Other demands included a doubling of output and the construction of refineries in Iraq. The companies agreed in principle to doubling output "depending on market conditions."[8] They also made it clear that they would not set up refinery installations in Iraq, nor relinquish anywhere near the entire 60 percent of the concession territory demanded by the government. They also denied the government any significant role in choosing the areas to be relinquished.

Before negotiations between IPC and Iraq resumed in the summer of 1960, another conflict emerged: a dispute over the government's attempt to raise cargo dues in the port of Basra. After undertaking a comparative study of port dues in Kuwait, Saudi Arabia, Bahrein, and Iran, the port authority announced an increase from 23.4 fils to 280 fils per ton for oil and other cargoes. The 23.4 fils rate had been set by an agreement in 1955 between the oil company and the Port Administration (run by British personnel), and ratified by the British-dominated cabinet under Nuri as-Said. IPC's response was to halt production from one field and curtail production in another. It rationalized its decision on economic grounds, asserting that higher costs made Basra oil noncompetitive. Since the oil industry has never based production decisions on the logic of competitive pricing, this meant that the international companies which owned IPC would simply turn to cheaper sources of supply in countries where their profits would be greater. This was July 1960. The next month came the second posted price cuts. The government went into the August negotiations with a set of far-reaching demands, old and new, which they were more than ever determined to achieve. Among them was the demand that the government be given 20 percent participation in the ownership of IPC and a position in the executive directorship of the company. Both of these provisions had actually been written into the old concession but were circumvented by the companies' tactic of maintaining IPC as a private rather than public corporation.

Negotiations continued intermittently for over a year before they were finally cut off in October 1961. While the internal stability of the regime and its open hostility to Nasser's Egypt pushed Kassem into taking a progressively more demagogic attitude toward the company, there is little doubt that the company shared a good part of the blame for the breakdown in negotiations and the long years of hostility that ensued. The company attitude as reflected in the negotiations was one of determination to maintain full and unhampered control of every aspect of the industry in Iraq, including production, pricing, industrial development, intermediate fees, and concession territories. One is forced to conclude that company behavior indicated a decision to make an example of Iraq and that there was a strong political flavor to this decision. Although ultimately Kassem was to have little success in uniting either the Iraqi people or the Arab world, the oil companies had every incentive to insure that he did not. The companies, in addition, felt themselves to be in a position of strength owing to the world crude surplus developing over the previous years. They all held other sources of supply and any decline in Iraqi production would only help to strengthen market prices.

While Iraq was not the first or only radical nationalist Middle East state to emerge in the 1950s, it was the only major oil-producing state to do so. Iran was back under the friendly, if not yet stable, rule of the Shah. Saudi Arabia was still under the rule of the reactionary and dissolute King

Saud, and in 1962, this rule was strengthened by the accession of Faisal. The main perceptible threat to royal rule in Kuwait came from Iraqi claims to that country following British withdrawal in 1951.

By contrast, Kassem's reassurances to the oil companies following his coup were strongly offset by his anti-imperialist rhetoric and more importantly, Iraq's formal withdrawal from the Baghdad Pact and simultaneous economic and technical aid agreement with the Soviet Union in 1959. In quick succession, Iraq withdrew from the sterling bloc, ordered British air force units out of the Habbaniya base, and cancelled the Point Four Agreement with the United States. Internal power plays led Kassem to give prominent, if temporary, cabinet roles to "Communist sympathizers."

Kassem closed the futile negotiations in October 1961 by announcing that the companies could continue to exploit existing wells as they wished, but went on to say: "I am sorry to tell you that we will take the other areas according to legislation we have prepared, so that our action will not be a surprise to you. Thank you for your presence here."[9] Two months later, the government issued Law 80, under which the companies were permitted an area of exploitation limited to little more than their existing facilities, or 0.5 percent of the original concessions. All previous rights in 99.5 percent of the concession area were withdrawn and assumed by the government. The lengthy explanatory statement that accompanied the promulgation of this law showed the extent to which abiding historical resentment of the colonial system that fostered the original concessions was as important as the immediate controversy in leading to expropriation.

The companies rejected the new law and demanded that the dispute be arbitrated. They retaliated by holding down production even though the investment and expansion program initiated in 1959 was largely completed. Iraq's production for 1962 increased by only 0.5 percent. By contrast, production in Kuwait, Iran, and Saudi Arabia, where each of the companies held concessions, increased by 11.5 percent, 12 percent, and 9.2 percent respectively. Furthermore, the government failed to get any bids from other companies on the expropriated area or offshore. By February 1963 there were indications that the regime was considering the arbitration the companies demanded, but in that month Kassem was killed in a new coup. The new Ba'ath regime moved toward negotiation, but made it clear that Law 80 was irrevocable. In February 1964, the Iraq National Oil Company (INOC) was established to facilitate state exploitation of the expropriated areas. The companies interpreted this as a further violation of their concession rights, but entered into negotiations with the Aref regime (which had replaced the Ba'ath in November 1963) that resulted in a draft agreement in June 1965.

Under the proposed settlement of June 1965, the companies would be granted a further 0.5 percent of the original concession area as permitted by Law 80, including the rich North Rumaila field which IPC had discovered but failed to exploit. Other disputes were settled along the lines being worked out in other countries through the OPEC negotiations. Minimum production quotas were established, and IPC agreed to enter a joint venture with INOC to explore and develop a large portion of the expropriated area. The conventional wisdom at the time viewed this as a reasonably good deal for Iraq and certainly the best it could expect under the circumstances. Nevertheless, there was sharp criticism of the proposed agreement, notably from Tariki, by then an oil consultant in Beirut.

Political instability was followed by a shift to the left in Baghdad. By the summer of 1966, the most recent cabinet was publicly questioning the desirability of the proposed agreement. An extensive dispute between IPC and Syria over the pipeline from Kirkuk to the Mediterranean (the outlet for two-thirds of Iraq's oil) erupted in 1966, further disrupting any movement toward an agreement. The June War of 1967 radically changed the whole situation. In the climate following the war, the Iraqi government issued Law 97 in August 1967, assigning to INOC the exclusive right to develop oil in Iraqi territory except that which IPC retained under Law 80. Law 123, issued in September, reorganized INOC as the base of the national petroleum industry that would serve as a core of the country's future industrialization. IPC, of course, protested this new legislation as a further infringement of its rights under the old concessions and threatened to prevent the sale of any crude from the expropriated area by legal means. The Ba'ath Party returned to power in July 1968. In June 1969, INOC entered into an agreement to develop the North Rumaila Fields with the Soviet Union which, in the words of one expert, "constituted the most significant development in the recent history of the Middle East oil industry."[10]

The basic strength of the companies resided in their unfettered ability to control production under the concessionary arrangements. Between the years 1961 and 1968, production in Iraq increased by only 165.3 million barrels per year, while the comparative figures for Iran, Kuwait, and Saudi Arabia were 600.5, 323.3, and 574 million barrels respectively.[11] The companies could adjust output among their several concessions in a way to best meet financial and political objectives. An individual country like Iraq, on the other hand, was totally dependent on the production decisions of the companies for its main source of government and development revenue. There can be little doubt that the financial constraints caused by the small increase in production in Iraq over these years contributed to the overall political instability of the country, and certainly prevented the various regimes from more substantial undertakings in the areas of land reform and economic development.

This is in sharp contrast to Iran, where the companies were most accommodating in increasing production when the Shah's regime was most dependent on increased revenues for its very survival. By 1962 even the Shah's nearly total control over the governmental, political, and military-security apparatuses was no match for the widespread disaffection and dissent among the population, especially the peasants. Iran's stability was rapidly deteriorating; the government was on the verge of bankruptcy, owing to widespread corruption of the ruling family and its lackeys. The country was on the verge of revolution. The Shah responded with his "White Revolution." Oil revenues jumped from 45 percent of total government revenues in the 1958-1962 period, to 50 percent in 1964, and 55 percent in 1967. Oil revenues were even more essential for foreign exchange, needed to finance the increased commodity imports that nearly doubled from $485.6 million in 1963-1964, to $967 million in 1966-1967. Special payments from bonuses by new offshore concessionaires and increased revenues through the OPEC negotiations in 1964-1965 rescued the country from severe foreign exchange crises in 1963-1964 and again in 1967-1968. These revenues were crucial to permit a liberalized commodity import program needed to keep the loyalty of the upper and middle classes.[12]

U.S. Policy in the Middle East in the 1960s

United States policy in the Middle East in these years, as before, aimed to protect U.S. strategic and economic interests, both centered on oil. The Soviet "threat" was used to justify the heavy dispensation of military aid to the northern tier countries, Greece, Turkey, and Iran. The main threat to American interests, however, was perceived to be the radical nationalism espoused by Nasser and the Ba'ath. The decade opened under the Kennedy administration with a "new" approach to achieve old goals. Kennedy himself made an attempt to establish a friendly personal relationship with Nasser and used the foreign aid program, particularly grain shipments, "to persuade Nasser to concentrate on making progress at home rather than trouble abroad." At the same time, Kennedy moved to institute what Israeli Prime Minister Ben Gurion called a "special relationship" with Israel in the first significant direct arms deal with that country, involving Hawk antiaircraft missiles.[13]

Kennedy did find the occasion to intervene militarily and politically in the Middle East with the coup in Yemen in 1962. As the civil war in that country developed into a struggle between Egypt and Saudi Arabia, the United States tried to pressure Nasser to withdraw his forces from the Arabian Peninsula and sent a squadron of USAF jet fighters (with U.S. pilots) to Saudi Arabia to prevent Egyptian incursions on Saudi airspace. Although military aid was stepped up, with the delivery of fighter planes and tanks, the largest quantity of military shipments and assistance came from Britain, following a magnanimous gesture on the part of the United States to allow the British to earn some badly needed foreign exchange. Joint military maneuvers were planned and carried out. The aim of the United States was to prevent the Yemen conflict from spilling over the border into Saudi Arabia at a time when internal dissension in the Saudi ruling family was at an all-time high.[14]

The struggle within the Royal House of Saud between King Saud and Crown Prince Faisal was brought to a head by the Egyptian military presence in Yemen. In 1962, Faisal maneuvered his way back into power as prime minister while Saud was whisked off for "medical treatment"; Saud was formally deposed in 1964. In the course of this struggle, the other princes aligned themselves with one side or the other. An attempt to form a third force of moderate and modernist princes got nowhere; some defected to Cairo. Faisal coopted some of their demands, such as sharply reducing the royal household budget which had been eating up the lion's share of the oil revenues, but by and large he suppressed their activities. More militant antiregime activities, including sabotage and propaganda by an Egyptian-backed Arab Peninsula People's Union, increased around 1964, but were met with a series of public beheadings. By 1965, a measure of repressive stability had been restored to Arabian Peninsula politics. Egyptian and republican forces had been stalemated in Yemen. Continued opposition to Faisal's rule resulted in several purges of the army and air force over the next several years, as late as 1969. By neutralizing and suppressing political conflict within the royal household, and by building up a modern police state apparatus to control any political activity, King Faisal was able to enforce the description of Saudi politics he had offered at a press conference in January 1960:

> We all know that our country is different, in its situation and its condition, from all other countries in the world. We have, praise be to God, no political parties, or sources of friction to menace our unique position. . . .[15]

United States satisfaction with Saudi developments was expressed in the conclusion of a report of a delegation from the House Foreign Affairs Committee issued in April 1965:

> After a generation in which oil revenues were not turned to proper account, Saudi Arabia, under King Faisal, now appears to be assuming a responsible attitude toward economic development. Strenuous efforts are being made to diversify the economy.[16]

Such efforts included a countrywide mineral search conducted by the U.S. Geological Survey and Army Corps of

Engineers, at Saudi expense, for the purpose of attracting even more foreign investment.

United States solicitude for the continued well-being of the House of Saud was gratefully acknowledged by Saudi oil policy, both inside and outside of OPEC. Saudi oil policy steadfastly supported the position of the international oil companies, while occasionally increasing the price of that support by way of financial demands on Aramco. In the case of Iraq's struggle for economic self-determination vis-à-vis the Iraq Petroleum Company, the reactionary Saudi role was expressed by Oil Minister Yamani in October 1965, when an Iraqi government-IPC agreement looked promising:

> Certainly we will ask for the same from Aramco, and I have already told them as much unofficially. There would, of course, be various ways of going about the matter. . . . But the main thing is that Aramco should be prepared to give Saudi Arabia what IPC has given Iraq.[17]

Obviously, the way to support the Iraqis, had that been Yamani's intention, would have been to make equivalent or farther-reaching demands on Aramco, independent of the outcome of the Iraqi talks. Instead, Yamani found for Jersey Standard and Mobil (Aramco parents with substantial interests in Iraq) every incentive to avoid making concessions in Iraq. A year later, Yamani gave this policy a more general formulation:

> I believe that we in Saudi Arabia have set an example worthy of emulation as regards the establishment of a truly fruitful relationship with the oil industry, sustained through cooperation and negotiation. I should like to differentiate between the right to take unilateral action and the actual exercise of this right. There is no need for Saudi Arabia, for the time being, to think of taking unilateral action. It is against our general philosophy of doing business and inconsistent with the friendly atmosphere which characterizes our relation with the oil companies at present. And I am sure that the oil companies operating in Saudi

Arabia have no interest whatsoever in shaking our faith in this philosophy by showing us that other means are more rewarding in safeguarding our oil interests.[18]

These sentiments were expressed in policy as well as in Yamani's words. Saudi Arabia sabotaged all efforts to establish a joint Arab oil policy—even on so innocuous a level as setting up an Arab Oil Company for the sole purpose of using the small amounts of royalty oil available to gain some experience in marketing and other "downstream" phases of the industry. In 1966, Yamani went so far as to decree that all papers submitted to the Sixth Arab Petroleum Congress by Saudi nationals or persons working in Saudi Arabia would be screened, and any paper "below the minimal standard" would be disqualified. One Tariki was enough, even in exile. Saudi "standards" were also brought to bear on any attempts within OPEC to strike a more demanding posture vis-à-vis the companies. One of Yamani's sidekicks, Muhammad Joukhdar, was Secretary-General of OPEC in 1967, and delivered a talk to institutional security analysts at the Wall Street Club in New York aptly entitled "OPEC as an Instrument of Moderation."[19]

Kennedy's attempt to establish close ties with Egypt had dissolved under the Johnson administration in the hostile political environment engendered by U.S. intervention in Vietnam, the Dominican Republic, and the Congo. In 1965, the United States moved to supply heavy armaments to Israel on a regular basis. Although internal stability in Saudi Arabia seemed to improve under Faisal after 1964, U.S. policy-makers were alarmed at the leftward shift in Arab politics toward the radical nationalism expounded by Nasser and the Syrian and Iraqi regimes. The radical oil technocrats, led by Tariki, were gaining a wider audience in the face of popular disillusionment with OPEC, and they used events like the Arab Petroleum Congresses to sharply attack Saudi and other OPEC spokesmen, using the same technical jargon. In

April 1965, Syria publicly attacked the sheikhdoms, calling the oil there the "rightful property of Arab people as a whole." According to Eugene Rostow, who served as Under Secretary of State for Political Affairs in the Johnson administration and was primarily responsible for the Middle East, the administration commissioned a secret, high-level policy study under the direction of former ambassador to Iran, Julius Holmes. Completed in early 1967, the study concluded that

> the rising tide of Soviet penetration, and the trends in Arab politics which that penetration encouraged and fortified, threatened major American and allied interests in the region . . . and that a continuation of the process, which could involve the Nasserization of Jordan, the Lebanon, Libya, Tunisia, Morocco, Saudi Arabia and the Persian Gulf, would present . . . a security crisis of major, and potentially catastrophic proportions.[20]

The June War of 1967 was of tremendous consequence in terms of changing the balance of political and economic power in the Middle East. Apprehension over the potential for widespread disruption of oil industry activities, as a result of popular outrage at the role of the United States, was well expressed by the *Middle East Economic Survey* just before the actual fighting erupted:

> If the worst comes to the worst, the present structure of the oil industry in the Arab world might be irreparably shattered, leaving a politico-economic vacuum to be filled by the Soviet Union and "neutral" western countries such as France and Italy.[21]

The immediate aftermath of the war seemed to bear this out. In Saudi Arabia there were demonstrations and strikes in the oil fields. Numerous Palestinian workers were deported and there were over 800 arrests. A strike at Aramco on June 25 resulted in some damage to American property. Radio Baghdad broadcast an appeal of the Iraqi oil workers' union calling on members to guard installations against spontaneous sabotage. In Libya, the oil workers' union kept the facilities there shut down for more than a month. Most of the Arab countries, including Saudi Arabia, suspended all oil shipments for about a week beginning June 7, and then resumed shipments but with an embargo on the United States and Britain. The oil companies diverted all available tankers to Iran, where output was stepped up to the maximum. United States Defense Secretary McNamara announced contingency plans to keep oil moving to U.S. armed forces in Vietnam in the event of a continued shutdown in the Persian Gulf. Abu Dhabi, where popular demands to join in the brief boycott had been ignored, finally shut down operations on June 11, after an explosion wrecked a British bank there. Abdullah Tariki proclaimed that this was the "chance of a lifetime" to nationalize Western oil interests.

By the end of the month, however, Arab oil operations were back to 50 percent of normal, with the main exceptions being Iraq and Libya—where oil workers continued a strike despite attempts by the government to renew operations. Saudi Arabia, though, was already complaining that the boycott of the United States and Britain was costing more than it was worth. Formerly moderate technocrats in Syria and Iraq began pressing more strongly for nationalization. Iraq passed a law which unilaterally resolved the dispute with IPC over the expropriated fields by authorizing the state company, INOC, to undertake production in the prized North Rumaila fields.[22]

Once the turmoil and shock of the Arab defeat in the June War had died down and decisions had to be made, only Iraq and Algeria among the oil producers, supported by Egypt and Syria, pressed in high-level Arab political meetings for strong and far-reaching measures that would affect the oil industry. After a mid-August meeting in Baghdad of all Arab ministers of finance and oil, Iraqi nationalization proposals were not even mentioned in their communiqué, which the *Middle East Economic Survey* characterized as "laconic."[23] The conference recommended the establishment of an Arab Petroleum Organization, long a goal of the nationalists, but this one would be restricted to those Arab states where oil was a major source of revenue. It also called for the creation of a special aid fund, pushed by Kuwait, and for higher Mediterranean postings by Saudi Arabia and Libya, due to the closure of the Suez Canal. An Iraqi plan for a total shutdown of oil production until Israel had withdrawn from the newly occupied territories was resolutely opposed by the oil-producing states. The stage was now set for the Arab summit conference in Khartoum.[24]

The new balance of power in the Arab world was well reflected in the course of the Khartoum summit, particularly in the "spectacular rapprochement" between Nasser and King Faisal. On the question of oil embargoes, the conference stipulated that

> nothing should be done to impair the financial capability of the Arab oil-producing states to back the unified Arab effort; and that the responsibility for deciding on appropriate measures should be left to the producing countries themselves. . . .[25]

The price for Egyptian capitulation to this and other Saudi demands (such as Egyptian withdrawal from Yemen) was an annual subsidy of $378 million from Saudi Arabia, Kuwait, and Libya, two-thirds of which would go to Egypt, and the rest to Jordan. The partial embargoes on the United States and Great Britain were formally lifted under a policy of using oil as a "positive weapon." Nasser's political subservience to Faisal is candidly and poignantly expressed in his remarks at the penultimate session at Khartoum:

> I think that the matter has now been clarified. I do not want to repeat what I said yesterday, and none of us wants to go around in circles. We want everyone to do what he can. We do not ask

anyone to give more than he is able to, or to do anything which, in his opinion, conflicts with the interests of his country. Yesterday we agreed with you that the oil embargo should be discontinued since you were unable to go on with it; of course, I am referring to our brothers in Saudi Arabia, Libya and Kuwait. And now I say we are not asking you to withdraw your deposits [from Western banks], since you are unable to do so or see no point in it.[26]

Saudi determination to hold the financial reins tightly is illustrated in this exchange concerning Syria, which had boycotted the conference in protest against the Saudi role:

Faisal: We are now dealing with those countries that suffered direct damage and in this connection my view is that these countries are the UAR and Jordan.
Iraqi President Aref: What about Syria, Your Majesty?
Faisal: Begging leave of the President, I would say that Syria was not directly harmed by the aggression. Its armed forces were not hit and its losses in men, money and materiel in the battle were not comparable with those of the UAR and Jordan.

Faisal simply ignored the attempt by an Algerian delegate to point out that the war had originated with Israeli threats and attacks against Syria alone.[27]

The shift in political power in the Arab world that the new Khartoum alignment led by Saudi Arabia represented was not well received in those countries where politically conscious forces were to be reckoned with. In Egypt the government-inspired press had to launch a campaign to "explain" the rationale behind the Khartoum program (oil as a "positive weapon") and thus justify Egypt's acquiescence. In Iraq, the leaders of the pro-Nasser Arab Nationalist Movement had to face the popular wrath of the ANM rank and file for their willingness to follow Nasser even unto "moderation and realism."

While the defeat of Egypt and Syria left Saudi Arabia in a dominant political position among the Arab states, the war and its aftermath also quickened the pace of political polarization around the oil question. In Iraq, events moved in a steadily more nationalist and assertive direction which culminated in the Iraqi-Soviet oil agreement in 1969. Within OPEC there was a postwar move to press for higher revenues by eliminating royalty discounts and raising posted prices on the Mediterranean to reflect the closing of the Canal. Just before the OPEC meeting of September 15, 1967, Aramco moved to head off possible trouble by eliminating the royalty discount for Saudi Arabian Mediterranean exports, and for Libya as well. No such concessions were made to Iraq. The political dimension of this move was outlined by the editor of *Middle East Economic Survey:*

> In the present political turmoil the major Arab oil producing countries have on the whole opted for a moderate course and, at the expense of considerable effort on their part in highly explosive circumstances, performed an invaluable service for the whole world in maintaining the flow of oil to international markets. In the process of accomplishing this, they have incurred very heavy financial burdens which, if not made up in some way or another, could result in heavy damage to their developing economies. It is therefore not surprising that the Arab oil producers should now be looking to the consuming countries and the oil companies for a broad measure of understanding when it comes to their current drive for an increase in their revenue from oil.[28]

Middle East Oil Profits in the 1960s

The production of crude oil in the Middle East continued to be an immensely profitable venture for the giant international firms throughout the 1960s. One indication of this was seen in the persistent and successful attempts of smaller independent companies to get involved in Middle East concessions and joint ventures even at the cost of reduced profit margins. The industry was thus squeezed from two directions: surplus production continued to weaken the monopoly price structure, particularly in Europe; expansion of the oil industry in Europe and elsewhere brought the entry of small independent and state-owned refinery and marketing operations eager to purchase the crude that the independents wanted to sell. The surplus crude situation in Europe had, no doubt, been exaggerated by the giant companies in order to keep posted prices down: although the majors' production world wide exceeded their refinery runs in 1966 by 1.8 million b/d; excess of production over product sales was only 800,000 b/d, not counting refinery wastage.[29] Nevertheless, the trend toward lower market prices was clear and unmistakable. At the other end, the producing countries were getting slightly more revenue per barrel at the expense of the companies. After the June War, the need to increase host country revenues as a form of "protection" became even more urgent.

It is impossible to calculate company profits with precision, since most of the relevant data is not available. This problem is accentuated by the growing discrepancy in the 1960s between posted and realized or market prices. I have computed company profits for two years in this period, 1963 and 1969, for operations in Iran, Iraq, Saudi Arabia, Kuwait, Libya, Abu Dhabi, and Qatar. The method I have used for 1963 is to multiply the total yearly production by the composite realized price per barrel as calculated by Adelman, based on European (Rotterdam) prices, minus refinery and transportation factors. This is $1.59 per barrel. From the total revenue I have subtracted a generous production cost estimate (including operating and development expenses) of 10¢ a barrel average, and the total oil revenues paid to host governments in that year. I then divided the total company income by the U.S. share in each operation (i.e., Iran-40 percent; Iraq-23.75 percent, etc.). Estimates of total oil company profits for 1963 are, in millions of dollars:

Iran	$423	(U.S. share-$169.2)
Iraq	$322.9	(U.S. share-$ 76.85)
Saudi Arabia	$363.7	(U.S. share-100 percent)
Kuwait	$596.4	(U.S. share-$298.2)
Abu Dhabi *and* Qatar	$ 78	(U.S. share-$ 19.5)

Total gross profits came to $1.7 billion, of which the U.S. share was $914 million. Allowing 10 percent for net investments in the Middle East, we can state that transfers of investment income abroad (repatriated profits) were in the neighborhood of $1.61 billion, of which the U.S. share was $841 million. Another $6 million was repatriated to the United States from Libyan operations of that year. These figures compare favorably with the more precise calculations of Issawi and Yeganeh for 1960, when total repatriated profits amounted to $1.57 billion, indicating that the drop in company income per barrel was more than compensated for by increased output.[30]

For 1969, we have the advantage of Adelman's estimate of company profit per barrel in 1969 at 30¢. This is probably too low, since it is based on Rotterdam prices which are the most competitive in Europe. There are indications that average profit was closer to 40¢ per barrel, but the trend in prices is clear, and it is probably safer to underestimate than overestimate. In any case, the companies were earning a far lower rate of return per barrel owing to a decline of realized prices to a (Rotterdam) low of $1.20 per barrel, and the elimination of various industry discounts by the OPEC negotiations. Multiplying output per country by 30¢, we have for 1969, in millions of dollars:

Iran	$347.4	(U.S. share-$139)
Iraq	$158.7	(U.S. share-$ 38)
Saudi Arabia	$347.4	(U.S. share-100 percent)
Kuwait	$301.8	(U.S. share-$150.9)
Abu Dhabi and Qatar	$103.2	(U.S. share-$ 24.8)
Libya	$339.3	(U.S. share-$305)

Total profits come to $1.6 billion, of which the U.S. share is just over $1 billion. After subtracting 10 percent for capital investment in production, the total is $1.4 billion, and the U.S. share, $905 million.[31]

The profitability of Middle East oil production is still considerable, and Department of Commerce figures for 1966 and 1970 indicate higher totals for U.S. oil investment income of about $1.16 billion for each of those years. In 1966, according to the Department of Commerce, U.S. oil investments in Asia and Africa (which means chiefly the Middle East and North Africa) accounted for 10 percent of total U.S. direct investments abroad, but 35 percent of total profits from U.S. direct investments abroad. Still, the most prominent and notable trend in these figures is the sharply falling rate of return per barrel. The result is that the total profit remained just about the same in the years considered, while output more than doubled from 2.6 billion barrels in 1963 to 5.5 billion barrels in 1969. Moreover, in 1969, the independent newcomers were getting a share of the total.[32]

6

Monopoly at Home: Prelude to the Energy Crisis

Today's energy crisis has its roots in the changes in the character and structure of the energy industry in the United States in the 1960s—changes that were prompted by the slow erosion of profits based on monopolization of crude oil reserves and production. One response was to diversify foreign operations, a trend which led to the emergence of Libya, Nigeria, and Indonesia as major producers by the end of the 1960s. The most important response, though, was the strategy of the oil companies to transform themselves into "energy" companies by obtaining controlling interests in competing raw energy resources: coal, uranium, and even exotic sources like geothermal energy.

"Total Energy" Companies

Evidence of the new strategy first became obvious in the coal industry. Gulf Oil acquired Pittsburgh & Midway Coal, the thirteenth largest producer, in 1963. Continental Oil bought out the giant of the eastern coal industry, Consolidation Coal, in 1966. Occidental Petroleum took over Island Creek Coal, the third largest producer, in 1968, and in the same year Standard Oil of Ohio took over Old Ben Coal, the tenth largest producer. Other major coal producers were taken over by large industrial firms like Kennecott Copper and General Dynamics. When the dust had settled, eleven of the fifteen largest coal companies were controlled by outside interests, mainly oil companies. Other oil companies moved into control of reserves. Standard of New Jersey, now Exxon, quickly bought at least 7 billion tons of reserves. Atlantic Richfield became the second largest holder of federal coal land leases, with 43,500 coal acres. Between 1962 and 1969, these top fifteen companies increased their share of growing U.S. coal production from 40.9 percent to 53.3 percent. The reselling and brokerage activities of these companies puts their actual control of the commercial market near two-thirds.[1]

The picture is similar in the nuclear industry. By 1970, eighteen of the twenty-five largest petroleum companies had interests in at least one phase of the mining and processing of uranium for nuclear fuel. Oil companies, on the basis of present known investments, account for 40 percent of all uranium reserves. The key stages of uranium refining and processing are controlled by two oil companies, Kerr-McGee and Atlantic Richfield. A clear trend is the integration of the industry along the oil pattern: from mining to refining, and even the manufacturing of nuclear equipment. Jersey Standard, Gulf, Atlantic Richfield, Getty, and Kerr-McGee (the leader) all have strong positions in at least four of the six phases of the nuclear fuel industry. The latest development in the nuclear field is Gulf's formation of a joint venture with Royal Dutch-Shell to form the third largest contractor for the construction of nuclear reactors. Gulf also recently purchased a 45 percent interest in the chief German nuclear reactor manufacturing firm.[2]

The major oil companies control most of the country's natural gas reserves and production and are heavily involved in other competing sources of petroleum such as oil shale and tar sands. With the appropriate technology, both of these resources may become the most important and richest sources of synthetic crude oil. One oil company executive describes the scramble for geothermal leases as "a new gold rush. No one is tipping his hand right now, but all the major companies have already leased private land and many are drilling" in the west and northwest.[3] The oil giants have expanded in major ways into nonenergy fields as well, particularly real estate. All the majors now have real estate divisions, and some, like Gulf, own whole towns.

The oil companies have not abandoned their historic dominance in the production and subsequent phases of the oil industry. According to a recent study by the Federal Trade Commission, the share of the top eight companies in the production of domestic crude increased from 44 percent in 1960 to over 50 percent in 1969, while that of the top twenty companies increased from 63 percent to 70 percent. A more important measure of concentration, says the FTC, is control of proved crude reserves. As of 1970, the top eight firms controlled 64 percent and the top twenty controlled 94 percent of crude reserves. Concentration in refining was even greater than in production, with the top eight companies controlling 58 percent of capacity in 1970, and the top twenty controlling 86 percent. Concentration figures for marketing are along the same order.[4]

The result of this concentration of energy resources in the hands of relatively few companies was a greater ability to manipulate and extort higher prices and profits from the public. This has been most clearly demonstrated in the coal and natural gas industries. Coal prices were stable between 1960 and 1965, while production increased 23 percent and steam coal consumption by electric utilities (a major market) rose by 40 percent. By 1969, under increasing demand from the electric utility sector, coal prices had risen 22 percent above the 1960 level. In 1970, following the merger and acquisition phenomenon described earlier, prices shot up by an average of 60 percent and over 100 percent in some areas, including the TVA. This incredible rise bore no relation to the market forces of supply and demand. Production was up 5 percent, and total steam consumption up only 1.6 percent. An economic analysis by a former staff economist with the antitrust division of the FTC concludes that "the sharp increases in coal prices in 1970 were made primarily to enhance the profits of the coal companies or their parents." The chairman of the board of TVA, a major coal consumer, has testified that one oil-coal company executive told him they would provide new supplies, but only at a price that would yield the same profit that they were accustomed to receiving on oil.[5]

These price increases were accompanied by a restructuring of the coal market from spot or short-term sales to long-term contracts of up to thirty years. By holding coal off the market, the companies forced their customers (utilities, etc.) to sign long-term contracts at prices reflecting the "shortage" on the short-term market. The companies also refused to supply coal under previously signed long-term contracts until the "supply problem" was eased and replaced by agreements to pay new, higher rates. The trend continues; the TVA's Power Annual Report for 1973 notes that coal costs continued to rise over the previous year. In uranium, too, there is now a "supply dilemma," according to uranium producers who, as we have seen, bear a certain resemblance to the coal producers, who bear a certain resemblance . . .[6]

In other words, the previous competitive and stable market relationship between producers and customers has been transformed by the concentration of resources in the hands of the oil companies. A noncompetitive sellers' market, like that in natural gas, is created whereby the producers (the oil companies) use their concentrated hold over reserves and production to command prices that have no relationship to production costs. One difference between coal and natural gas is that in the latter industry prices are regulated by the Federal Power Commission, or at least they were until the Nixon administration took office in 1968.

Natural Gas Prices: How the "Energy Crisis" Began

The so-called energy crisis in the United States is nothing more or less than a well-coordinated attempt by the oil companies (now the "energy companies") to extort higher prices and profits from the consuming public for all energy and fuel resources in order to maintain their profit margins, once dependent on total control of low-cost crude oil in the Middle East and elsewhere. The best description of the energy crisis is this remark by Allan Hamilton, the treasurer of the biggest energy giant, Exxon (Jersey Standard). In a statement worthy of the late and great John D. Rockefeller himself, Hamilton said: "Unless and until the real nature of the crisis is understood, and profit levels become such that the industry is confident that its investments will bear fruit, the supply of energy required will not be forthcoming."[7]

The energy crisis did not begin in 1970 or 1971 (when Middle East producers demanded higher taxes), but in 1968, when the oil industry decided to smash the one vestige of public control over private power, the regulation of natural gas prices by the Federal Power Commission (FPC). The history of public regulation of natural gas has been and, it seems, will be brief. Congress passed the Natural Gas Act in 1938 without a dissenting vote in either house. It covered the interstate sale and transportation of natural gas and required that rates charged be just and reasonable, nondiscriminatory, and publicly posted. In 1954 the Supreme Court ruled that under the Natural Gas Act the FPC had jurisdiction over sales by producers to interstate pipelines. Industry attempts to have the Eisenhower administration introduce legislation to exempt these sales from FPC jurisdiction failed when large industry campaign contributions became a matter of public record.

The FPC attempted to determine rates for cost-plus-fair-return on a company-by-company approach, and so overwhelmed the commission staff that it would have taken a tripled staff almost one hundred years to clear the dockets. The result was that very little actual regulating got done, and contract prices were approved without much regard to their reasonableness. In order to carry out its mandate and halt spiraling gas prices, the FPC initiated in the early 1960s the *area rate method* by which production costs for each geographical area are used to set the price for all gas produced in that area. By 1965, the FPC had completed its first area rate determination covering the Permian Basin, with prices based on composite industry data on costs of finding and producing gas from a well in that area, plus a 12 percent rate of return. Industry challenged this procedure, but the Supreme Court, in 1968, upheld the FPC. Also in 1968, the FPC issued further rate schedules, the most important being that for southern Louisiana. In 1968 Richard Nixon took over in the White House, and in 1968 the "energy crisis" began.

Natural gas is frequently found in conjunction with oil. It is an extremely clean and efficient fuel, but transportation difficulties have traditionally caused it to be treated as an unwanted by-product of oil production. Much of it was simply burned away at the well; industries situated near the wells, especially oil refineries, could, did, and still do use it as a prime fuel. Only after World War II did government-built pipelines like the Big Inch make it feasible to transport large quantities of natural gas to consuming areas like the northeastern United States. Because of its cleanness, efficiency, and relative cheapness, its use escalated until it accounted for 33 percent of the energy fuel used in the United States.

The natural gas industry, which is to say the oil industry, contended that government price regulation at the producing end on the basis of cost-plus-15 percent did not provide enough financial incentive (i.e., profits) for the companies to explore and develop new reserves in order to meet the growing demand for natural gas—a demand enhanced by environmental legislation such as the Clean Air Act. To support their contentions they pointed to the fact that annual production now exceeds additions to reserves, so that the reserves-production ratio has declined from a range of 15 to 19 percent in the mid-1960s to 11.3 percent in 1971. Year end reserves which stood at a high of 289.3 Tcf (trillion cubic feet) in 1967, had declined to 247.4 Tcf in 1971. The industry claims that in addition to not providing an adequate rate of return, price regulation makes the cost of natural gas *artificially* low and thus encourages extravagant and wasteful use. They have enlisted the support of virtually the entire administration, from the White House to the Interior Department and the Federal Power Commission, in their campaign to raise gas prices to their "natural" or "market clearing" level: i.e., until they are as high per energy unit as oil and coal.

The evidence suggests that the only thing *artificial* about natural gas prices is that they are based on the costs of production and hence cannot contain the spread of monopoly profits in other "competing" fuels controlled by the oil companies. The public, including the FPC, is totally dependent on the industry for the declining reserve figures, through the American Gas Association (AGA). These industry figures are not even open to inspection by the FPC on the grounds that they are confidential and proprietary. The frivolity of this position is obvious when one notes that the AGA personnel who compile the data are on loan and salary from the various gas producers: only the public but not "rival" companies are deprived of the data.

All available independent evidence suggests that the natural gas crisis is phony. The reserve figures in question took an inexplicable nose-dive beginning in 1968, a drop that bore no relation to drilling effort. In 1969 there was more successful well footage drilled in south Louisiana than in any year since 1962, and this was true in other prime producing areas as well. Annual reserve additions over the years have been composed of roughly 30 percent new discoveries and 70 percent extensions and revisions. In 1968, the extensions and revisions were made to show a net negative revision of 642 billion cubic feet. In 1969, the figure was again far below normal. Had the relationship between new discoveries and extensions and revisions continued at the normal 30-70 rate, reserve additions would have been nearly double the 4.5 Tcf shown for 1969. For example, a contract between producer Gulf and pipeline Texas Eastern in 1963 based on the prevailing price range of 19 to 26¢ per thousand cubic feet (Mcf) was brought by Gulf to the FPC for revision in 1968 on the grounds that Gulf had overestimated its reserves in that field by 60 percent. Gulf suddenly ended up with a field of 1 Tcf instead of 2.5 Tcf, and its reserves were suddenly down by 1.5 Tcf—on paper at least.[8]

Another indication of the real nature of the natural gas shortage is that the ratio of exploratory wells to development wells has been extraordinarily high over the "shortage" years. Development drilling is many times more prolific, and is what turns probable reserves into proved reserves. Furthermore, development drilling is done on areas already under lease and is ten to fifteen times cheaper than exploratory drilling. These facts indicate that in some key areas such as southern Louisiana there are large potential reserves waiting to be developed on already discovered reservoirs at relatively low cost or risk, once the "price is right." Additional evidence can be found that the AGA reserve figures are wholly unreliable. According to FTC antitrust chief James Halverson, the estimates of proved reserves for a given lease that are reflected in in-house reports . . . were primarily used for tax purposes, have been found to be lower than the estimates of proved reserves that are used for other in-house purposes such as decisions whether to build a drilling platform on a tract or to sell reserves to a pipeline company.[9]

The differences, according to Halverson, range as high as 200 percent.

The industry contention that higher prices will improve the supply crunch is belied by yet another set of statistics. Comparing averages for the five-year periods 1963-1967 and 1968-1972, we find net production rose 29 percent, weighted average initial contract prices for each period rose 33.5 percent, and new gas prices rose 13.8 percent, but reserve additions fell by 50 percent, according to FPC staff economist David Schwartz. Schwartz, in testimony to the Senate Antitrust and Monopoly Subcommittee in June 1973, also pointed out that of the 117 leases sold in the 1970 offshore sale by the Interior Department, only 9 (7.7 percent) are currently in production and 38 (32.5 percent) are classified as producing plugged-in. Almost 60 percent of the leases have not been classified yet.[10] This would indicate that the industry anticipates higher prices.

And well they might. Natural gas policy is made in the White House, the Interior Department, and the FPC. Never have the foxes had such complete run of the chicken coop. Nixon's chairman of the FPC, John Nassikas, testified in June 1973 that

I predicted on November 13, 1969, at the first policy hearing on natural gas before the Senate Interior Committee, that there was a deepening gas crisis, that we had to do something about it, and that the policies of the FPC should be designed to elicit more supply of gas, better allocate the resources, and to amend its policies of pre-existing Commissions to meet these objectives.[11]

These sentiments have been echoed ever since by the president, former Treasury Secretary Connally, Interior Department officials, and FPC Commissioners, especially those appointed by Nixon. In the face of congressional resistance to flat deregulation of natural gas, the FPC has moved obliquely to deregulate by administrative fiat, most notably in the Belco case, in which a price-hike of 73 percent was granted. This decision was made against the recommendation of the FPC staff, and was authored by FPC Commissioner Rush Moody, formerly a lawyer for a Houston firm that represents one of the parties to this case, Texaco. The fact that another petitioner, Tenneco, was arguing for a hike in the price charged to its own pipeline subsidiary indicates the extent to which "market forces" prevail in the natural gas industry. FPC general counsel at the time was Gordon Gooch, also from a Houston firm with oil companies prominent among its clients. Gooch left the FPC in July of 1972, to become a key fundraiser for the Committee to Re-elect the President. The same law firm—Baker, Botts, Shepherd, and Coates—sent Stephen Wakefield first to the FPC and most recently to the Interior Department as Deputy Assistant Secretary in charge of energy. The administration has given every encouragement to the gas producers to withhold gas supplies from contract in the expectation of fabulous price-hikes that are totally unrelated to costs. The industry has shown its appreciation by its prominence on the lists of individual and corporate (illegal) contributions to Nixon's campaign in 1972, totaling more than $5 million.[12]

The industry is not satisfied with 15 percent profit, and has shown itself capable of anything to raise those profits. Coastal States, a major Texas pipeline and producer, virtually cut off supplies to the city of San Antonio in the winter of 1972-1973, until its long-term contract was renegotiated. At present the twenty-five major producers, almost all large oil companies, control over 75 percent of current natural gas production. The level of monopoly concentration is even greater with regard to reserves, the relevant figure for the future. The FPC staff, in protesting the Commissioners' decision to grant the 73 percent hike in the Belco case, estimated that the top four firms control over 50 percent of the relevant market and the top eight firms, 70 to 80 percent. The proportion of total volume of natural gas under contracts controlled by the Big Five due to expire over the next six years ranges from 56.6 percent in southern Louisiana to 99.6 percent in the Permian Basin.[13]

It has been estimated that every 10¢ per Mcf increase over the cost-plus-12 percent formula will result in windfall profits of $110 billion for the producers. One aim of the industry is to bring prices up to a level where coal gasification will become "economical." The process of making gas from coal has been controlled by the oil industry since the days when Jersey Standard entered a cartel with the German firm of I.G. Farben in 1936. By the early 1950s a federally sponsored coal gasification project, and an oil shale project, had succeeded in reducing costs to a level close to that prevailing for regular petroleum products. At that point, the plants were closed down by the Interior Department at the behest of the oil industry.[14] Once the industry has total control of gas prices, it plans to strip mine the vast coal reserves bought up over the last decade to manufacture synthetic gas.

Shortages for Fun and Profit

The natural gas situation gives us a capsule view of the extortion campaign being waged on all fronts by the oil industry through its $3 million media campaign and the "free time" provided by industry spokesmen masquerading as public servants in Interior and the FPC. The shortage of heating oil and gasoline over the last few years indicates that industry efforts to jack up prices are intensifying on all fronts. Just months and weeks before the heating oil "shortage" in the 1972-1973 winter, the major oil companies assured the Office of Emergency Preparedness that the industry "had the capacity to meet the fuel requirements of the coming winter and that it would do its best, within its usual marketing patterns, to meet the needs of individual customers." This assurance was backed up by the results of an Interior Department survey of refinery prospects. Texaco told the OEP in July that "there is sufficient refining capacity available in the U.S. to meet anticipated demand for clean products over the balance of this year. . . ." On January 22, 1973, Texaco informed the state of Massachusetts that it would run out of low-sulphur residual fuel in ten days, and threatened to stop delivery of oil to most customers by the end of the week unless pollution requirements were waived.[15]   Other cities across the country, notably Denver, were forced to shut down schools and other public institutions for lack of fuel. Requests for wholesale fuel oil price-hikes were granted with uncomfortable regularity by Nixon's Price Commission, but the industry sages told us to be prepared for more shortages in 1974.

Washington residents were treated to the spectacle of Nixon flying in from Key Biscayne or San Clemente at the end of a two-week air pollution crisis in September 1973 to announce that cities should relax clean air standards. Standards now in effect in most cities do not meet federal standards. Even these federal standards are dangerously inadequate according to a study conducted by the Environmental Protection Agency that is being kept secret under White House pressure.[16]

With its remarkable ability to "seasonably adjust" crises, the industry experienced a gasoline "shortage" in summer. Like the heating oil "shortage," the root of the problem lay with refining capacity, not crude supplies. Refining, like every other phase of the industry, is dominated by the top twenty companies, which control 87 percent of capacity and a greater share of actual production. The industry, which has not built a single refinery or made a major plant expansion in this country in more than a decade, blames environmental opposition. The true story, however, was hinted at by the *Oil and Gas Journal* when it assessed the higher profits of the industry in 1971: "In general, the better profits stemmed from high operating volumes, improved prices for chemicals, a lowered rate of spending for exploration and capital expansion, and more efficient operations in crude-production and refining."[17] An investigation by the Senate Government Operations Subcommittee asserts that refinery operations were purposefully cut back in 1972, in the face of increased demand for products, thereby driving up prices further. In other words, a higher rate of spending for finding, producing, and refining more oil will only come when higher returns are guaranteed. Interestingly, within several weeks of a Nixon energy message, plans for fourteen new refineries or additions were announced.[18]

It is probably true that an unanticipated factor in the gasoline shortage was the sharp rise in gas consumption for automobiles using the mandated pollution control devices. This fact has not prompted any call by the industry for a more rational transportation system than one based on Eldorados and Toronados which get seven miles per gallon. Transportation is by far the largest consuming sector for petroleum. A confidential White House study has predicted that the country could conserve 7.3 million barrels of oil a day, which is well over half the total import needs forecast by the industry.[19]

This lack of investment planning on the part of the industry, if that's what it was, has had some payoff: the shortage of refined gasoline (and the shortage of crude to some extent) gave the majors the perfect rationale for cutting off

independent marketers and thus increasing their control of
the marketing sector. In order to compensate for the fall in
return rates at the crude production end, the majors are
trying to increase profits at the transportation, refining, and
marketing level. One tactic is the regional market withdrawals
that have been announced by practically all the majors.
ARCO has sold its Florida stations to Mobil; Gulf has pulled
out of the West Coast and Exxon out of Illinois. In the
opinion of the Federal Trade Commission:

> All the majors can increase their regional market concentration
> simultaneously by pulling out of markets where their share is
> lower than their national average and selling their operations to
> those majors who remain. Their strategy will only work, however,
> if the majors can retain or expand their regional market shares.
> To do so they must prevent the further entry and expansion of
> independent marketers.[20]

A Crisis of Capital?

Most conventional predictions of the U.S. energy supply
outlook assume that U.S. crude oil production has peaked at
around 11 million barrels per day, and that even Alaskan oil
can merely offset declining production in the lower forty-
eight states by 1980. It should be noted that with regard to
oil, as with gas, we are completely dependent on the industry
for reserve figures. The industry has cried wolf before, but
until the 1970s production steadily increased to meet
consumption. Indeed, the historic dilemma of the U.S. oil
industry has been to restrict production in order to bolster
prices. In oil and gas, the reserves are probably great and
much more remains to be discovered. Much of it lies on
federal lands or federally controlled offshore areas. Consumer
group suggestions that a public corporation be forced to

explore and exploit this public domain have not been welcomed by the industry or the administration.

Once the debate moves away from questions of actual physical resources, which almost all agree are available at some cost, the energy crisis is framed in terms of financial resources.

> If the energy industries are expected to expand their goods and services to meet national goals approved by the majority of this country, then they must be granted access to the funds necessary to carry out their assignment. These funds must be paid for either by the consumer through higher prices in the marketplace or by the taxpayer through the tax structure of the government.[21]

The message is the same whether it comes from Standard Oil, Chase Manhattan, or the Department of the Interior: The era of cheap energy is over; pay up or else. "The key problem," says Chase Manhattan's energy division chief John Winger, "is a shortage of capital to produce energy, not the lack of energy resources."[22]

There are, however, some who think that the real problem lies in the opposite direction:

> Namely, that the present producing companies . . . will continue to advance their control over available capital in the United States with the result that they will achieve an even higher concentration of capital pre-eminence among total United States industry than is presently enjoyed today.[23]

The capital crisis argument, curiously enough, is just about as old as the natural gas "shortage." The evidence suggests it has about equal validity. Of the top 25 largest industrial corporations in the Fortune 500, 9 are petroleum companies. The petroleum industry has traditionally been exceptionally reliant on internal financing, that is, profits rather than bank loans or bond issues. Over the last decade there has been an increase in the reliance on external financing, but it is still well below the industrial average. Moreover, the oil industry has had no particular difficulty in securing capital from the giant banks and insurance companies that dominate finance capital in the United States today. It is no coincidence that the head of Chase Manhattan is a Rockefeller, or that all of the major banks have energy divisions, or that a large portion of the business loans issued by the giant New York banks in the 1950s and 1960s were made to large energy corporations. All of these banks (as well as the largest insurance companies) sit on the board of directors of the major energy companies, hold large quantities of those companies' stock, hold significant portions of their bond issues, and manage their trust funds. These same bastions of finance capital, under the rubric of "institutional investors," control more than 40 percent of the stock on the New York Stock Exchange, and have a demonstrated ability to determine the relative success or failure of stock issues.[24]

The major energy companies have shown a steady increase in available capital over the years since the "energy crisis" began. Capital generated internally, as well as cash and marketable securities, increased substantially, and the companies have demonstrated no inability to secure bank and institutional financing. Their average capitalization ratio of 85 percent equity to 15 percent debt is substantially above most U.S. industries.[25] An increasing tendency of the large companies toward acquiring the assets of smaller companies has been accompanied by an increasing movement to joint ventures. Both courses effectively eliminate any competition between the companies for capital. Any extension of the tax privileges endemic to the industry will only accentuate the concentration of capital in the oil (energy) industry at the taxpayers' expense. On top of research and investment subsidies expected to top $10 billion over the next five years, economist Walter Heller regards the profit jump resulting from the latest price-hikes as "downright spectacular." The cash flow of the companies will increase by $13 to $15 billion after taxes. In an economy entering a recessionary phase with overall growth approaching zero, this represents a significant transfer of capital resources.[26]

7

The Road to Teheran Is Through Tripoli

The energy crisis rhetoric with which we are now so familiar had its roots in the campaign of the U.S. energy (oil) industry to raise prices and profit margins on all energy resources and in particular to lift federal price controls from interstate sales of natural gas. A recurrent but subdued theme in this campaign, as in the perennial fight to retain the mandatory oil import quota system, was the presentation of independence from foreign energy supplies as a virtue. Explicitly or, more often, implicitly, the chief potential culprits were the Middle East oil-producing countries, invidiously lumped together as "those crazy Arabs." The proportion of Middle East production and reserves to "free" world figures came to seem outrageously unfair in light of the notorious irrationalities and tensions that periodically plunged that region into war. There was, at the same time, a curious complacency about Middle East oil in the popular consciousness that probably derived from the same "crazy Arab" image. This attitude perceived a lack of solidarity among the producing states and a collective incompetence to pull off anything like a credible threat to Middle East oil access. The fiasco of the 1967 embargo, along with the demonstrated ability of the oil companies to avoid any serious supply disruptions by juggling supplies and markets, seemed most reassuring.

The industry vigorously pursued its long-run strategy of diversifying crude oil supplies and increasing control of non-oil resources in the face of slowly declining rates of return on Middle East operations. It displayed confidence that the growing surplus of crude supplies and production capacity that was causing so much trouble in maintaining high prices would also provide the kind of supply insurance that would deter any attempts at nationalization by a Middle East producer. The tighter supply situation brought about by the June War had translated itself into higher prices on the European market, prices which were maintained by the Suez Canal closure and resulting high tanker rates. The rise in anti-Western political tempers as a result of the Israeli victory was offset by the clear inability of Nasser's Egypt to have any political impact on the Gulf and Arabian Peninsula states, an important consideration in the face of the proposed British military withdrawal from the Gulf area. Increased revenues (through increased production) for friendly regimes like Saudi Arabia and Iran were a small enough price to pay for unrestricted access to the low-cost oil of the Middle East.

Developments in the Gulf After the June War

Oil company-government relations and policies continued more or less along a prewar pattern in the Persian Gulf area. Political polarization and mass tensions had reached a new high during and immediately after the war, but only had direct consequences for the industry in Iraq, where the concession takeover of 1961 (Law 80) was made more or less irrevocable. For their part, the companies continued to punish Iraq for its economic nationalism by keeping production (and therefore government revenue) abnormally low. According to a secret U.S. government report, the companies (IPC) actually drilled wells to the wrong depth and covered others with bulldozers in order to reduce productive capacity.[1] Faced with the alternative of capitulating to company terms regarding past disputes over concessions and revenues, Iraq headed toward full national control of its oil industry (with Soviet and other socialist countries' help) through the Iraq National Oil Company (INOC). The Ba'ath Party coup of July 1968 did nothing to change the course of relations with the companies.

The prewar pattern of company relationships with Iran continued also, at a more intense level. Familiar Iranian pressures for increased revenues through increased production were accelerated in the spring of 1968 by Iran's demand for a 20 percent increase, designed to finance the five-year development plan then going into effect. Iran modified its demand to an increase of 16.5 percent, as compared with the industry's planned overall Middle East increase of 8 percent for the year. Thus the Iranians were, in effect, demanding 63.7 percent of the companies' planned 1969 increment. This would be clearly unacceptable to the Arab oil producers. Some of the consortium companies, the small "independents," were interested in getting all the low-cost Iranian oil they could. The majors, with responsibilities to other monarchs and with large stakes in the Libyan bonanza, had reason to keep Iranian increases at a minimum. As talk of a showdown in Iran increased in March 1968, Under-Secretary of State Eugene Rostow called a special meeting of representatives of the consortium companies and "advised us as to the delicate status of the whole Middle East situation."[2] Rostow, according to this account, strongly urged the companies to placate rather than antagonize the Shah, and left the clear message that it was in the United States' national interest to increase his revenues. A settlement was finally announced in August 1968 which boosted revenues by 22 percent, but with an overall production hike of only 12 percent. The added revenue would come through producing more light, or premium, crudes rather than heavy crudes, at a higher posted price. The crisis was settled for one year, but left few rabbits in the companies' hat to meet the next year's demand.[3]

In Saudi Arabia similar pressures for increased production existed, although they did not take the form of government threats, as in Iran. Publicly, Oil Minister Yamani was content to introduce the notion of participation, by which countries like Saudi Arabia would acquire direct equity interest in the operating companies. The companies themselves were astute enough not to let production languish and invite government pressure. Moreover, the competitive market pressures resulting in overproduction led to even more production, since any restraint on the part of one company would most likely result in the loss of customers or markets without significantly reducing the downward pressure on prices. Howard Page, former Middle East coordinator for Exxon, recently described why Aramco let its crude-short partner, Mobil, overlift its Aramco quota at reduced prices:

> I mean, I have never seen such a competitive market. The market was falling on its face, and as it fell on its face, why Mobil had a better argument to get a lower price for their overlift than before, and we recognized that if we didn't agree with them why they would go to BP and other people and get large quantities of oil and then Aramco would lose that outlet, and we were all interested in keeping the Aramco outlet reasonably high, because this was the most important concession in the entire world and we didn't want to take any chances of losing it.[4]

While Mobil's 10 percent equity share of Aramco left it with not enough crude, Socal's 30 percent share left it with a surplus, and as a result its liftings had dropped to 19 percent by 1967. An aggressive sales campaign for long-term customers raised Socal's independent sales of Saudi crude (that is, outside the company's own integrated channels) from 42,000 b/d in 1960 to 354,000 b/d in 1971, from 11 percent to 22 percent of its gross liftings.[5]

Competition between Iran and Saudi Arabia in the Gulf was not limited to production rivalry alone. With the announcement in 1968 of British military and political withdrawal from the Gulf in 1971, Iran took to asserting itself as the local power to be reckoned with. A potentially serious conflict arose in the Persian Gulf where both Saudi Arabia and Iran, with overlapping claims in the area, let out separate concessions. In early February 1968, after an Aramco rig moved into the disputed area from the Saudi side and an Amoco-Iranian joint venture moved in from the Iranian side, an Iranian gunboat came alongside the Aramco rig and arrested the American and Saudi crew. They were soon released, but the Shah abruptly canceled a planned visit to Saudi Arabia. Under-Secretary of State Eugene Rostow made a quick unscheduled visit to Teheran to try to smooth things out. Drilling by both sides was stopped in the disputed area, and by the end of the summer a settlement had been reached. A serious clash was thereby averted and the gunboat incident was kept under wraps. The incident (as the Shah probably intended) did demonstrate that Iran was out to dominate the Gulf area politically and militarily, and any minor obstacles would be dealt with summarily.

Another more long-standing issue concerned the status of Bahrein, which Iran claimed. The Shah repeatedly voiced opposition to the British-planned Federation of Arab Emirates because of his claim on Bahrein and some smaller but strategically located Arab islands in the Gulf. The United States encouraged the peaceful settlement of these disputes and armed both sides. The first deliveries of Phantom jets to Iran were made in 1968 and in the same year a Saudi naval expansion program was initiated. There was no doubt, though, that the big money was on Iran.[6]

On the Arab side of the Gulf, Saudi Arabia and Kuwait backed the British in their effort to erect a federation or union of the various Trucial States. The emergence of the radical National Liberation Front regime in Aden at the end of 1967 and the continuing guerrilla war in the Dhofar province of Oman posed a new, indigenous revolutionary threat in place of the real or imagined Egyptian threat which ended with the June 1967 defeat. Iranian cooperation could be secured, but only by Arab deference. In the words of the Teheran newspaper *Ayandegan,*

> we are ready to collaborate with the sheikhdoms to safeguard the security of the Persian Gulf, but they must not forget so long as Iran is there the Persian Gulf will never become an Arabian Gulf.[7]

The new oil action in the area was centered directly on the Gulf itself: offshore joint ventures with Iran, Kuwait, and Saudi Arabia, and more favorable, traditional concession terms both on and offshore from the British-controlled sheikhdoms. The Iraq Petroleum Company, under the old Red Line Agreement, had first dibs on Qatar and Abu Dhabi, which went into production in the 1960s. In areas relinquished by the IPC, and in new areas from the other sheikhdoms, European and Japanese companies, along with one major company, Royal Dutch-Shell, were securing new concessions. In November 1967 *Fortune* wrote:

> The Gulf is a haven of generous concession areas, where producers can rapidly amortize their initial capital costs and where no one is coercing them with threats of a shutdown of production or expropriation . . . The Abu Dhabi fields, moreover, provide the [IPC] consortia members with a useful hedge against the uncertainties they are facing in Iraq.[8]

It was not for nothing that the British had sunk $30 million into their military base at Bahrein, or that United States and other Western oil companies placed a premium on the billions of dollars they stood to make from Persian Gulf crude oil.

The conservative Arab oil states, under the leadership of Saudi Arabia, moved to insulate themselves politically from the radical upsurge in the rest of the Arab world following the June War by creating the Organization of Arab Petroleum Exporting Countries (OAPEC). Such an organization had long been demanded by the radical-nationalist oil technocrats like Tariki, unhappy with OPEC's timidity and moderation. What they wanted was an organization what would include all Arab states, or at least those like Syria and Egypt which, as transit states, had some stake in oil politics. OAPEC, though, was restricted to states for which oil production was the main source of income, This limited membership to founding states Saudi Arabia, Libya, and Kuwait, and potentially to Iraq, which rejected the initial invitation to join, and to the tiny Gulf states. As the *Middle East Economic Survey* observed:

> The purpose of this restrictive condition is clearly to ensure that all countries admitted into the organization would be equally anxious to maintain a purely economic approach to the development of the oil sector to the exclusion, as far as possible, of dangerous political crosscurrents.[9]

Iraq remained in opposition to the conception and orientation of OAPEC and initiated contact with Algeria and Egypt to set up some coordination of their respective national oil companies as a possible counterweight to OAPEC, but nothing much developed from this. Egypt itself, the traditional purveyor of political crosscurrents in the Arab world, was subdued on the oil question, owing to the Saudi and Kuwaiti subsidies and Egypt's dependence on its small but expanding domestic oil industry. The Egyptians were counting on domestic oil production to meet domestic consumption and earn badly needed foreign exchange as well. This program consisted mostly of joint ventures with American independents like Standard of Indiana.

The most radical and unsettling political force in the post-June War years was the Palestinian resistance movement, and its main body, Fatah, received substantial financial support from Saudi Arabia and other oil states. Radical Palestinian groups, however, made no secret of their emnity for "Arab reactionaries" who, along with Zionism and imperialism, were seen as responsible for the Palestinian predicament. This hostility went beyond rhetoric in June 1969 when the Popular Front for the Liberation of Palestine (PFLP) sabotaged a portion of the Trans-Arabian Pipeline running through the occupied Golan Heights. The Saudis labeled the action "criminal" and somewhat fancifully alleged that it was a plot hatched by the PFLP and the Israeli Communist Party (Rakah) at the Intercontinental Hotel in Vienna! Cairo's *al-Ahram* called the attack "incomprehensible"; the semi-official Algerian newspaper *Al-Moujahid* asked, "Is history to record Khartoum as the Arab Munich?"[10] This was the first of several Palestinian attacks against TAPline and oil facilities in Israel like the Haifa refinery.

Despite the ominous portent of the Palestinian attack and the potential for conflict among the Gulf states, Western control of Middle East oil in 1969-1970 seemed more secure and stable than could have been predicted in June 1967. Saudi Arabia, and then OPEC as a whole, was making studies and speeches about "participation." The future of the industry in Iraq was still a question mark. But the main problem faced by the industry was the familiar one of oversupply and weakening prices, particularly for Persian Gulf crude faced with the long tanker haul to European markets. Royal Dutch-Shell's chairman offered this assessment in early 1969:

> The picture as I see it is on the whole a reassuring one. Despite the situation created by the withdrawal of British forces from the Middle East after 1971, despite the Russian enigma, despite the economic and political pressures from the producing countries, I think that the system will remain broadly in balance—although like a mobile the whole thing may swing a bit this way or that with the wind from time to time—and that the oil will continue to flow.[11]

The North African Klondike: Libya and Algeria

Shell and the other giant oil companies were never the ones to sit passively by and "swing a bit this way or that." Diversification of crude oil sources has always been a strategic precept for the industry, particularly in the face of crises and political tensions. Throughout the 1960s, with a situation of increasing supplies and weakening prices, the companies continued to spend millions looking for oil in Indonesia, West Africa, Alaska, and the North Sea, in what *Business Week* (in 1969) called "the wildest—and most widespread—oil rush in history . . . in the face of an oversupply of crude so massive that if not one additional barrel of oil were found the world could maintain its current consumption for more than 30 years."[12]

Back in the 1950s, with a special impetus from the Suez crisis, the industry moved to diversify supplies by sidestepping the Suez Canal and concentrating its exploration efforts in Libya and Algeria. In Algeria, then waging a bitter war for national liberation from French colonialism, the oil effort was dominated by several French state-supported companies, although some small American companies like Getty and Sinclair also ended up with concessions. Production began at 10,000 b/d in 1948 and had reached nearly 1,000,000 b/d ten years later. Most of it was channeled through special postindependence arrangements to the French market. Algeria did not become a significant factor in world oil by virtue of its productive capacity, which was relatively marginal and mostly under French control, but because it took a strong political stand against the companies, beginning in the late 1960s.

Libya was another story. Most of Libya's political significance today is rooted one way or another in its emergence in the 1960s as a major crude oil producer. The possibility of developing crude oil reserves on Libyan territory was considered good even before the country's independence, although no significant moves were made by its assorted occupiers to exploit this potential. Soon after independence in 1951, however, the British-sponsored regime of King Idris proceeded to open the country up to foreign oil capital in a move designed to avoid total dependence on the direct (rental) and indirect (foreign aid) payments of the United States and Britain for military base rights. The country's first petroleum law was issued in 1955, drafted by advisors to the government solicitously provided by the major oil companies.

The results, not surprisingly, "reflected the aims of those who shaped it."[13] In its broad outlines the law provided for concessions under terms then prevailing in the Middle East, with profits divided 50-50 between company and government, and a royalty of 12½ precent. The fine print contained provisions though that would turn Libya into "an Arabian Klondike."[14] Libyan concession holders could deduct a 20 percent depreciation charge on all physical assets and choose between a 20 percent amortization of preproduction expenditures and a 25 percent depletion allowance on gross income. More far-reaching was the provision that tied income (for government tax purposes) to actual market prices rather than the posted price formula that prevailed in the Persian Gulf area. As market prices dropped in the 1960s and the other Middle Eastern countries forced the oil companies to hold posted prices steady, company profit margins from Libyan oil zoomed.

The Libyan law did represent something of an innovation in that the concession areas were divided into many separate tracts for company bidding rather than auctioned as a whole to a single consortium or company (as in Saudi Arabia, etc.). Within three months of the passage of the petroleum law, forty-seven separate concessions had been let to fourteen different companies or groups of companies. Six of the seven majors were represented. Standard of New Jersey (Exxon) became the first to discover large quantities of oil, in 1959, and was followed by the Oasis group (American independents Marathon, Continental, and Amerada). Export production began at 20,000 b/d in 1961, reached 1.5 million b/d by 1966, 2.6 million in 1968, and more than 3.5 million in 1970. These production rates nearly equaled those of long-established producers Iran and Saudi Arabia.

The political role of Libyan oil was as important as the economic rewards it provided the companies in determining the incredible growth of Libyan crude production. Libya was seen by Western oil strategists as an alternative to the "unstable" political environment that characterized the rest of the Arab world in the 1950s and 1960s. Preproduction expenditures and activities climbed sharply in 1957 and 1958 after the nationalization of the Suez Canal, and the political turmoil that ensued with the American and British military intervention following the 1958 revolution in Iraq. Another spur was the discovery of a large crude deposit in Algeria, near the Libyan border, in 1956. Geological and geophysical crew-months multiplied from 70 in 1955 to nearly 1,000 in 1959. The number of exploratory wells climbed from zero to 343 in the same few years.[15]

The division of the country's oil-producing areas into many concessions and the calculation of income on market prices meant that many companies, especially the so-called independents (smaller American companies and European state companies), had tremendous economic incentive to produce as much as possible, without many of the cartel restraints that held most Middle East production in check according to the carefully worked out schedules of the Big Seven companies. Libya at once became not only the source of bonanza profits (even by the standards of the oil industry) but also fueled the political and economic dynamics in the oil world that would backfire on the companies following the revolution of 1969.

With the commencing of large-scale crude production and export in the early 1960s the Idris regime made some half-hearted attempts to remedy the financial loopholes in the 1955 law. In 1961 the notorious depletion allowance was abolished. The government next tried to bring its tax policy in alignment with the prevailing posted price system. An amendment to this effect was enacted, but the companies got around it by insisting that marketing expenses be deducted from income. By "marketing expenses" the companies meant all the discounts and rebates from the posted prices that enabled Libyan oil to gain a large and growing share of the European oil market. In effect, nothing had changed. In 1964 Libya got 90¢ a barrel from Exxon, whose corporate interest dictated strict adherence to the posted price system. The Oasis independents, on the other hand, paid an average of less than 30¢ a barrel—slightly more than bare royalty and rental charges.[16]

The need to change Libya's oil policies regarding the posted price was dictated not only by the country's own financial interest but also by pressures emanating from the major companies, on the one hand, and the other oil-producing countries, operating through OPEC, on the other. The emergence of OPEC itself was rooted in the attempts of the producing countries to maintain the posted price as a basis for the 50-50 profit split independent of the falling market prices. "There is little doubt," writes one observer, "that much of the loose legislation which cost Libya so much during its early years was due to the large sums of money which changed hands between certain companies and the oil ministry." Pressures for Libya to join OPEC and adhere to the posted price formula were handled in like manner: "Each time OPEC approached the government, one of the independents opened a new private bank account in Switzerland for a Libyan oil official, who then used his influence to veto membership."[17] Even the extensive venality of the Idris regime could only delay but not prevent the inevitable. Libya joined OPEC in 1965 and, with the support of that body, finally legislated the prevailing system of posted price taxation.

James Akins, presently U.S. ambassador to Saudi Arabia and before that the State Department's resident energy expert, recently characterized the Idris regime as "one of the most corrupt in the area and probably one of the most corrupt in the world. Concessions were given, contracts were given on the basis of payments to members of the royal family."[18] Joining OPEC did nothing to alter this fundamental character of the regime. Some flavor of the wheeling and dealing with which Libyan concessions were granted can be gleaned from civil suits now before two U.S. courts in which disgruntled participants in several of these schemes are suing their erstwhile partners. One of these involves the phenomenally successful Occidental concession with average gross earnings of around $1 billion per year. The other court case involves a less successful venture and is notable because one of the main figures achieved some attention as the unpaid advisor for the preparation of former President Nixon's tax returns.[19]

The oil industry strategy of developing reserves outside of the Persian Gulf area and particularly west of the Suez Canal was vindicated in the years following the June War. Idris, like his reactionary colleagues in the Gulf, paid mere lip service to the idea of cutting off oil shipments to countries aiding Israel. But that strategy was shown to be shaky when the oil workers and port workers of Libya went on strike after the June War and closed down Libyan oil operations for almost two months before they were finally suppressed.[20]

In 1968, Libyan oil production shot up an astounding 49 percent, while the increase in Middle East production generally increased by only 12 percent. At 2.8 million barrels at the end of the year, Libya had overtaken Kuwait as the third largest Middle East producer, behind Iran and Saudi Arabia. Production was up another 22.9 percent in 1969, bringing daily production to within 40,000 barrels of Saudi Arabia's. One reason for the increase was that Libya was the only crude source for a number of aggressive independent companies, notably Occidental and the Oasis group. Prices for Libyan crude, meanwhile, remained solid, owing to the great transportation advantage over oil from the Persian Gulf, and to an additional low-sulphur premium. Libyan oil was fetching $1.89 a barrel in Europe at this time, while Iranian light crudes were hovering down in the all-time low range of $1.20–$1.30. This advantage for Libyan oil on the European market resulted in a lot of Gulf oil heading east, primarily to Japan, further weakening prices there. Continued expansion of the Libyan industry, even on reserves that were only a fraction of those in the Gulf countries, was planned to reach some 5 million b/d by the early 1970s, when it was expected to level off. The major companies, pushed by their independent competitors in Libya, foresaw an eventual balancing of supply and demand and a consequent strengthening of European prices only after Libyan expansion had run its course. Discussing the problem the companies were having in meeting the Shah's production demands in Iran, *Middle East Economic Survey* wrote in May 1969:

> For the next two years it is estimated that at least 2/3 of the growth in Eastern Hemisphere oil demand will be supplied from North and West Africa, leaving an aggregate annual output increase of at most 5-6% for the Middle East producing countries east of Suez. In around 3 years' time, with Libyan and Nigerian production expected to level off, the outlook for substantial volume increases from the Arabian Gulf area throughout the 70's looks very bright.[21]

The vast quantities of oil being produced in Libya and other areas of the Middle East greatly exceeded the integrated capacities of the major companies. Thus, Occidental lined up long-term purchasers among independent European refining and petrochemical complexes, notably in Italy. For Standard of California, with its huge Saudi jackpot as well as stakes in Libya and Iran, one answer was to market low-sulphur Libyan crude to suppliers of U.S. utilities formerly dependent on Venezuelan (high-sulphur) crude. Standard, for example, made a deal with one such distributor, New England Petroleum Corporation (NEPCO), to supply Libyan crude for a 250,000 b/d deep water refinery that NEPCO had built in the Bahamas. NEPCO then contracted with an Italian ENI subsidiary, Snam Progetti, for the refinery construction. ENI got Standard to guarantee the financing of the refinery by agreeing to sign a separate long-term contract for crude deliveries to Italy from Standard's Aramco concession.[22]

Pushed by the dynamics of competition between the major and independent oil companies and unrestrained by any political interference from the complaisantly corrupt regime of King Idris, the market for Libyan oil constantly expanded but with relatively stable prices that were the envy of the rest of the industry. By 1969, Libyan production averaged 3.1 million b/d; in 1970 it peaked at 3.5 million b/d at mid-year. More than 80 percent was marketed in West Germay, Italy, Britain, France, and Holland, all of which (in that order) had grown heavily dependent on Libyan supplies. All the concern in 1967 about European dependence on Persian Gulf supplies had been resolved at the cost of a much heavier dependence on this single country. When the "winds of change" came upon the Middle East oil industry, they blew not from the Gulf, where everyone's attention was riveted, but from the hot North African sands of Libya in the west. The consequences of rapid and intense exploitation of Libyan oil resources included a strategic vulnerability the industry had not experienced in any other country at any other time.[23]

The Libyan Revolution

On September 1, 1969, a group of young army officers and soldiers seized control of the government and brought down the monarchy in a bloodless coup. Within a few days, Tripoli, Tobruk, and other points around the country were secured with no reported violence or resistance. The code words for the coup were "Palestine is ours." It was led by twelve officers who became the Revolutionary Command Council. All were lieutenants and captains and were under thirty years. On September 8, Captain Mummar Qaddafi, who had been in charge of the Benghazi part of the operation, was promoted to Colonel and made Commander-in-Chief of the armed forces. He emerged as the leader of the coup and the chairman of the RCC. The ideology of the RCC as it was expressed in the coming months had a strong populist component based on the rural and humble origins of many of its members. It was anti-Communist as well as anti-imperialist. In Qaddafi's view, history is shaped by contending nationalisms and religions, not classes. "We are socialists within the frame stipulated by the Quran," he says. "We are socialists indeed, but first and above all we are Muslims."[24]

The RCC immediately began to implement its anti-imperialist policies by abruptly canceling a $1.2 billion air defense system contract with Britain and by asking American Peace Corps volunteers, whom Idris had brought in to replace Egyptian teachers and technicians, to leave. The minimum wage was doubled and the contract labor system utilized by the oil companies to sidestep unions and minimum wage demands was outlawed. The "Libyanization" of the economy was initiated with a decree that all businesses (oil production companies excepted) had to be owned by Libyan nationals—which effectively meant the expansion of public ownership, due to the lack of indigenous bourgeois capital. Most of these moves were eclipsed in the eyes of the Western industrialized countries by the direction the new regime would take in achieving long-standing Libyan revenue demands on the oil companies. Although the revolutionary regime would move cautiously through most of the first year, the *Middle East Economic Survey* articulated the apprehension of those with a stake in the status quo:

> [The coup] could well signify a momentous realignment of the balance of forces in the Arab world. Up to now the inner dynamics of inter-Arab politics have been regulated by a precarious equilibrium between the so-called "progressive" or "revolutionary" states on the one hand and the traditional monarchies, which include most of the major oil producing states, on the other, with the political leverage of the former being to a certain extent offset by the financial strength of the latter. Now, with the abrupt transfer of Libya into the "revolutionary" camp . . . the previous balance between the two groups has been drastically upset. And this is likely to have important long-term implications on a number of fronts—not the least for the remaining oil-producing monarchies, for the oil industry, and for the Arab struggle against Israel together with the related strategic interests of the big powers.[25]

On September 8 the RCC announced that Mahmud Suleiman Maghribi had been appointed prime minister. Maghribi, thirty-six years old, had worked for Esso as a petroleum lawyer, and later headed up the oil workers' union. He had been in prison since July 1967 for his role in the oil workers' shutdown of Libyan export operations following the June War. While Maghribi served in this position only through mid-January, when Qaddafi assumed the Prime Minister's office, his appointment indicated that the new regime would make unprecedented demands on the companies. Through January, though, no moves had been made. The old monarchy's demand for a ten-cents-per-barrel hike in the posted price was assumed to stand. Any expansion or elaboration of this long-standing point of contention, along with more recent Libyan demands that the companies restrain production to the most optimum level, would come after the new regime had acclimated itself to power. At year end, Libyan production stood at a record 3.1 million barrels per day. The oil companies announced in November that Eastern Hemisphere profits had risen 21 percent, bolstered by rising European and Japanese demand, high tanker rates, and increasing attempts by East European countries to line up Middle East crude supplies.

Up to this point none of the producing countries, either singly or collectively through OPEC, had made any attempt to increase their revenues by increasing the price of oil. On the contrary, by agreeing to the special discounts on posted prices demanded by the companies in return for royalty expensing, OPEC countries implicitly acknowledged the companies' assertion that weakening market prices would not permit such a move. Iran, and the other countries, were pushing for increased production; Iraq was moving, virtually alone, toward developing a national oil industry that could supplant the international companies. Only Algeria had attempted to raise prices. In March 1969 the Algerian government legislated a 30-cent per barrel hike in its posted price, but a low production level and its virtual monopolization by French companies and the French market tended to diminish the impact of this move. As late as June 1970 the Shah of Iran specifically disclaimed any intention of demanding higher crude prices. OPEC gave verbal endorsement to the Algerian and Libyan demands at its conference in December 1969 but took no action to support those demands. Dissident technocrats like Tariki argued that coordinated production restraints among the producing countries would provide the appropriate setting for such price increases. Such proposals, however, foundered as companies and countries competed with one another for more markets, more production, and more revenues.

The atmosphere and the setting began to change as early as January 1970 when representatives of the national oil companies of Algeria, Libya, Egypt, and Iraq met in Baghdad to organize a possible counterweight to the Saudi-dominated OAPEC. The meeting produced an agreement to share information and coordinate marketing activities. Following the meeting Algeria formally called for a revision of its 1965 oil pact with France. On January 16 in Libya, Qaddafi became Prime Minister, Jallud became Deputy Prime Minister, and a civilian, Izz al-Din al-Mabruk became Petroleum Minister. A three-man committee was set up to negotiate with the companies. On January 20 Mabruk opened negotiations with the twenty-one companies operating in Libya with a speech that displayed the political and economic sophistication and the tone of firm understatement which the Libyans would perfect over the coming months:

> We do not wish to dig up the past, nor to bring it back to mind. What we wish to emphasize, with absolute clarity and frankness, is that the new revolutionary regime in Libya will not be content with the previous passive methods of solving problems. . . . The just demands we seek here are not intended to bring about any basic changes in the existing structure of the world oil industry, nor specifically in the price system. This does not mean to say that we approve of the existing system or believe that it is an equitable one. Libya will continue to support the collective efforts undertaken by OPEC to alter conditions in this respect. . . . What we ask of you gentlemen, is to recognize the changed circumstances in our country and, accordingly, let your actions be guided by flexibility and reasonableness.[26]

There was no indication over the next few months that Mabruk's warning had fallen on anything but deaf ears. Negotiations with Esso and Occidental, representing the majors and the independents respectively, found the companies still resisting the old ten-cent hike demanded for years under the old regime. In March one of the Libyan negotiators, the

Director of Technical Affairs in the Ministry, threatened the companies with unilateral action if they did not come to terms, and refused to confirm or deny rumors that the government was asking for a hike in the range of 40 to 50 cents. Petroleum Minister Mabruk stated in April that nationalization was not being considered at that time. In April former Prime Minister Maghribi was put in charge of negotiations; he announced that they would not be allowed to drag on indefinitely and that the ten-cent raise now being offered was not enough. In May the government warned the companies not to lessen their exploration activities, then running at about half the rate for the previous year. At the same time a Soviet geological survey team arrived in Libya, raising the specter of a Soviet role in Libya similar to their role in Iraq, offering the advice, technical assistance, and financing for which the producing countries had traditionally been dependent on the companies.[27]

Demand in Europe for petroleum products continued at a rate higher than anticipated. A shortage did develop in heavy fuel oil, mainly due to increased demand for low-sulphur fuel on the U.S. East Coast. This aggravated the tanker shortage, bringing tanker rates between January and May 1970 up to the 1967 postwar level. Then, in early May, the Trans-Arabian Pipeline, which carries 500,000 barrels a day of Saudi crude to the Eastern Mediterranean, was accidentally ruptured by a construction bulldozer in Syria. Syria refused to allow the company to make the relatively simple repairs, using the occasion to press for a substantial hike in its transit revenues. Tanker rates accelerated even faster, along with European market prices.[28]

Libya was coordinating its negotiating policy to an unprecedented, if low-keyed, degree with Algeria, and to a lesser extent, Iraq. A special meeting of Libyan and Algerian negotiators in late January issued a joint communiqué announcing an agreement to coordinate efforts to raise posted prices. A meeting in late May of the oil ministers of Algeria, Libya, and Iraq produced an agreement to stand together in their separate struggles with the companies. Specifically, the meeting called for: (1) setting a limit to "lengthy and fruitless negotiations"; (2) unilateral implementation of posted price increases, should negotiations break down; and (3) establishing a cooperative fund to support any country in the face of company attempts to cut off production and revenues. The final communiqué also stressed the need for "authentic national industry in all the principal phases of oil operations. . . ."[29] Following the meeting the *Middle East Economic Survey* perceptively noted that all three countries had "revolutionary" regimes, and all were involved in separate confrontations with their operating companies. "There is no need to stress," wrote the *Survey* editors, "the strategic pull conferred upon the three countries by their virtual monopoly of this most vital source of supply for Western Europe." Along with Syria, which kept TAPline shut down, these countries controlled more than 90 percent of Mediterranean oil exports, then amounting to more than 5 million b/d.

In early June the Libyan regime indicated its seriousness by imposing port dues on export operations. It also initiated the first of a series of production cutbacks that only added to the temporary and localized, but nonetheless real, shortage of oil in Europe. The first reduction was directed at Occidental, forcing the company to reduce production in one field from 800,000 b/d to 500,000 b/d. There is strong evidence that this initial cutback was motivated (as the government claimed) by Occidental's wasteful and greedy exploitation of this field in excess of the field's optimum capacity. It is equally likely that subsequent cuts were prompted by the effect they had on the companies' willingness to make "realistic" offers, although the government denied that they were in any way intended to affect price negotiations. But affect them they did. Following the first Occidental cutback of 300,000 b/d in June, the Oasis consortium (three American independents) was trimmed by 150,000 b/d in July. The government, in a separate maneuver, held up the launching of the massive Esso LNG (liquified natural gas) facility pending an agreement on prices. Not coincidentally, Algeria was at that time pressing for higher natural gas prices from its customers. In early July all internal marketing and distribution of petroleum products was taken over by the Libyan National Oil Company (Linoco). Amoseas (Texaco and Standard of California) was cut back by 30 percent or 100,000 b/d. The *Middle East Economic Survey* calculated in mid-July that the total Mediterranean "shortfall" stood at 1.07 million barrels per day, causing "fairly severe dislocation" in European refinery supply contracts and a "phenomenal rise" in Persian Gulf-Europe tanker rates.[30]

The readiness of individual countries to take advantage of this situation of imposed scarcity that they had helped create was quickly apparent. Syria, for one, continued to put off repairs on the TAPline. Some critics have charged that Libyan subsidies accounted for this. While this cannot be ruled out, it should also be noted that the increase in prices for Syrian crude exports, small as they were, easily compensated for the loss of TAPline revenue. In late July Algeria unilaterally raised the posted price for exports to France from $2.08 to $2.85 per barrel. The *Middle East Economic Survey* regarded this move as "a decisive turning point" in the history of the world oil industry, citing its specific effects on Algerian-French relations and the yardstick it would provide for other Mediterranean exporters. Moreover, wrote the *Survey*, it

constitutes a strong challenge to the whole of the traditional concessionary system and the internationally integrated financial structure of the oil industry whereby the oil companies, for reasons of history and convenience, have generated the bulk of their profits from the production stage of their operations.[31]

A month later, in late August, the Libyan government cut Occidental's production by another 60,000 b/d, bringing the total drop in production to 760,000 b/d. The government used the occasion to call again for "realistic" price offers from Occidental and Esso.[32] The companies held to their 10-cent offer even in the face of State Department advice that a 40-cent hike would be more reasonable.[33] Responsibility for the negotiations was transferred from the three-man committee and put at the cabinet level under the direction of Deputy Prime Minister Jallud. The industry took this to mean that the long economic and technical discussions were over, and that the dispute was nearing a decisive point. It was presumed that the government would take some action or announce some agreement on September 1, the first anniversary of the revolutionary regime.

The announcement came on September 4. Occidental Petroleum, the Los Angeles independent almost wholly dependent on Libyan crude for its new European markets and the company most severely affected by the production cuts, had agreed to raise its posted price by 30¢ per barrel, with an annual increment of 2¢ per barrel over the next five years. In lieu of the retroactive payments demanded by the Libyans, Occidental agreed to raise its tax rate to 58 percent (thus taking full advantage of the foreign tax credit provisions of the U.S. tax code). A more favorable rate for premium low-sulphur qualities was also established. The increase in Libyan revenue amounted to 17¢ per barrel; Occidental's production was allowed to rise to 700,000 b/d. Occidental promptly passed on the increases to its customers via the price escalation clauses in its supply contracts. European market prices had already risen to levels that exceeded the Libyan tax increase, a fact reflected in the sale of

Libyan royalty crude to Austria at $2.90 per barrel in late August.[34]

Following Occidental's "capitulation" to Libyan demands, the Oasis independents were invited to sign a similar accord—which they did two weeks later. Shell, one of the majors with a stake in Oasis, refused to sign, so its share of 150,000 b/d was cut off. Shell did finally come around a month later. Immediately following the Occidental agreement, on September 7, representatives of the major oil companies, headed by Wall Street lawyer John McCloy, met in Washington with State Department officials to enlist the government's backing in a showdown with Libya. The State Department's view was that an appropriate increase in Libyan postings would be in the vicinity of 40¢. By those calculations, then, the State Department correctly regarded the Occidental settlement as something of a bargain and presumably advised the companies accordingly. Subsequent to this, Esso and British Petroleum, the majors with the largest stake in Libya, tried to make virtue of necessity and unilaterally raised their posted prices by 30¢ per barrel. They followed this move a few days later by raising the posted prices at the Eastern Mediterranean pipeline terminals by 20¢ per barrel. Industry sources generally agreed that this new "plateau" of prices was likely to be a "more or less permanent fixture; or at least one can say with certainty that product prices will never fall back to the low levels of the past decade."[35] For the industry, these clouds of crisis were lined with silver and gold. Back in the United States the Interior Department's Assistant Secretary Hollis Dole (now an oil company executive) announced that U.S. consumers "aren't going back to the bargain basement prices we have so long been accustomed to paying for energy. Not this year, or next. Not ever."[36]

The repercussions of this Libyan victory were not long in coming. Iran raised its tax rate to 55 percent, and Saudi

Arabia announced that it was opening price talks with Aramco. In December, Syria agreed to begin talks with TAPline representatives for higher transit revenues in return for permission to repair the pipeline. In Iran the Shah announced that his country would indeed press for higher revenues per barrel, following the Libyan success. The twenty-first OPEC conference was held in Caracas on December 18. Algeria reopened price negotiations with the French companies, and Boumedienne paid a four-day visit to Libya at the end of December. Venezuela increased its tax rate to 60 percent and retroactive to January 1970. At year end in Libya production stood at 3.1 million b/d, still 500,000 b/d less than just before the first production cuts were enacted. Libyan Oil Minister Mabruk described the new environment as "very encouraging indeed." "A totally new situation has arisen in the oil market," he said. "Libya has given the lead and now it is up to OPEC to take positive action in respect of prices."

Teheran and Tripoli Negotiations: Confrontation and Collusion

The twenty-first OPEC conference met in Caracas, Venezuela from December 9-12, 1970, and its resolutions were published on December 18. They included (1) the establishment of a minimum income tax rate of 55 percent; (2) the elimination of disparities in posted prices in member countries; (3) a uniform increase in prices to reflect the improvement in the international petroleum market; (4) the adoption of a new system of adjustment of gravity (sulphur content) differential, and (5) the elimination of all remaining discounts. In addition, the conference decided on a negotiation strategy of three regional groupings—the Gulf countries, the Mediterranean exporters, and Venezuela and Indonesia. The

Gulf negotiations would be held first, and a producers' committee consisting of the oil ministers of Iran, Saudi Arabia, and Iraq was appointed. The conference also declared that its members were ready to legislate the new terms if negotiations did not proceed to a satisfactory conclusion. The resolution also set up a firm timetable, mindful of the companies' delaying tactics that had dragged out the royalty expensing negotiations for more than five years. The Gulf producers' committee was directed to

> establish negotiations with the oil companies concerned within a period of 31 days from the conclusion of the present Conference and report to all Member Countries through the Secretary General the results of the negotiations not later than 7 days thereafter.
>
> Within 15 days of the submission of the committee's report to Member Countries, an extraordinary meeting of the Conference shall be convened in order to evaluate the results of the committee's and the individual Member Countries' negotiations. In case such negotiations fail to achieve their purpose, the Conference shall determine and set forth a procedure with a view to enforcing and achieving the objectives as outlined in this Resolution through a concerted and simultaneous action by all Member Countries.[37]

Libya did not even wait for the Teheran negotiations to begin before summoning oil company representatives on January 3 and presenting a new set of demands based on the Caracas meeting. Libya took the position that the September-October settlements merely redressed an iniquitous state of affairs but did not represent a significant advance over existing industry price and tax terms. The new Libyan demands included: (1) a 5 percent hike in the tax rate—as in the Gulf countries (the earlier hikes in Libya being in settlement of retroactive claims)—with retroactive claims to be settled either by a cash payment (with a 10 percent discount), or five-year installments plus interest, or a tax rate above 55 percent; (2) a post-1967 freight differential and a post-May 1970 freight differential later reported to be 39¢ a barrel and 30¢ a barrel, respectively; (3) monthly rather than quarterly tax payments; and (4) increased investment in oil and non-oil areas amounting to at least 25¢ per barrel of oil exported. A general increase in the posted price was not specified but would presumably follow the rate set in the Gulf negotiations. This unilateral move by Libya was probably motivated by an increasingly sophisticated sense of negotiating strategy aimed at throwing the companies off balance, in addition to the particular fear that the Gulf countries, dominated by Iran and Saudi Arabia (neither of them famous for their militance versus the companies), would settle for too small an increase.[38]   Libya presented these demands on January 11 to Occidental and Bunker Hunt, two of the most vulnerable independents.

The companies moved to coordinate their strategy and secure the backing of the United States and other Western governments. Meetings were held in Washington by representatives of the United States, Great Britain, France, and the Netherlands—all parent countries to the major companies. Japan and other OECD (Organization for Economic Cooperation and Development—essentially the rich capitalist industrialized countries) members, who had a vital stake in the outcome as consumers, were merely "kept informed."[39] Simultaneously, Akins from the State Department, along with a representative of the Justice Department, flew to New York on January 11 where industry lawyer John McCloy was presiding over an ongoing meeting of top oil executives in his office at One Chase Manhattan Plaza. Akins was kept in an anteroom and from time to time was brought drafts of an agreement that would permit the companies to form a united front in their negotiations without fear of antitrust prosecution and to supply any independent company cut off by Libya with alternative sources of crude. One result was that the industry issued a joint message to OPEC dated January 13 and signed by Standard Oil (New Jersey), Standard of California, British Petroleum, Gulf, Mobil, and Texaco. It was made public on January 16, by then having the signatures of the other majors, a number of American independents, and one West German company. The letter expressed "great concern" at OPEC's "continuing series of claims" and proposed "an all embracing negotiation" between all the companies on the one hand, and OPEC on the other. This proposal, of course, was in direct contradiction to the "regional approach" adopted by OPEC. The companies' letter offered a revision of posted prices, an annual "worldwide inflation" adjustment, and a further transportation premium for Libyan and other "short haul" crude. In return, the companies demanded no tax increase beyond the now-standard 55 percent, no retroactive payments, and no obligatory reinvestment. The companies further stipulated that a settlement should be binding for five years.

The second concrete result of the New York oil executives' meeting was the Libyan Producers' Safety Net Agreement, which called for the companies to jointly supply other sources of crude to any company cut off by Libya. This was designed to forestall Libya's ability to threaten the more vulnerable independents with a partial or total shutdown. On January 15, McCloy, representing the oil companies as a group, flew to Washington to meet with Secretary of State Rogers and other top officials to secure high-level approval and backing for these measures. Meetings were continuing in Washington among the parent nations of the giant oil firms. None of this activity was known to the public.

The Teheran negotiations had been scheduled to begin on January 12, in the midst of all this oil industry-government conniving. Oil Ministers Amouzegar, Yamani, and Hamadi from Iran, Saudi Arabia, and Iraq respectively, were there. The first meeting lasted only ninety minutes and broke up in bitter acrimony. The companies had sent what they described as a "fact-finding" team that was not even empowered to set a date for substantive negotiations, let alone actually discuss the issues. The oil ministers were furious at this cavalier treatment and refused to discuss specific price demands until the companies indicated their readiness to negotiate seriously. An extraordinary OPEC ministerial meeting was set for one week later, January 19, in Teheran, amid talk of unilateral legislation of OPEC demands. The head of the OPEC team, Iranian Finance Minister Jamshid Amouzegar, told a press conference that OPEC members were "one for all and all for one; we are united and will take simultaneous action."[40] He pointed out that the companies had already hiked the selling price of oil by 55¢ over the last several months—20¢ could be attributed to higher transportation costs, 10¢ went to the producing countries, and 25¢ remained as additional company profit. Saudi Minister Yamani said after the January 12 breakdown, "I am afraid that they are going to have to pay a heavy price for this for it will hurt them as well as the innocent consumer."[41]

The most charitable explanation for the companies' sabotage of the January 12 meetings is that they were trying to buy time while a strategy was worked out in New York for dealing with the Teheran negotiations and the new Libyan demands. It could also have served the purpose of testing the solidarity and determination of OPEC to stick to its demands and, more importantly, to its tight schedule. It was, however, highly provocative and contributed greatly to the sense that a "producer-consumer confrontation of unprecedented magnitude" was brewing.[42] In light of what was to follow, there is much to suggest that the companies *were* interested in promoting an atmosphere of confrontation, and that a less abrasive means of delaying the January 12 meeting a few days was well within the collective means of the oil industry.

At the January 15 oil company-government meeting in

Washington, after the air of crisis had been communicated from Teheran to the American and European public, the oil companies, through McCloy, suggested that it would be appropriate for the President to send a high-level official to Iran, Saudi Arabia, and Kuwait to convey the United States' "concern" that an impasse in the negotiations not develop and that there be continued access to Gulf oil at reasonable prices. McCloy suggested that Under-Secretary of State John Irwin, formerly a Wall Street colleague, was the appropriate emissary. Irwin had just enough time to pack his bags, pick up the letters from the President, and be briefed by the oil executives, to leave the next day, January 16, for Teheran. On the same day, the companies' collective message to OPEC (described earlier) was made public. These two moves were presented to the public as evidence of company reasonableness in the face of the OPEC demands, and of the seriousness with which the United States took the threat of a potential oil supply disruption. The fact that the United States depended on the Middle East for only about 3 percent of its oil supplies at the time, while U.S.-based companies were aware that higher payments to OPEC could also mean higher profits for them, did not dissuade the industry from portraying Irwin as representing the "interests of the oil-consuming nations."[43]

The OPEC team took the companies' collective message as an offer to negotiate seriously. The extraordinary ministerial meeting scheduled for January 19 was postponed and instead, negotiations were scheduled to open that day. This was just about the time that Irwin was making his rounds with his "concern" tucked in his briefcase. It should be noted that up to this point there had been no specific threat to cut off oil supplies. The "concerted and simultaneous action" mentioned in the original OPEC resolution of December and repeated by Amouzegar after the January 12 breakdown was purposefully vague, but its context clearly seemed to refer to the decision to unilaterally legislate price hikes if the negotiations were not successful. After the Libyan experience of selective and partial cutbacks the possibility of such actions by the Gulf producers could not be ruled out. However, the different nature of the concession patterns in the Gulf, in addition to the history of "moderation" and production maximization there, indicated that such moves would only be taken as last resorts. It is difficult to see, therefore, what purpose Irwin's trip had other than contributing to the environment of imminent crisis or showdown in order to justify a new, higher plateau of costs and prices that would boost company profits as well as government revenues. The main controversy at the time of his trip was a procedural one: would the OPEC team represent only the Gulf producers, as OPEC insisted, or all the OPEC exporters, including Libya and Algeria, as the companies wanted. Irwin, it seems, was quickly convinced by Iranian officials that not only would OPEC insist on the regional approach, but that this was to the companies' advantage, since all-embracing negotiations would give the Libyans a more direct role in setting Gulf prices. Irwin was assured that the Gulf countries would not attempt to leapfrog Libya even if that country would get a better deal later. In any case, Irwin communicated this to Washington right away, which informed McCloy, who told the company executives. When the first negotiations ended on January 21 it was reported that the negotiators had gotten around representation by agreeing on regional negotiations coupled with the Gulf states' assurances of five-year agreements, regardless of prices set by other regions.[44]

Following the initial high-level negotiations in Teheran on January 19 and 21, economic and technical experts from both sides began several days of discussion on January 24. According to subsequent reports, the companies did not finally agree to treat the negotiations as exclusively applicable to Gulf states until the next high-level session on

January 28.[45] In the concrete price negotiations that followed, the companies offered an immediate increase of 15¢ per barrel rising to 22¢ by 1975. According to the companies, OPEC's initial demands amounted to an immediate 49¢ hike rising to 87¢ over five years. It seems that the countries quickly settled on a 35¢ increase with the companies raising their offer to 20¢. There was every indication that a settlement would be reached without a confrontation; but on February 2, the day before the OPEC ministerial meeting, negotiations broke up with the companies claiming that the remaining issue was not so much the price increase but their demand for assurances of stability. The OPEC meeting took a predictably dim view of this and resolved to legislate the price changes on February 15 unless the companies signed before then. The resolution also specified that any company not complying with legislation would be embargoed although oil would continue to be made available to consuming countries. As an enticement to the companies, the OPEC meeting expressly limited its support of Libya and Algeria to the framework of the Caracas resolution and the pending Gulf settlement. OPEC also rejected company insistence that production controls or export limitations be eliminated during the five-year life of the agreement.

Whatever the companies lacked in official assurance that the Gulf states would hold to the agreement for five years should have been offset by the new resolutions, which pretty much left Libya and Algeria on their own as far as getting *better* terms than Teheran. Nevertheless, it was not until February 12, three days before the February 15 deadline, that the companies' negotiating team met with its OPEC counterpart. The reasons for the delay have not been forthcoming. It provided the perfect setting for a hairline finish. The Venezuelan Oil Minister, Perez La Salvia, observed in New York on February 11 that the companies wanted what looked like an imposed settlement to justify the highest possible market prices, and asserted that there would be no cutoff.

The Teheran agreement was announced on February 14, one day before the OPEC deadline, taking full advantage of the anticipation and sense of impending crisis that had been carefully nurtured over the preceding six weeks. There is nothing to indicate that the terms were any different from those offered at least two weeks earlier. The financial terms called for a hike in the posted prices from 35 to 40¢ per barrel, depending on the quality of the crude; an elimination of previous discounts, worth 3 or 4¢ per barrel; and a scheduled increase of 5¢ per year over five years, plus a yearly increase of 2.5 percent on the posted price as an inflation allowance. The tax rate was fixed at 55 percent. Government revenues would immediately increase a total of 27¢ per barrel, up from the present average of $1 per barrel, rising to 54¢ per barrel by 1975. The total increase in revenue to the Gulf states was calculated to be $1.2 billion in 1971 and $3 billion in 1975.[46]

Following the Teheran settlement, attention turned to Libya and the Mediterranean region. Almost lost in the publicity was the fact that Syria had reached a settlement with TAPline which nearly doubled Syria's annual income on transit rights, making it $8.5 million, and renewed the flow of some 500,000 barrels per day of Aramco oil to Mediterranean ports. Libya was delegated by OPEC to head the Mediterranean talks, although the procedure was different from the one followed in the Gulf. Essentially, Libya would try to get the best deal it could from the companies, and those terms would then be applied by the rest of the Mediterranean exporters, including those of Saudi Arabia and Iraq. A two-week limit on negotiations was set. Libya was guaranteed a hike in posted prices of at least 54.5 cents per barrel, based on an across-the-board application of the Teheran terms.

Before the OPEC Mediterranean procedure was set, Libya's Jallud commented on the Teheran agreement by saying that it did not meet "our minimum demands." The Algerians, maintaining close contact with Libya, announced a "complete identity of views" on the subject.[47] Algeria proceeded to warm things up by resolving its continuing controversy with France through a 51 percent takeover of the French companies. Algeria and Libya were unhappy because, in their view, the Teheran agreement secured too little a financial gain, in the light of the unique (and possibly temporary) bargaining strength of the producers, and because the increase was arbitrary and not related to market prices.

During the negotiations the companies remained headquartered in London. Libya refused to bargain collectively with the companies, insisting that head negotiator George Piercy of Standard (New Jersey) only represent his company. The companies got around this by formally complying in making separate but identical offers. Of the companies' first proposal, Jallud said, "When we received the cable from the companies, we laughed, and laughed and laughed and laughed."[48]

After the laughter stopped there were several Libyan rejections of company final offers, along with repeated extensions of deadlines. The Libyans imposed a high degree of secrecy on the details of the negotiations. The *Middle East Economic Survey* estimated the demands to entail a posted price-hike of $1.20 per barrel, including freight premiums, from which the government share would be about 73¢. By the middle of March the companies' offer had reportedly reached 55¢ per barrel. When the settlement was finally announced at the beginning of April, the posted price had risen 90¢. There were the other standard features from Teheran: 55 percent tax rate; five-year guarantee; 2.5 percent inflation allowance. Tax payments were to be made monthly instead of quarterly.

The Libyan demand for a specific reinvestment formula was left to be worked out on a company-by-company basis. Negotiations between Iraq and Saudi Arabia and the companies over the East Mediterranean postings of Iraqi oil continued until the summer but for all intents and purposes the landmark negotiations of 1971, in pursuance of the resolutions of the twenty-first OPEC Conference in Caracas, had been concluded.[49]

Assessment and Evaluation

Oil industry advisor Walter J. Levy wrote a few months after the negotiations that as a result of the Tripoli-Teheran-Tripoli settlements, "the economic terms of the world trade in oil have been radically altered," and that "clearly, a very real challenge to the historical structure and operation of the internationally integrated oil industry is emerging."[50] Since this was the standard refrain in the industry and popular media, it is worth a closer assessment.

There can be no question that the nature of the 1971 OPEC price negotiations was substantially different from previous negotiation attempts throughout the 1960s. On one level, this simply represents the fact that OPEC collectively had learned its lesson: that negotiations are a tactic used by the companies to interminably delay and circumvent the issues at stake—hence OPEC's insistence that these negotiations not be subject to such tactics. In pursuing this it is moreover clear that the OPEC countries had learned the value of suppressing political differences among themselves in pursuit of a commonly beneficial goal. There is no doubt that the 1971 negotiations represented an historic shift of political power from the Western industrialized countries and companies to a small number of Third World countries that happen to be the world's major oil producers. The 1971 negotiations represented the first serious OPEC attempt at collective bargaining and the results, not surprisingly, were unlike those of any previous encounter between companies and governments.

At the same time it is important to recognize the limited nature of their gains and the particular set of circumstances that permitted even this limited success. In the first place, OPEC had virtually nothing to do with the development of the shift in political and economic power represented by the negotiations. The shift in power was a cumulative one, developing over the years out of larger historical forces and in particular out of the contradictions in the world oil industry that eroded the monopoly power of the largest companies. Specifically, developments in the Libyan oil industry, and the readiness of the Libyan (and Algerian) revolutionary regimes to take advantage of those unique (and probably temporary) conditions, set the stage for the success of the OPEC negotiations. In other words, the basically conservative political orientation of OPEC under the domination of Iran and Saudi Arabia remains despite the 1971 developments. In fact, given the Libyan achievements in September-October 1970, OPEC could have done no less than it did do, though it certainly could have done more. As another observer has already noted, "One is compelled to conclude that the producing countries' collective bargaining power was really a function of changing external conditions rather than an independent force capable of development on its own."[51]

While it is possible to agree with Levy that "a very real challenge" to the industry is "emerging," it must be said that the 1971 OPEC negotiations revealed the *potential* of that challenge rather than any real shift in power or structure. While the achievements of the negotiations—the price-hikes, the inflation index—should not be underrated, they all remained strictly financial in nature, increasing the revenue of the host governments but leaving the companies free to control industry operations and extract whatever profit they could in the markets they still controlled. Although the OPEC resolution stated the demands in terms of "the general improvement in the conditions of the international petroleum market," the actual increases were quite arbitrary and not specifically related to those changed conditions.[52] In his press conference of January 24 the Shah of Iran complained, as he had many times before, that the producing countries got only $1 a barrel for oil that is sold as petroleum products at something close to $14 per barrel, the rest going to the oil companies, various middlemen, costs of transportation and refining, and consumer country taxes. After the Teheran settlement the countries in the Persian Gulf were getting between $1.25 and $1.30 per barrel. This is hardly an indication of momentous change, even without considering the post-Teheran market price-hikes. The initial Libyan rise of 30¢ per barrel must be modestly assessed against the opinion of the State Department's Akins that a 40¢ hike would have just brought Libyan crude up to a par with Persian Gulf crude in the main markets.

One alternative would have been that suggested by the Algerians: price Middle East crude on the basis of U.S. product prices less transportation, refining, and other costs. Such an option would have done more than simply provide a bigger chunk of revenue—it would have concretely related crude prices to "the general improvement in market conditions" and the corresponding improvement in company profits. Instead, the OPEC countries settled for a flat 5¢ increase per year plus a fixed 2.5 percent inflation escalation, both of which were related only arbitrarily to the actual state of inflation, market prices, and company profits. By locking these provisions into a firm five-year contract, OPEC gave the companies a free hand over a significant period of time to raise prices sharply, blaming it all, of course, on the OPEC "challenge."

The companies hesitated not a minute to raise their crude and product prices to cover the tax hikes and then some. In Britain for example, the 28¢ tax hike was covered by a price rise of 42¢ The overall results were characterized by Kenneth Hill, a prominent Wall Street financial analyst, as "an unexpected boon for the worldwide industry. . . . As a result earnings in the downstream function in the Eastern Hemisphere are for the first time in recent memory quite attractive. . . ."[53] All the talk of crisis and shortage was temporarily shelved around the end of the year, when "increased earnings . . . were the order of the day" among the major oil companies, with a "phenomenal 36% jump in overseas earnings" reported in 1971.[54]

Petroleum product prices in the United States rose in this period also, although the low level of U.S. imports of OPEC oil made the U.S. price structure somewhat independent of OPEC negotiations. Almost all OPEC oil was to be sold in Europe and Japan, although the bulk of it through U.S. companies and their affiliates. The benefit to the U.S. balance of payments from oil company remittances also helped to offset the negative impact of higher prices for imported crude. In a more sinister vein, there has been some suspicion that the Nixon administration "saw increased oil prices as a quick and easy way to slowing down the Japanese economy, whose exports were bothering America mightily at the time and which would be hurt more by rises in oil prices than any other nation."[55] The other consequence of the Teheran settlement was that for the first time in the public mind Middle East oil and Middle East politics became inextricably linked with the "energy crisis." This bolstered the efforts of the domestic "energy industry" to generate higher prices and profits, government research and development subsidies, and lent political legitimacy to increasing monopoly control of energy resources, markets, and capital.

Financial analyst Hill's characterization of the Teheran

settlement as an *"unexpected* boon" now deserves a close
look. As far back as 1968 Shell president D.H. Barran noted
that:

> Pressure from the producing governments on costs is something
> that we can live with provided we are not at the same time denied
> freedom to move prices in the market so as to maintain a com-
> mercial margin of profit.[56]

At about the same time a prominent British oil economist
discussing the concept of participation, opined that the
companies would be willing to enter an arrangement which
gave the *appearance* of partnership while denying any effective role to the government partner over operations, "and
especially over offtake, prices and even investment. . . ."[57]
The main fear of the companies, she wrote, was the specter
of competitive production increases, with further weakening
of prices and company flexibility in the markets. "Clearly
the only alternative would be for the governments as a group,
perhaps through OPEC, to join with the companies as a group
in an effective cartel to plan the amount and distribution of
oil supplies." Such a development, she concluded, was
impossible in present cirucmstances.

It is tempting to look at Teheran 1971 as just such a development, for this was its net effect. This helps explain the
otherwise baffling and provocative moves of the companies,
with the assistance of the U.S. government, and the acquiescence of the other industrialized consumer countries, to
create an atomsphere of confrontation and crisis that was
totally unwarranted by the objective situation. Beyond this,
however, there is no direct evidence of collusion between the
major companies and the OPEC countries. The OPEC
demands were genuine, albeit prompted by the unilateral
initiative of Libya and Algeria. Once the need for a tax hike
became unmistakably clear, the companies could mold and
channel those demands into specific forms that did *not*
threaten the structure of the industry and at least temporar-

ily enhanced its stability and profitability. In this way they were aided, of course, by the fundamentally conservative and pliable politics of the dominant OPEC countries, Iran and Saudi Arabia, and were able to short circuit more serious challenges from Libya, Algeria, and Iraq.

8

Controlling Oil Resources: Nationalization and Participation

Once the crude oil price issue was temporarily resolved (by the middle of 1971) in a manner favorable to both the producing countries and the international companies, the question of control of the basic decisions concerning the use and exploitation of the OPEC countries' oil resources regained prominence. The questions of political and economic control and self-determination that lay behind pressures for nationalization were not entirely absent from the price negotiations of Teheran and Tripoli. Company control over price and production levels had been circumscribed by the growing political power of the countries, individually and collectively, but the companies retained control of markets, transport facilities, technology, capital resources, and title to almost all the oil produced. Operational control of the industry was still in their hands.

The question of prices and revenues can be viewed as part of the larger question of overall control. OPEC itself originated from just such a perception of the significance of the unilateral company posted price reductions in 1959-1960. In practice, OPEC tended to abstract to the greatest extent possible financial questions from the openly political issues surrounding control. This tendency provoked the emergence of the anti-OPEC group of oil technocrats and economists grouped around Abdullah Tariki and Nicholas Sarkis, whose political program for the Arab oil producers was based on the need for the countries to nationalize the producing companies. They argued that financial benefits would be greater than under the existing concession system but stressed that the question of political control was central and should be resolved even in the face of a temporary loss of revenue.

The one notable attempt at nationalization of Middle East oil facilities was the Iranian nationalization of British Petroleum in 1951. That attempt was defeated by the combined efforts of the United States and British governments, their secret services, their oil companies, and their political clients, including all of the other Middle East oil producers. In place of British Petroleum the Iranians got a consortium made up of all the giant oil companies with a sprinkling of American independents and the Shah. The oil companies and local anti-nationalist forces never failed to remind the nationalists of later years of the penalties of such "extremism." The story had been different, of course, with the nationalization of the Suez Canal, where Nasser's success was in proportion to Mossadeq's failure. Subsequently, Egypt, Syria, and later Iraq nationalized significant sectors of their economies, as did a number of other newly independent Third World countries. Elsewhere, in Latin America, the Mexican nationalization of its petroleum industry proved enduring, despite the open attempts of the international companies to reverse it. The experience of other countries was not as reassuring. As a result, the political constituency for nationalization was located by and large outside the regimes of producing countries. The exception was Iraq, but even there the takeover of the concession area in 1961 was not made irrevocable until the late 1960s.

Elements of national control were introduced to the Middle East in small doses with the formation of joint ventures between the host countries' national oil companies and small foreign independents. We have already noted the details of this development, beginning in 1957 in Iran, and the effect of this marginal production on the industry price structure, which accentuated growing contradictions. The most concrete contribution these ventures made to the trend toward national ownership was the development of an experienced local cadre of oil technicians and administrators. While the amount of oil produced under these joint-venture arrangements has remained relatively insignificant, it was variations of these joint-venture arrangements that set the style for all new concessions after 1957 except for Libya and the British-ruled Persian Gulf emirates.

After the June War, even this changed. Only companies willing to consider "partnership" with the local national oil company were eligible to be considered for new concession areas. On a broader level, popular pressure for nationalization increased due to the primacy of U.S. oil interests in the area and the U.S. role in supporting Israel. Only in Iraq did this take immediate and concrete form, but the trend was unmistakable. In a major strategic move to head off and defuse this pressure, Saudi Oil Minister Yamani introduced the concept of "participation" as an explicit alternative to nationalization. In June 1968 he announced that the Saudi government, through its national oil company, would seek some form of partnership with Aramco:

> So many new-style agreements in the area are based on the partnership principle, and the June War, with all its psychological repercussions, has made it essential for the majors—and not least Aramco—to follow suit if they wish to continue operating peacefully in the area.[1]

Later that same month at its sixteenth conference OPEC formally took up the question with a resolution endorsing "participation" as applicable to the old concessions on the grounds of "changing circumstance." As defined by Yamani and others, participation meant that the host country would hold equity, initially at a level around 20 percent, in the producing companies like Aramco and IPC. Whether this would involve any more than a 20 percent holding of the company's stock and a seat or two on the board of directors was left vague. Since Saudi Arabia already had seats on the board (which had proved more or less honorific and without decision-making substance), and since many of the producing companies were nonprofit—they sold crude oil at cost to the parent majors—participation surely had to mean something more. As it was, participation remained on the OPEC agenda over the next several years as the subject of many studies, and its main proponent, Yamani, preferred to define it simply in terms of what it was not: namely, nationalization.

Defining Nationalization and Participation

Ambiguity has continued to characterize much of the discussion of the comparative merits of nationalization and participation, so much so that in many instances the terms have come to be used interchangeably. It seems necessary at this point to clarify matters. Nationalization refers to the takeover by a nation state of the assets and resources of a private individual or corporate entity. In the case of Middle East oil, these private entities have been, without exception, foreign-owned. The foreign individual or corporation may continue to play a role in the operation of the nationalized entity, as a purchaser or even as an operator under contract for a set fee or share in the product. Ownership of the capital and means of production, and the power to make decisions concerning output, investments, prices, and so on are in the hands of the state or its agent, such as a national oil company. The use of the term nationalization assumes no judgment concerning the representativeness of the state in question, or whether such national ownership will be exercised in the popular interest. Our use of the term does rule out, however, mere nominal or cosmetic use of the term, such as in Iran under the Shah's regime. The consortium set up in 1954 left formal ownership of the resources and means of production in the hands of the state. Real control, and ownership of the means of production, remained with the foreign companies. There was no nationalization of oil in Iran, *pace* the Shah, at least until 1973.

Participation means, on the face of it, a state-company partnership in which ownership, control, and rewards (profits) are shared proportionately. Unless the share is specified, it is a rather meaningless term. One hundred percent participation is, by our definition, nationalization. Fifty-one percent nationalization is, by our definition, participation; 51 percent participation is significant in a way that 25 percent participation, for example, is not. Majority participation, or even nationalization, acquires significance with exercise of ownership and control that is real and not nominal. Frequently the significance of nationalization or participation can be measured only on the basis of the proportion of state decisions which go against the interest of, or would not be taken by, the foreign owner. A clear example of this is the Iraqi decision, in 1969, to develop fields confiscated from IPC in 1961 with Soviet technical and financial assistance. Any political or economic evaluation of the merits of one country's arrangement, or policy, must rest on a concrete and specific examination of that policy and the circumstances. Nationalization and participation acquire political value only through exercise and implementation. The 51 percent participation of Libya in many of its oil-producing arrangements can represent a more important, more radical, and more progressive relationship than the 60 percent or even 100 percent participation/nationalization of Saudi Arabia if that regime continues to see its interests tied to U.S. power and hegemony.

It is necessary to concentrate on the circumstances surrounding nationalization and participation decisions and to understand the interests and dynamics behind these steps as well as their consequences. In this process it is well to keep in mind the legacy of struggle over the issue of national control of resources and economic self-determination as regards Middle East oil and the conflict within and among the various regimes over these questions.

Nationalization has been one of the most explicit means that a dependent Third World country could use to break the dominant hold of foreign capital and economic interests over a national economy. Nationalization may or may not be successful in such an endeavor; that depends on the country's ability to mobilize all its popular political and economic forces to successfully operate the nationalized industry or develop alternatives for the economy. In the case of an industry as strategically important as oil, a country must mobilize its domestic and international forces to withstand retaliatory attacks—military, political, or economic—of the interests opposed to expropriation. The experiences of Iran and Cuba are example enough. Nationalization has historically been seen as the only way to radically break the hold of foreign monopoly capital over an underdeveloped economy, and to change radically the distorted foreign-oriented structure of the economy. In the oil industry, at least, nationalization has been seen as the one effective way of changing the structure of the industry and modifying or eliminating the dominant role of the giant firms.

Participation, on the other hand, was advanced by Saudi Arabia (through OPEC) in specific reaction to the threat and intent of nationalization. Sheikh Yamani was most explicit about this in June 1969 in a speech entitled "Participation Versus Nationalization":

> For our part, we do not want the majors to lose their power and be forced to abandon their roles as a buffer element between the producers and the consumers. We want the present setup to continue as long as possible and at all costs to avoid any disastrous clash of interests which would shake the foundations of the whole oil business. That is why we are calling for participation.[2]

Yamani based his case for participation and the need to strengthen the majors on the surplus conditions of the late 1960s. Only the international companies, he argued, could maintain crude prices above a purely competitive level. If one country nationalized its oil the companies would simply increase their offtake from other countries, as happened with Iran in the 1950s and with Iraq all through the 1960s. At this point there was intense competition between the producing countries, notably Iran and Saudi Arabia, for increased production as a source of greater revenue. Even if all the OPEC countries acted together in nationalizing producing facilities, Yamani said, competition and price cutting among the national oil companies would continue and accelerate a downward spiral of prices and government revenues.

Yamani's argument utilized a convincing scenario, but its plausibility was shaped by Saudi Arabia's persistent veto of any attempt to set up a production plan under OPEC auspices. Each country would produce a specified amount of crude according to a schedule which would be keyed to market demand, thus preventing surplus crude from reaching the market, and maintaining prices at a profitable level. There are a good number of political and technical impediments to the success of any such plan, with enforcement dependent on the solidarity of the countries involved and the temptation to shade prices for a marginal sale ever present. The potential for cooperation was also there, but Saudi Arabia used its weight as largest producer to prevent production plans from getting past the drawing boards. By preventing one step that would be a necessary component of nationalization, Saudi Arabia made nationalization all the more difficult for the other oil-producing countries.

Algerian Nationalization

The basic momentum for nationalization, and thus indirectly for the participation alternative, came from Algeria, Iraq and, after the September 1969 revolution, Libya. In early December 1968 Algeria took over 51 percent of the Getty operating assets and set up a new operating company with Getty in which the state oil company, Sonatrach, had the controlling interest. Tariki and Sarkis hailed this new arrangement as the only kind of participation worthy of the name and suggested that Saudi Arabia, if it was serious about asserting some measure of national control over its resources, would at least move to implement a similar arrangement with the small Getty operation in its neutral zone territory. Algeria quietly continued to build its national oil industry capacity by forming a number of 51 percent ventures between Sonatrach and small nonintegrated drilling, exploration, and other specialized firms, some European and most American. In 1969 Sonatrach spent about $70 million on exploration through these joint ventures, compared with some $30 million in exploration expenditures by the French companies that dominated Algerian production quite profitably. In June 1970 Algeria nationalized Shell, Phillips, and other small operators. Later that year Oil Minister Belaid Abdessalam outlined the strategy behind Algeria's oil moves in a speech to a meeting of Arab oil economists:

In short, the path . . . is to take in hand the national oil industry and, as much through the many production and service activities related to it as by the cash flows it creates, act so as to integrate it fully into the economic life of the country, and by means of a systematic commercialization of the products derived from it make it a permanent source of income. In other words, it is a question of "sowing" oil and gas in order to reap factories, modernize our agriculture, diversify our production and create an organized national economy oriented towards progress.[3]

This was in October 1970 during the months leading up to the Teheran price negotiations. Speaking of the existing structure of the industry (dominated by the major companies) and the political differences among the producing countries over the question of national control, Abdessalam observed that

the forced relationships and iniquitous state of affairs which have been imposed on us cannot be modified by negotiations and talks which last for years and result at best in minor changes to the status quo. As the experience of our various countries shows, negotiations are only one of many tactical means whose effectiveness depends basically on our will to regain our usurped national rights, on the actions we take to achieve this and on the new situations and advantages which we ourselves create to dominate the game of negotiations and change forced relations into legal ones and lopsided relationships into equal ones between partners who respect their mutual interests. . . . the problems raised cannot really be restricted to the readjustment of prices and the redistribution of financial charges. They also call for a reorientation of our international economic relations. . . . In the present context and in view of the revolution taking place, our desire to reorganize the system of exploitation of our natural wealth to ensure our own control, in order to use it primarily for our development, is in no way less legitimate than the major concern of the consumer countries to ensure the security of their supplies.

In February 1971, in the midst of the historic price negotiations, Algeria took 51 percent control of French oil interests that controlled 70 percent of Algerian crude production. France retaliated by organizing and promoting a boycott of Algerian oil and other exports; World Bank loan applications by Algeria were rejected with United States support; and the Sonatrach-El Paso liquified natural gas (LNG) project was temporarily held up by the White House, evidently in response to the French campaign. In April 1971 Algeria unilaterally raised the posted price on its oil to $3.60 a barrel and promulgated a new oil code which abolished the concession system. Compensation settlements were reached with the French companies over the summer of 1971, and the French embargo was said to have modified the Algerian position in these negotiations. The settlements left Algeria in direct control, through Sonatrach, of 77 percent of its crude oil production of about 1 million b/d, 100 percent of its natural gas production, and majority or total control of all other phases of the industry. In December 1971 a U.S. refiner with a huge refinery complex in Puerto Rico, Commonwealth Oil, made an agreement with Sonatrach that covered "one of the largest transactions ever made between a private company and a state-owned concern"[4] providing for an $8 billion purchase of 380 million tons of crude over 25 years (about 400,000 b/d). The major oil companies had historically refused to deal with state-owned companies for fear of sullying their commitment to free enterprise. It was therefore significant when Exxon made a large purchase of Sonatrach crude in January 1972; several months later Gulf Oil Company did the same.[5]

In the process of struggling to free itself of French economic control in oil and other economic sectors, Algeria has relied heavily on U.S. business interests as well as assistance from the Soviet Union and the socialist countries. For a while John Connally represented Algerian legal interests in Washington, and when he became Treasury Secretary former Secretary of Defense Clark Clifford took over. In March 1969

an Algerian trade delegate in the United States noted that top
U.S. consulting firms had earned over $20 million over the
previous five years from Algerian contracts and pointed out
to his U.S. audience:

> As a developing nation, Algeria will continue to provide a growing
> and important new market for American goods, industrial equip-
> ment, services and know-how. Moreover, Algeria may well pro-
> vide a key to American expansion into the markets of the
> Mediterranean area because of its geographical position.[6]

The Arthur D. Little Company of Cambridge is said to have
advised Sonatrach "at every turn, and played a crucial be-
hind-the-scenes role in the nationalization negotiations with
Compagnie Française des Pétroles and with ELF-ERAP [a
second French company]."[7]

While these developments are important to any assessment
of the Algerian economic development effort, their political
connotations are secondary to the fact that Algeria has been
able to establish a high degree of national autonomy in the
economic sphere, and in so doing has managed to serve as a
model to some countries in similar positions (like Libya) and
to put pressure on the old concessionary arrangements even
under the reactionary Gulf regimes. The fact that it has made
a number of commercial agreements with U.S. firms is no
more indicative of reactionary tendencies than Iraq's agreements with the Soviet Union and other socialist countries to
help develop its hydrocarbon resources are unqualifiedly
progressive.

Nationalization in Iraq

Iraq took the first step on the long road to developing an
independent and autonomous national oil industry with the
promulgation of Law 80 in 1961, which expropriated all of

the IPC concession area except those fields in production. The expropriated area included some fields, notably North Rumaila, where IPC had discovered oil but had taken no steps to produce it. Over the next several years there was sharp political struggle in Iraq over further courses of action. The IPC companies, of course, vowed to use all their resources to prevent any settlement which did not include rights to the North Rumaila fields. We have already noted the 1965 government-company agreement that would have given IPC full rights to North Rumaila in return for acknowledging nationalization of the rest of the expropriated concession area and recognition of the full rights of Iraq National Oil Company (INOC) to develop them through joint ventures or other means. This settlement was favored by the technocrats running INOC but was opposed by the leading radical nationalist political organizations, including the Ba'ath and the ANM (Arab Nationalist Movement), and by Tariki. The political instability that characterized Iraq throughout the 1960s was most intense in the 1965-1967 period, including more than one attempted coup, and in 1968, a successful one. The moderate forces in the government and INOC were not able to muster sufficient support for the agreement, and as a result it was never ratified. This represented a success for the nationalist forces who insisted that INOC take sole responsibility for developing North Rumaila. The June War, and the anti-American and anti-British political pressures it created, eliminated any possibility that Iraq would give North Rumaila back to IPC. In early August 1967 Law 97 was promulgated, specifically barring the restoration of North Rumaila to the Anglo-American companies and assigning exclusive development rights to INOC. INOC was expressly forbidden to grant any concessions. A joint venture between INOC and the IPC companies was not ruled out, but the political atmosphere made such an outcome extremely unlikely.

This move was followed a month later with the reorganization of INOC under Law 123. The company's responsibilities were expanded and its autonomy reduced. The post of Managing Director, usually staffed by a top technocrat, was abolished and decision-making power was invested in a board of directors nominated by the Minister of Oil and approved by the Cabinet. Final authority was vested with the Cabinet. In establishing this high degree of political control over the national oil company, an explanatory memorandum accompanying Law 123 noted that the company is expected to work

> for the development and expansion of oil exploitation in Iraq in all phases of the petroleum industry including the production of crude oil and petrochemicals and refining, export and marketing operations; and to undertake all domestic and foreign operations required to promote the growth of the national income and achieve a self-sufficient and balanced economy.[8]

Iraq looked mainly to France and the Soviet Union for assistance in implementing its national oil policy. In November 1967 INOC signed a contract with the French company ERAP for the development of some very promising areas (not North Rumaila) expropriated under Law 80. Under the terms of the contract, all exploration and development costs were to be put up by ERAP as a loan, repayable only after commercial production began. Fifty percent of discovered reserves was to be under the exclusive control of INOC; the remaining 50 percent was to be developed cooperatively. Of this cooperatively developed oil, ERAP would buy 30 percent: 18 percent at cost of production—plus 13.5 percent royalty plus one-half the difference between that sum and the posted price—the other 12 percent for the cost of production plus the royalty. INOC was to have full control over the remaining 70 percent of production to sell at whatever prices it could get on the market. If these prices are not satisfactory ERAP has to market a substantial portion of INOC oil through its channels at a per-barrel fee of one-half cent on the first 100,000 b/d and one and one-half cents on each barrel above that amount. This agreement came in for sharp criticism from the technocrats formerly in charge of INOC on the grounds that it would result in greater profits for ERAP and less revenue for Iraq than under the traditional concession system. Western oil economists have made similar charges. It is impossible for us to compute here the merits of this criticism. Whatever validity it had in the days of market surplus and weak prices is surely negated by the strong seller's market that developed around 1971. In any case, such arguments do not and cannot measure the independence and autonomy established by INOC that would have been absent in any concession with the IPC companies.

To aid in the development of the North Rumaila fields, where oil had already been discovered, Iraq turned to the Soviet Union. This was due in part to IPC threats to sue any Western company involved in producing or even purchasing "their" oil. A Soviet delegation visited Iraq in late 1967 and signed a letter of intent that in July 1969 was translated into a definite program by a Soviet-Iraq agreement. It was designed to prepare and put the North Rumaila fields into operation at an initial production rate of 100,000 b/d beginning in early 1972, rising to nearly 400,000 b/d by 1975. Implementation of the government agreement came through a contract between INOC and the Soviet Machino-export Organization, providing for drilling rigs, geophysical and geological teams, pipeline construction from the fields to the Gulf port of Fao, and ancillary services. The operation was financed by a $70 million Soviet loan at 2.5 percent interest payable in crude oil at market prices, and a further $72 million in credits from Machinoexport. This deal, combined with previous loan and barter arrangements with socialist bloc countries, represented a pronounced orientation of Iraq's foreign economic policy toward the socialist countries and a "proportionate erosion of the Western political and economic position in the country."[9] Production from the North Rumaila fields began in April 1972, more or less on schedule, at a cost of $55 million. INOC predicted further investments of $153 million over a ten-year period, with earnings of $3.8 billion and claimed to have outlets for some $550 million worth of crude on the basis of post-1969 barter deals alone.[10]

Iraqi success in developing the North Rumaila fields with Soviet assistance did not make any easier the settlement of the long-simmering compensation dispute with IPC (for the 1961 Law 80 expropriation). IPC claims on North Rumaila crude hindered Iraq's ability to market that oil in the West. IPC undertook little or no expansion of its Kirkuk and Basra facilities, with a continued restraint on the growth of Iraqi government revenues and sought in negotiations to tie up the bulk of INOC's North Rumaila exports in a twenty-year purchase contract. In November 1971, just months before North Rumaila came on stream, President Bakr warned IPC that Iraqi patience with their obstructionist tactics was wearing thin. Talks began anew but made no progress. In February there were reports of a renewed impasse. In March and April Kirkuk-Mediterranean exports were down to half of normal. IPC claimed that this was due to the extra premium on Mediterranean liftings which made them more expensive than Gulf exports to the European market. The Iraqis pointed out that the 44 percent decline in Kirkuk production paralleled an increase in Nigerian production, which was largely controlled by IPC parents like Shell, and that there had been no reduction in Aramco's Mediterranean output via TAPline. The government issued a two-week ultimatum in mid-May, insisting that IPC restore Kirkuk production to normal levels and rejecting the proposal of a 35¢ discount per barrel from the posted price. The Revolutionary Command Council gave IPC three options: (1) hand over excess production at cost to INOC; (2) relinquish idle producing capacity to INOC; or (3) turn over the Kirkuk fields to INOC. The political essence of the dispute was confirmed when IPC refused to make a satisfactory response by the end of May. The company's assets were nationalized—on June 1, 1972.[11]

Iraq moved to neutralize Western opposition by offering France the opportunity to buy Compagnie Française des Pétroles' 23.75 percent share of the nationalized oil (280,000 b/d) at tax paid cost (the same as it had been paying under the concession), citing "the just policy pursued by France towards our Arab causes and more specifically towards the Palestinian cause."[12] Additional purchases could be made at competitive prices. The tactic was successful and resulted in a sharp increase of French trade and credits for Iraq as well as CFP's serving a mediatory role in the compensation talks with IPC. The company postponed taking action against buyers of the nationalized oil, and Iraq vigorously pushed sales to state companies in Brazil, Japan, Ceylon, and elsewhere, including more deals with Eastern European countries. Political support for Iraq was strong in some parts of the Middle East, especially in Algeria and Libya. The Shah publicly declared himself "totally out of sympathy" with Iraq's move, proclaimed his belief in "genuine cooperation with the foreign oil firms," and asserted that Russian involvement in Iraq "put a potential stranglehold on the West's oil supplies."[13] Iran and other countries failed to implement an OPEC resolution against raising production to supplant what IPC lost.

The Iraqi move came in the midst of the participation negotiations. It increased the political pressure on the companies for a nationalist solution to the question of control and threatened "the speedy collapse of the whole concession system."[14] Although Iran helped the companies redress the balance a bit by pulling out of the participation talks a few weeks later, there is little doubt that the Iraqi action resulted in more pressure on the companies to come to some agreement with Yamani. A settlement with IPC was finally worked out at the end of February 1973, removing legal obstacles to INOC sales of Kirkuk and North Rumaila crude. Together with 25 percent participation in the remaining Basra concession, the settlement left Iraq in full control of 75 percent of its crude production, which then had a total capacity of 2.8 million b/d. Compensation to IPC for the takeover was set at $300 million, payable in crude, but was effectively offset by company payment of $345 million in back claims.[15] The price was not cheap: more than a dozen years of economic stagnation, political instability, and confrontation.

Participation: Takeover or Sellout?

The pursuit of national control over oil resources in Algeria and Iraq represented the very development that Yamani was trying to head off with his participation scheme. After 1968, participation was on the agenda of every OPEC meeting, with the only result being a series of technical studies on implementation. In 1970 the participation issue was overtaken by the price demands initiated by Libya, which culminated in the Teheran and Tripoli agreements of 1971. The shift in political power that characterized the price negotiations had the overall effect of strengthening the hand of the OPEC countries in general. The success of the Algerian nationalization program in particular led Sheikh Yamani to refer to participation as a "national demand" in July 1971.[16] New price talks to deal with compensation to the OPEC countries following the devaluation of the dollar parity and participation were on the OPEC agenda in September. On the latter question OPEC followed the price negotiation tactic of setting up a regional approach through which the Gulf states, under the leadership of Sheikh Yamani, would reach a basic settlement. Other oil producers would conduct their own negotiations. Much of this procedure was irrelevant, however, because in fact it was only the Gulf states that were interested in Yamani's brand of participation. Algeria had already moved, and Libya announced after the OPEC meeting that it would conduct its own negotiations and demanded an immediate 51 percent, as compared with Yamani's 20 percent. In December 1971 Libya had already begun negotiations with Italy's ENI, and that same month nationalized British Petroleum's operation in retaliation for Britain's collusion in the Iranian takeover of three small Arab islands in the Persian Gulf.

The oil companies offered little public argument in late 1971 against the participation notion, except to protest for the record that it violated the five-year stability clause of the Teheran price agreement—as if participation had not been on the OPEC agenda for more than three years. In fact, the companies had to display some opposition to the idea for fear of undermining Yamani's already flimsy credibility as a "nationalist." The companies formed a joint consultative board, and secured an extension of the U.S. Justice Department's antitrust waiver. Even before the negotiations began in February 1972, it was acknowledged by oil company sources that the crucial problem was not participation but the compensation value of the assets to be transferred.[17] The companies surely wanted to delay participation as long as possible and then secure the best possible financial terms they could. There is no serious question that they had reconciled themselves to the "principle" well before negotiations began.

Nevertheless, it was deemed necessary to rig a confrontation scenario. Yamani, in an interview with the London *Evening Standard* prior to negotiations, was appropriately threatening:

> The atmosphere is so nice with us now in the moderate countries that the oil companies could cooperate with us smoothly and reach an agreement rapidly in a friendly manner. But it is changing. The price is going up. The price for them. We moderates cannot underestimate the effect of national pride among the Arabs in, say, Libya and Iraq. If the Libyans nationalize all their oil resources and the Iraqis follow suit, I don't think Kuwait, Saudi Arabia and the Gulf States can keep quiet. There will have to be changes. I have got to take these matters into consideration.[18]

A few days later in an interview with the *Washington Post* in his "handsomely furnished Lausanne apartment," Yamani referred to Libya's nationalization of British Petroleum:

> The others had to do something politically speaking for their own public opinion. We must start quickly on the participation process to assure the companies any participation in the concessions.[19]

The confrontation charade followed quickly. Despite reports that the crucial question was compensation, the companies at first declined to accept publicly the participation demand. Yamani arranged for King Faisal to warn the Aramco companies that participation was "imperative,"[20] and scheduled an extraordinary OPEC meeting with the message that he would recommend an embargo of noncomplying companies. Aramco announced agreement "in principle" to Yamani's demands on the eve of the OPEC meeting.[21] Talks between Yamani (representing the Gulf states) and Aramco (representing the companies) continued intermittently over the next few months over the questions of compensation and crude oil allotment to the national oil companies.

On the question of compensation, OPEC's position was

that it should be based on the net book value of their assets for tax purposes. This prospect horrified the companies. They maintained that they were entitled to compensation for the value of the oil underground and the profits they would forego on such oil through participation. Total net assets of the companies operating in the Gulf on concessions (including the Iran consortium) were estimated at just under $1.5 billion in 1972.[22]

The other issue concerned the disposal of the oil available to the producing states. The basic proposition from Yamani was that the countries would take very little of their 20 percent share of the oil, and that the bulk of that share would be sold back to the companies at a price halfway between the concessionary tax paid cost and the posted price. Yamani's fear of "flooding" the market was matched by company interest in keeping their hands on as much of the total production as possible. Hence the benefits of participation for the countries became almost solely psychological and financial but did not provide any meaningful control over the oil industry.

The Iraqi nationalization in June provided some genuine pressure vis-à-vis the participation talks. The Shah tried to offset this radical pressure by backing out of the participation talks several weeks later on the spurious claim that since Iranian oil had been nationalized under Mossadeq, there was nothing to participate in: Iran already owned its oil. The Shah proceeded to work out a tentative deal with the consortium companies in which Iran would forego any formal participation in the capital ownership and management of the operation in return for a production increase to eight million b/d and increased deliveries to the National Iranian Oil Company. The concession, due to expire in 1979, would be extended an additional fifteen years. The only shred of national control Iran stood to gain was the end of consortium restrictions on NIOC sales. Previously such sales had been restricted to Eastern Europe, where the majors had no market position. The ratification of this arrangement was delayed until after the participation deal was conducted.

There is no evidence of Saudi hostility to this unilateral Iranian move. In mid-July Saudi Deputy Oil Minister (and Faisal's son) Prince Saud visited Iran for a "very useful exchange" with Iranian Oil Minister Amouzegar.[23] The next month there were more threats orchestrated by Yamani. Faisal issued another statement stressing the need for the companies to reach a participation agreement or face unilateral legislation. Prince Saud headed off to Washington "to inform the United States Government of the precise significance of King Faisal's recent proclamation."[24] The *Middle East Economic Survey* lent its authoritative voice to the sense of impending crisis, referring to the upcoming OPEC meeting in Beirut:

> This time it will surely be the very last chance for an amicable settlement. The companies have secured their flanks in Iran, but they—particularly the U.S. companies—will not be able to assure themselves a stable future in the area until they have also reached a settlement on the other side of the Gulf which will satisfy the aspirations of Saudi Arabia and the other oil producers of the Arabian Peninsula. Only then will the chill north wind of nationalization blowing from Iraq seem less menacing.[25]

Negotiations continued at an apparently lackadaisical pace. A few weeks later Yamani took the opportunity to spell out "the aspirations of Saudi Arabia" in a speech at an oil conference organized by the *Financial Times* of London.[26] Nationalization and participation, he asserted, are two alternative and mutually exclusive ways of "appeasing patriotic sentiments" in the Arab world. Without the alternative of participation, nationalization is an irresistible political demand that would lead to the necessity of a production program among the OPEC producers, which Saudi Arabia has consistently rejected for unspecified "practical and realistic reasons." One way in which participation would help avoid this is by channeling producer country investments in refining, marketing, and other "downstream operations" in the industrialized countries. "The investments we make in the process would absorb our surplus income, help the consumer countries and lighten the burden on the international companies." In response to a question, Yamani distinguished his program from the 51 percent takeover by Algeria. Participation is a "substitute for nationalization" and "it would continue to be participation because it would be difficult, if it is implemented the way we intend, to go on to a second stage of nationalization." At another point Yamani said that "I wouldn't even go to 51 percent if it were not for our political reasons for having participation as a substitute for nationalization."

Yamani's solicitude for the well-being of the major companies and the consuming countries was even more pronounced a week later in Washington in an address to the annual conference of the Middle East Institute. There Yamani proposed that the United States give tariff preference for Saudi oil imports, direct investments in refining, and other oil industry activities. In return, Saudi Arabia would commit itself to a production level of at least 20 million b/d (i.e., equal to total Middle East production in 1972), of which the United States would have an assured and abundant supply. In passing reference to participation, Yamani affirmed that "participation would bring about important changes in the oil industry, strengthening the existing structure and protecting it."[27] From Washington Yamani went to New York to hammer out the details on the participation accord with Exxon negotiator George Piercy. A draft agreement was announced on October 5, and quick acceptance by the Gulf states was predicted.

The one surprise in the agreement produced by Yamani and Piercy was that initial participation would be at 25 percent rather than 20 percent. This would rise to 30 percent in 1979, and 5 percent more each year thereafter, reaching 51 percent in 1983. Compensation would be on the basis of "updated" (for inflation) net book value, which comes to about four times the actual net book value the countries had demanded. The actual amount of oil to be made available to the national oil companies was a miniscule 6.25 percent initially, rising to 12 percent by the third year, and to be fixed three years in advance thereafter. The countries would receive an additional sum for any national crude marketed by the companies, adding anywhere from 10 to 30 cents to the average cost of the crude to the companies. Management and operation of the production facilities would continue in the hands of the companies.

This last point, in the eyes of many nationalist critics of Yamani's brand of participation, certified the fact that participation is merely financial and does not even begin to deal with the question of control. One of the more systematic critiques of the New York agreement was advanced by petroleum economist Nicholas Sarkis in a speech to a Kuwaiti nationalist organization in November.[28] Sarkis began by noting that the major oil companies' initial public reservations about participation were now buried by their unconditional support of the New York agreement. Reservations were now expressed by the producing countries, notably Kuwait. To many the New York agreement looked like the old concession system under a new name. It contained no provision for the foreign company to cede any part of the operational control to the national partner, in contrast to the Algerian model. The small amounts of oil allotted to the national companies for direct sale would prevent them from affecting the hold of the major companies on the markets. The requirement for a three-year advance schedule of national company liftings was seen by Sarkis as but a further obstruction, especially when coupled with a standing company offer to buy back the allotted oil at prices approaching or higher than average market prices. In light of the fact that in 1972 Iran and Saudi Arabia accounted for 64 percent of Middle East crude output and are planning for an even greater proportion of production over the coming decade, "it is obvious that any 'special' agreement between these two countries and the foreign companies or the consuming countries would inevitably lead to obstructing the direct marketing operations of the other Gulf countries." Even when the countries' shares rise to 51 percent in 1983, the New York agreement will give them no control of the market, and thus no determining role in fixing prices and production levels. To refuse to exercise the inherent national right of setting production levels is to allow the companies to continue to make those determinations "to favor the docile countries and penalize the others."

Sarkis also criticized the participation accord for making no modification of the outrageously long concession periods, like twenty-seven years for Saudi Arabia and fifty-four years for Kuwait. In this respect the participation agreement strengthens rather than modifies the existing arrangement. He criticizes the compensation figures as outrageously high (about four times the net book value). This, plus the investments the countries will have to make in the production operations, and the downstream investments in refining and transportation in the industrialized countries, represents a substantial channeling of national revenues into an industry still controlled by foreign interests. In Sarkis' view, the most outrageous part of the participation agreement is that it provides for investment by the Middle Eastern countries in the industrialized world, but "does not include any provision obligating the foreign partner to invest part—if only very small—of its profits in developing downstream activities in the host countries." The participation agreement also contains elaborate provisions detailing the rights and privileges of the companies and provides for the continued economic development of the industrialized consuming countries, but does not say a word about the role of oil in the economic development of the producing states.

Libya's Participation

While Sarkis' political analysis of the participation agreement correctly notes Yamani's aim of interlocking the interests of the Gulf states with those of the major companies, it does not appreciate the quickening dynamics of Middle East oil politics. Saudi Arabia and Abu Dhabi ratified the agreement in December 1972, but the Kuwaiti National Assembly, with its vocal nationalist bloc, defied the government and refused to ratify the agreement. The fact that independent purchasers were already lining up in December to purchase the crude made available to the national companies under the agreement indicated that the companies would indeed market more crude than the participation agreement allowed them. Venezuela pointedly declined to sign the OPEC endorsement of Yamani's efforts and, more critically, Libya proceeded to push for an immediate 50 percent participation, beginning with ENI and the small Texas company, Bunker Hunt. In a press conference after the OPEC endorsement session in Riyadh, Libyan Prime Minister Jallud opined that "the Gulf countries could have obtained better terms."[29] He went on to specify Libya's objectives in participation: (1) company guarantees of stepped-up investment and exploration; (2) Libyan control of operations, including an increase in the number of trained cadre; and (3) adequate crude supplies to develop direct relations with consumer countries outside the market structure of the majors. He specifically criticized the compensation terms of the Gulf agreement, saying that Libya would pay only net book value less depreciation, and the lack of provision for investment of oil revenues in the producing countries rather than abroad.

The course of Libyan participation negotiations with the various companies proceeded under a shroud of secrecy, with the smaller and more vulnerable companies being, as in the past, the first targets. Agreement with ENI for 50 percent Libya also ordered several production cutbacks for Occidental, Oasis, and Esso, although not as serious as the ones during crude involved (100,000 b/d to start) made it a minor news item.[30] The same demands were put to Bunker Hunt and the Oasis group of American independents (plus Shell). Negotiations with Bunker Hunt broke down in late December 1972. Libya agreed to arbitrate the dispute several months later but then nationalized Bunker Hunt in early June 1973 as a "slap" at American "arrogance."[31] Oasis had earlier offered a deal along the lines of Yamani's agreement, which was rejected. Libya also ordered several production cutbacks for Occidental, Oasis, and Esso, although not as serious as the ones during the price negotiations two years earlier. The major companies were described as presenting a solid and aggressive front toward Libya, and it was later reported that at their behest (again through Wall Street lawyer John J. McCloy) the U.S. government had sounded out the main European countries about a boycott of Libyan oil in the event of a takeover. This effort met no success.[32] In mid-August Occidental and Oasis (except Shell) announced their "acquiescence" to the Libyan terms,[33] which provided for a 51 percent takeover with the companies staying on as operators with each being under the supervision of a three-man committee (composed of two government representatives and one from the company). Compensation would be strictly on the basis of net book value. The companies would have first preference on the Libyan share of production at market price, calculated then by the Libyans to be $4.90 per barrel. This price was implicitly verified by the purchase of Libyan royalty crude by independent companies at $4.90 (30¢ above the posted price) in the same month.[34]

The other companies, mostly majors, refused to "acquiesce," claiming that to do so would jeopardize the carefully constructed Gulf agreement, particularly with regard to government share of crude for direct marketing and prices.[35] The Occidental and Oasis deals left Libya in direct control of more than 40 percent of its crude production. Another third, about 780,000 b/d, was in the hands of the majors. On September 1, anniversary of the 1969 revolution, Libya unilaterally decreed the 51 percent nationalization of the remaining companies.[36] In his press statement on September 2 explaining that decision, Prime Minister Jallud asserted: "We stated it clearly to the oil companies: we are revolutionaries, not princelings from the Gulf. . . . It is clear that effective participation means that we have the final say."[37] The companies refused to acknowledge nationalization. Some, like Socal and Texaco, were cut off. Others, including Exxon, were not. Following the October War, in February 1974, Socal and Texaco were nationalized fully. Shell was nationalized at the end of March. Mobil and Exxon came to an agreement on the 51 percent pattern in April and May 1974.[38]

Evaluating Participation

Libyan militancy on the question of control took place against a background of constantly escalating market prices and, after October 1973, war with Israel and the embargo and production cuts that accompanied it. The most significant moves, however, were made well before the war broke out, and can be analyzed separately. As early as February 1973, the instability of the price structure laid down by the Teheran-Tripoli agreements and supplemental devaluation and dollar parity talks became evident. An agreement was made between Abu Dhabi and a Japanese shipping company with some refinery interests, Japan Line, for the sale of participation crude at a price of $2.38, more than 10¢ over the company-quoted market price. A month later Abu Dhabi made further sales at $2.53. In the face of objections from the Western press, government, and company officials over escalating prices, Oil Minister Otaiba retorted: "A month ago the Japan Line deal was unique; now it is history."[39] Within OPEC, following the failure of talks over the February dollar devaluation, there was sharp pressure from Algeria, Libya, and Iraq to scrap the Teheran agreement as "structurally obsolete."[40] The same pressures were renewed at the OPEC meeting in mid-May and again in September. Saudi Arabia and Iran prevailed in their efforts to maintain posted prices and participation buy-back prices at low levels until the price-hike that coincided with the war.[41] In May Saudi Arabia sold its participation crude at record prices to a variety of independent companies. Of the Saudi sales by volume, 27.4 percent went to Japanese companies, 24.7 percent to U.S. companies (independents like Ashland of Kentucky), 18.2 percent to European independents, and the rest to the national oil companies of Third World consuming countries like Brazil.

The tight supply situation that lay behind the rising market price of crude in 1973 was a continuation of the circumstances that emerged in 1970-1971. Continued expansion of demand in Western Europe, Japan, and notably, the U.S. East Coast (for low-sulphur crude) was met with expanded production only in the Gulf, notably Saudi Arabia. Libya and Kuwait, among the largest producers, clamped on production limits for conservation. Other countries like Algeria and Indonesia experienced difficulty in meeting planned production rates. Actual shortages were limited to specific items (e.g., residual fuel oil) in specific places (e.g., the U.S. Midwest) that could be fairly attributed to lack of refinery capacity and other instances of company mismanagement or manipulation. Nevertheless, the lack of surplus crude, coupled with the price increases and increasing talk in Arab circles of using oil as a "weapon" in the struggle with Israel, created considerable anxiety in many countries and small firms dependent on the majors and added impetus to the growing number of direct sales between crude producers and consumers, circumventing the channels dominated by the majors. The new availability of participation crude, and the growing amount of royalty crude directly available, provided the opportunity for these contacts to flourish. Prices rose under these pressures.

What emerges from developments in Middle East oil under the rubrics of nationalization and participation is a picture not unlike that described earlier with regard to the Teheran-Tripoli price agreements. In the face of small and incremental, but fundamental, shifts in economic and political power from the major companies to the producing countries—shifts brought about largely by the development of contradictions within the oil industry, of which national control was only one—certain producing countries moved to heighten these contradictions and thereby strengthen their own national economies at the expense of the companies and the (largely capitalist industrial) consuming states. Algeria, Iraq, and Libya (and outside the Middle East, Venezuela) were in the forefront of this struggle. Other countries, notably Saudi Arabia, used these developments to achieve greater material benefits and political security by allying themselves with the major companies and with the existing industry structure, by becoming "partners" in a system still oriented to the benefit of the consuming, industrialized countries at the expense of the underdeveloped raw material-producing countries. Iran, for example, was in a position to make and secure more far-reaching demands than Saudi Arabia, and after the participation accords did negotiate a new arrangement in which its national company (NIOC) assumed control and operating responsibility for consortium operations, with the companies assuming the role of long-term contract purchasers and service-contract operators. The financial terms were precisely tied to prices in the rest of the Gulf. In pursuing this course, Iran let it be known that it was prepared to make arrangements with Japanese and other independent and state-owned companies, by-passing the majors altogether, if the companies did not accept its terms. In the end, though, the Shah was willing to limit the use of this alternative relationship as leverage for bettering his terms with the majors. He saw his basic interest in remaining allied with the major companies, preserving as much as possible the existing industry structure, and relying on the continued military and political support of the United States.

This was even more true of Saudi Arabia, where Faisal's interest has clearly been to solidify and consolidate the position of his regime internally, to play a "protective" role on the Arabian Peninsula, and a leading role in the Arab world. Where participation for Libya is an assertion of political and economic independence from Western companies and governments, for the Saudis it is a means of repudiating such goals and establishing a "special relationship" with the United States. "Participation," said Yamani at one point, "was the result of King Faisal putting his full weight behind the issue. He established direct and personal contacts with many countries, and particularly with the White House of America. . . ."[42] Elsewhere he elaborated on the proposed "special relationship":

> We are thinking in terms of economic cooperation and shared interests in the financial sphere. As is always the case in inter-national relations, whenever mutual interests exist between two countries in the economic field very strong political relations immediately develop . . . Certainly participation will increase our revenues, but . . . financial gain is not what we are after. Our real

objective is to acquire control over our natural resources *and to help the oil industry remain on a sound footing and maintain the present structure.*[43]

The United States reacted cautiously to Yamani's proposal, fearing that acceptance would create further competitive scrambles among other consuming nations for similarly secure supplies, resulting in still higher prices. In May of 1973 Akins of the State Department (and soon to be appointed ambassador to Saudi Arabia) told the Senate Foreign Relations Committee that the United States had established an informal agreement with Saudi Arabia regarding Saudi investments in the United States.[44]

The role of U.S. companies in promoting, rather than submitting to, participation has been unclear. There had been a growing consensus on the part of American multinational corporations that alternative institutional arrangements (like joint ventures) had to be developed in order to cope with the rising nationalist demands of Third World countries. A decade ago Chandler Morse wrote that behind such moves

> there will usually be found a perception of political writing on the wall and an estimate of net advantage in the long term. The major problem, therefore, would seem to be that of formulating the political writing. . . . What is needed, consequently, are systematic efforts to help the raw-material countries equip themselves with the sociopolitical goals, values, institutions, and processes that are prerequisite to the efficient and socially beneficial operation of private enterprise, whether domestic or foreign.[45]

In Sheikh Yamani, Hisham Nazer, Mohammed Joukhdar and other Saudi technocrats trained in the United States, Morse's prescription has worked remarkably well. In the swiftly changing dynamics of Middle East oil in the 1970s, though, the technocrats have succeeded only in lessening and blunting the sharp changes demanded by the radical nationalist states. In every case, whether it be prices or participation, the direction and the pace of change has been set largely by the radicals. The other countries, and with them OPEC as a whole, have been forced to increase their demands in order to survive politically. The same pattern is repeated on yet another issue—industrialization. We have pointed out the contrast between the radical nationalist orientation—developing an independent and autonomous national economy that utilized hydrocarbon and other resources to increase the material well-being of its population—and the program enunciated by Yamani in calling for heavy investments in the already industrialized and economically dominant countries, particularly the United States. Even this has now changed under pressure of the course followed by Algeria, Iraq, and Venezuela. In the months before the October War and the oil production cutbacks, the main issue on the OPEC agenda, under Venezuelan leadership, was how to tie continued crude production to a firm commitment by the industrialized consuming countries to provide aid and assistance in building up diversified, industrial national economies and even guaranteeing markets for the industrial and commercial exports of these economies once they were developed. This represented a sharp change from the simple revenue-maximizing policies of the Saudis. By the middle of 1973 even the Saudis were following that line, specifying that continued increases in crude oil output would depend on large-scale technical assistance to help industrialize the Saudi economy.[46] It was this question of industrialization, a question that goes to the heart of the relationship of power and dependence existing between the Third World and the West, that was emerging with a momentum all its own when war broke out between Egypt, Syria, and Israel on October 6, 1973.

9

The Oil Weapon, the October War, and the Price Explosion

The oil weapon, that is, the implementation of production cutbacks and destination embargoes by the Arab producing states in order to pressure the United States and its allies to force Israel to withdraw from Arab territories seized in 1967 and 1973, must be seen in two separate but interrelated contexts. One is the Arab-Israeli conflict and the dynamics of inter-Arab politics centered around that conflict. The second is the drive by the oil-producing countries, aided and abetted by the major oil companies, to take full advantage of the temporary and localized shortages and generally tight balance of supply and demand in world markets to register a hefty increase in the price of crude oil and its products.

The war, of course, and the implementation of the oil weapon acquired a momentum of their own which effected qualitative changes in the structure of the world oil industry and in the present alignment of political power in the Middle East. In addition, they accentuated significant developments in the competitive economic and political relationship of the United States with the European industrial capitalist countries and Japan. We shall attempt to explicate the background, the course, and the consequences of the oil weapon in all of these areas.

The Oil Weapon Before the War

The close affinity of politics to oil and politics to war have made it almost inevitable that war in the Middle East and oil in the Middle East should be closely intertwined. In 1948 this consciousness was evident in the deliberations and memoranda of U.S. political and military strategists as they attempted to formulate a policy toward the Zionist-Palestinian conflict. More than one Arab leader publicly threatened to cut off oil exports to those countries supporting Israel. As it turned out, only the shipments of the Iraq Petroleum Company to Haifa were significantly (and permanently) affected. Whether a more concerted campaign could have been mounted is highly doubtful, given the low level of technical competence and political collaboration in and between the Arab states, and the limited dependence of the industrialized countries at that time on Arab oil. In any case, the Saudis refused to entertain it from the beginning, so the threat remained dim and distant. Threats were raised again in each succeeding war (in 1956 and 1967), and in both cases limited embargoes were implemented, with little effect. In 1956, the United States was able to compensate for this by making deliveries to Britain and France from domestic production. In 1967, the low level of U.S. imports from the Arab states and the ability of the major companies to reroute supplies rendered the embargo a farce. In subsequent years radicals blamed this failure on the perfidy of reactionary oil producers like Saudi Arabia, which reciprocated by blaming the incompetence of the radicals. The result was that the idea of an embargo or cutback fell into disrepute. The rich producing states wrapped a cloak of virtue around their ever expanding production by declaring that oil had to be used as a "positive weapon." Oil was seen as a source of ever growing revenues that might then be dispensed, at the ruling families' discretion, to build up the military and economic strength of those Arab states that behaved themselves politically.

Debate about the potential of the oil weapon continued nonetheless. Given the conditions of surplus production and capacity that prevailed through 1970, it appeared axiomatic that no embargo could be effected without the support and participation of at least one of the two major producers, Saudi Arabia and Iran, which between them accounted for about two-thirds of Middle East production. Besides being non-Arab, Iran had a tacit strategic alliance with Israel, facilitated by the United States. In 1967, Iranian production was increased and used to supplement the Arab cutbacks. Iran, in short, could be counted on to sabotage and neutralize any Arab oil offensive.

Saudi Arabia, for its part, had to maintain a modicum of solidarity with the rest of the Arab states in the struggle against Israel, which was covered with the oil-financed subsidies to Egypt and Jordan following the Khartoum Conference in 1967. The oil weapon was, for Saudi Arabia, a lever for modifying and controlling the political behavior of the radical Arab states, notably Egypt. To those who called for the use of Saudi oil production to pressure the United States into a less pro-Israeli position, the constant Saudi refrain was that oil and politics don't mix. Oil Minister Yamani offered a unique interpretation of the oil weapon in November 1972:

> I must say that we do not believe in the use of oil as a political weapon in a negative manner. We believe that the best way for the Arabs to employ their oil is as a basis for true cooperation with the West, notably with the United States. In this way very strong economic ties are established which will ultimately reflect on our political relations.[1]

The changes in the structure of the oil industry which accelerated after 1970, notably the disappearance of surplus production and the increase of government control over production levels, renewed the momentum for a more militant stance than Yamani's. The rising level of low-sulphur crude imports to the U.S. East Coast, though still a small proportion of U.S. supplies, accentuated the slowly growing reliance of the United States on Eastern Hemisphere sources. "Studies" by the oil industry lobby in the United States, such as the National Petroleum Council, all stressed the growing shortage of domestic energy supplies and the accelerating dependence on foreign supplies. The Middle East, and in particular Saudi Arabia, was seen as the most likely if not the only source for the coming decade. While much of this was self-serving propaganda in the drive for higher prices, profits, and subsidies, the general thrust of the argument was assailed by very few. The voracious appetite of the United States for energy resources could only be met over the next decade by Middle East imports. This in turn focused a good deal of attention on the stability and potential disruption of that source.

The close and intricate relationship between the United States and Iran and Saudi Arabia led U.S. policy-makers to see the main threat to an uninterrupted oil supply as coming from the radical oil states—mainly Iraq and Libya—and from the radical political movements working to overthrow the reactionary monarchies. United States policy was directed toward suppressing those movements and strengthening the monarchial regimes to guard against the possibility of a Libya-type coup. Military and security assistance was stepped up to the "key" states in the area, Iran and Saudi Arabia, "in the spirit of the Nixon Doctrine." "Support for indigenous regional collective security efforts" and massive amounts of U.S. arms have been channeled into those countries.[2] In both countries internal security and intelligence forces have been beefed up to notorious efficiency, making the police the most "modern" part of the government apparatus. U.S. military and intelligence personnel number in the hundreds in Saudi Arabia and thousands in Iran. "Social science" research into ethnic and tribal divisions has been extensive, especially in Iran, where it is put at the disposal of the Shah's regime.

Arms transfers from the United States to Iran in grants and sales amount to more than $3.5 billion, most of that over the last two years, and it is growing at a rate of nearly a billion a year. Iran has actively sought the role of policeman in the Gulf and Indian Ocean and has provided troops as well as arms and aid to the embattled British-backed sultan in Oman, which sits at the mouth of the Gulf. The liberation struggle led by the Popular Front for the Liberation of Oman in Dhofar, the western province of Oman, has been the most visible threat to the continued rule of the reactionary sultans throughout the Gulf. Saudi Arabia has also provided some support to Oman but has concentrated its financial and mercenary assistance at the other end of the Peninsula, propping up the stagnant and paralyzed North Yemen regime while conducting hostilities against the radical National Front regime in the People's Democratic Republic of Yemen to the south. Mercenary assistance, financed by Saudi Arabia, Britain, and the United States, is also supplied in Oman and the Gulf by "retired" Jordanian and Pakistani military and intelligence personnel.

This mercenary strategy known as the Nixon Doctrine assigns (and where necessary finances) regional defense responsibility to a local "power" and is no more than a refurbished version of the Indochina strategy attempted under the direction of former Defense Secretary McNamara. In many other respects, too, the Nixon Doctrine displays an important continuity with the developments initiated under McNamara. Many of McNamara's weapons programs—the giant C-5A transport plane, the new generation of amphibious landing craft, the portable airfields that can transform any reasonably flat area into a staging area for combat aircraft—have only become operational in the last few years, providing the potential for quick and powerful military intervention that did not exist in Vietnam.[3] Indeed, the October War provided an important crisis utilization of the C-5As in the arms airlift to Israel. The Pentagon, and particularly the navy, have wasted no time using the "energy crisis" to support their arguments for whopping budget increases. This has been a key factor in the recent navy acquisition of a "communications station" in the Indian Ocean at Diego Garcia designed to provide "a range of involvement options from no involvement to whatever involvement is deemed necessary."[4]

In this setting, another key country is Israel itself. Far from jeopardizing American accessibility to Arab oil, according to this line of thinking, Israel's military capabilities help contain and deflect radical Arab political and military potential that might otherwise be directed against the weak and unstable collection of Gulf states. This view has been pushed by Israeli defense strategists and seems to have prevailed among American policy-makers until the October War. One of the most articulate, although idealized, expressions of this approach was offered by the Senate's "energy expert," Henry Jackson, in May 1973:

Such stability as now obtains in the Middle East is, in my view, largely the result of the strength and Western orientation of Israel on the Mediterranean and Iran on the Persian Gulf. These two countries, reliable friends of the United States, together with Saudi Arabia, have served to inhibit and contain those irresponsible and radical elements in certain Arab states—such as Syria, Libya, Lebanon and Iraq—who, were they free to do so, would pose a grave threat indeed to our principal sources of petroleum in the Persian Gulf. Among the many anomalies of the Middle East must surely be counted the extent to which Saudi Arabia and the sheikhdoms—from which, along with Iran, most of our imported oil will flow in the years ahead—will depend for regional stability on the ability of Israel to help provide an environment in which the moderate regimes in Lebanon and Jordan can survive and in which Syria can be contained.[5]

A corollary to the Nixon Doctrine—the buildup of local police states to act as regional guarantors of the status quo, backed by a capacity for rapid and massive armed intervention—is reliance on the regimes of the favored client states to act consistently with United States interests. In terms of Middle East oil, this means assuming that the governments of Iran and Saudi Arabia will not undertake to interrupt oil supplies. It means confidence in the word of Sheikh Yamani that oil and politics do not mix, and in Saudi Arabia's ability to prevent any coordinated Arab action.

With the consistent failure of various military and political strategies to force the Israelis into a negotiated withdrawal, and the equally consistent readiness of the United States to provide Israel with arms and financial assistance, pressure began to build again for an Arab oil policy that could be used to pressure the United States. The OPEC success in the price negotiations, particularly Libya's careful use of production curbs and Yamani's apparent success in the participation negotiations raised the obvious question of why such tactics could not be used to secure a political objective.

The Saudis had managed after the 1967 war to isolate their oil policies from the contagious political emotions connected with the Israeli conflict. Saudi Arabia's new role as the one country with the capacity and willingness to expand production to meet growing Western, and particularly American, oil needs signaled the end of the isolation of oil from politics. In 1972, the Economic Council of the Arab League commissioned a study concerning the strategic use of Arab economic power in the fight with Israel. The report, finished at the end of the year, took the view that while Arab interests in the long run would be best served by developing independent and autonomous industrially based economies, in the short term a more restrictive oil production policy would both conserve wasting resources for future use and bring a significant degree of pressure on industrial consuming countries to alter their support for Israel and hostility toward the Palestinian cause. The report did not call for an embargo or cutback but a slower expansion rate than that desired by the consuming countries. This tactic would allow maximum flexibility for proportionate degrees of escalation and de-escalation that the "total cutoff" approach lacks.[6] This particular tactic of a freeze or constraint on production expansion was endorsed by a top Arab oil economist and former OPEC Secretary-General Nadim Pachachi, in June 1973, when he predicted it would be "quite sufficient to cause a world-wide supply crisis in a fairly short period of time."[7]

By this time Saudi Arabia was already coming under considerable and increasing pressure to limit its rate of expansion; Saudi Arabia's isolation from the Arab-Israeli confrontation was ending. Another factor in this process was the Egyptian shift to the right under Sadat. In order to consolidate the power of the new and old bourgeoisie represented by the Sadat regime, some concession to the patriotic sentiments of the Egyptian masses was necessary. Some means of restoring Egyptian sovereignty over the occupied Sinai was essential to undercut the insurgent political movement of students and workers that had regularly brought Egypt to the edge of political crisis since the 1967 war and especially since Nasser's death. With the changing balance from a buyers' to a sellers' market in oil, and with increased U.S. declarations of the importance of Middle East oil in that balance, politically conscious Arabs saw the irrefutable need for an oil policy that would match the rhetorical militance of the politicians. For Egypt the logical partner by geography and political temperament appeared to be Libya, and this was matched by Qaddafi's stress on the need for some specific form of Arab political union or federation. Qaddafi's staunch fundamentalist Islamic and anti-imperialist politics, though, forced Sadat to look to Saudi Arabia as an alternative. Faisal's regime, looking for a role that would give it pan-Arab legitimacy and would limit the influence of radicals like Qaddafi, was inclined to cautious cooperation.

Qaddafi's stress on confrontation and popular mobilization was countered by Faisal's emphasis on close ties to the United States. This meant that Egypt's links to the Soviet Union had to be sharply reduced. Faisal apparently had some role in Sadat's decision to expel the Russian military advisors in 1972 as a way of inviting the United States to use its leverage with Israel to secure some territorial concession. There was no perceptible change in U.S. policy. Sadat went ahead with plans for an Arab federation comprised of Egypt, Libya, and Syria. In early 1973, after renewed clashes in Egyptian cities between militant students and workers and security forces over foreign and domestic policy, Sadat dispatched his special national security advisor, Hafez Ismail, to Washington to sound out any possible shifts in U.S. policy. Ismail's visit coincided with that of Golda Meir, and Washington's response was to leak word of a new shipment of Phantom jets to Israel. It appears likely that the Egyptian decision to go to war in October was made in the aftermath of these events.

The Saudis, now, had some stake in the outcome, having helped to persuade Sadat that a diminished Russian role would lead to a change in U.S. policy. In April, Yamani paid one of his frequent visits to Washington, where he publicly linked oil and Israel for the first time. It is politically impossible, he reportedly told U.S. officials, for Saudi Arabia to expand production at the desired rate in the absence of a change in U.S. policy toward Israel. In a story accompanying an interview with Yamani, the *Washington Post* wrote that

the Saudis are known to feel under increasing pressure from the radical Arab states and the Palestinian guerrillas to use their oil as a political weapon for pressuring Washington into forcing Israel into a compromise settlement with the Arabs.[8]

Soon after this King Faisal called in the president of Aramco to stress that Saudi Arabia was "not able to stand alone much longer" in the Middle East as an American ally.[9] Aramco cabled the details of Faisal's remarks to the American parent companies, which began taking out newspaper ads and appearing at congressional hearings to warn of the need for an "evenhanded" American policy in the Arab-Israeli dispute. On May 15, 1973, anniversary of the establishment of Israel, four oil-producing states engaged in a minor but symbolic stoppage of crude oil production: Algeria, Kuwait, and Iraq for one hour and Libya for a full twenty-four hours. In June, Libya nationalized the small American independent, Bunker Hunt, invoking U.S. imperialism and support for Israel as a primary reason. Washington responded to the Saudi initiatives by offering Saudi Arabia a squadron of Phantom jets hitherto reserved in the Middle East for Israel and Iran. Saudi Arabia indicated that this offer was not acceptable as a substitute for changing U.S. policy toward Israel. In July, the Palestine Liberation Organization endorsed the tactic of freezing oil production at present levels. Faisal let it be known that the debate inside Saudi ruling circles was between those who wanted to limit production increases and those who wanted to freeze them. Saudi Foreign Minister Saqqaf predicted that Arab oil would be denied to "those who help our enemy," but indicated that a decision might not come for as long as a year.[10]

The Saudi campaign to use oil as a weapon was consistently restricted to the arena of communiqués and press interviews. Production for the first seven months of 1973 was up 37 percent over the previous year; July production was up an incredible 62 percent, to 8.4 million b/d,[11] with production scheduled to hit a whopping 10 million b/d by the beginning of 1974. The aim of Saudi publicity about the "oil weapon" was to secure an equally public and superficial indication of change in U.S. policy that would validate the Saudi-United States relationship and preclude the need to implement the oil weapon. Above all, the campaign was designed to neutralize mounting public pressure throughout the Arab world for some move, especially after the U.S. veto in the UN of a resolution condemning continued Israeli occupation of Arab territories captured in 1967. In August Prince Abdullah, Faisal's brother and head of internal security forces, the National Guard, was in Britain and France shopping for arms and advisors, where he urged that:

> All the Arab countries, whether oil producers or not, should act to prevent the debate on the use of oil from being transferred to the street. . . . I deem it imperative for the Arabs to draw up a higher Arab oil policy whose aim should be to consolidate relations between friends and to isolate enemies. . . . We must differentiate between the uncompromising enemy and the potential friend. Only reason and a historical sense of responsibility can provide the right answers to these questions.[12]

This finely balanced public maneuvering continued through August and into early September when, at the conference of nonaligned nations in Algiers, it was predicted that some joint Arab oil stand would emerge. Standard Oil of California, one of Aramco's parents, sent a letter to its stockholders stressing that U.S. policy must understand "the aspirations of the Arab people" and support "their efforts toward peace in the Middle East."[13] This aroused a predictable flurry of protest from some American Jewish organizations. The Libyan takeover of majority interest in Occidental and Oasis in mid-August and its nationalization decree of September 1, 1973 heightened the sense of brewing confrontation. Nixon's subsequent press statement invoking the specter of Mossadeq, along with well-publicized desert warfare training operations of the U.S. Marines, did nothing to de-escalate the politics of oil.

A meeting of Arab oil ministers in Kuwait on September 4 was expected to produce policy suggestions for the Arab heads of state meeting in Algiers, but the traditional politics espoused by some of the participants showed that nothing had really changed in the Saudi approach. No recommendations came out of the conference. It was reportedly split between those states advocating seizure of controlling interest in U.S. oil operations, led by Libya and Iraq, and "those urging that oil shouldn't be used as a political weapon."[14] Saudi Arabia reportedly stressed the need for a wait-and-see approach. Nixon reciprocated the next day with a statement that "both sides are at fault" in the Arab-Israeli conflict, prompting the *Wall Street Journal* headline: "Nixon Mutes Support for Israel as Arabs Appear to Get Results With Oil Threats." The day after Nixon's press conference, the Saudis were delivering messages in various Western capitals playing down the widespread press reports of possible output restrictions directed against the United States.[15] Later in September there were reports of two new Kissinger initiatives: one was a secret "peace plan" which would have settled most of the particular territorial claims between Israel and the Arab states in Israel's favor; the other was an equally secret decision to pursue the "special relationship" with Saudi Arabia proposed a year earlier by Yamani. The policy under the new Secretary of State was to be one of "compatability rather than confrontation."[16]

The Oil Weapon and the October War

On October 6, 1973, Egypt and Syria launched full-scale military attacks against Israeli occupation forces. There were immediate calls from several quarters for the use of oil as a political weapon, including the Palestine Liberation Organization and the radical nationalists in the Kuwait National Assembly. Only Iraq, among the oil-producing Arab states, took action by nationalizing the U.S. (Exxon and Mobil) interest in the Basra Petroleum Company. Although the amount of oil lost to the companies was not more than several hundred thousand b/d, it was a serious long-term blow because it cut them off from a relatively cheap and expanding source of crude. In addition, the Iraqi action set a militant tone for the current debate about the "oil weapon." But it was more than a week before any further steps were taken.

When the war broke out, the Gulf states via OPEC were preparing to enter negotiations with the companies to raise the price of crude. For almost a year the more militant producing countries had been pressing inside OPEC for a revised price schedule that would take account of the galloping rate of inflation, currency revaluations, and rising market prices for crude oil and its products. Saudi Arabia and Iran had stood resolutely against these pressures as long as they could. Algeria and Libya finally, in September, unilaterally raised their posted prices to the five-dollar range. Even Yamani had to admit that "the Teheran Agreement is either dead or dying and is in need of extensive revision."[17] Negotiations with the companies had been scheduled to open in Vienna on October 8.

The negotiations opened on schedule against the background of war, but were apparently conducted in an unreal isolation from the fighting. Press reports suggested that the producers were asking for a posted price-hike of 35 to 50 percent. Negotiations, formal and private, stretched over five days before they were broken off on October 12. The companies asked for a two-week recess; the countries refused, and scheduled a meeting among themselves for October 16 in Kuwait. At the Kuwait meeting the countries finally did what the radicals had been advocating for several years—determine crude oil prices unilaterally. The posted price for Gulf crudes was hiked about 70 percent—from $3.01 to $5.12 for Arabian light. This meant a boost in government revenue from $1.77 to $3.05 per barrel.

In the war zones, full-scale fighting continued uninterruptedly, eliciting more and more calls for the oil producers, and particularly Saudi Arabia, to use the oil weapon. On October 10, President Sadat sent a special emissary to Saudi Arabia and then to the other Gulf oil-producing states, impressing on them the need to support Egypt and Syria by pressuring the United States to refrain from resupplying Israel with arms. It appeared that Faisal had given Sadat assurances in August, when Sadat visited Riyadh, that Saudi Arabia would support an Egyptian war effort with its oil. When, after four days of fighting, there was no movement by the oil producers (except Iraq's national-ization), Sadat felt the need to increase public pressure on Faisal. By the end of the first week, the only interference with Middle East oil supplies had been the disruption in the export of about 1 million b/d of Iraqi oil through Syria, due to heavy Israel bombings of port and refinery installations.

The Arab oil ministers, following their OPEC price meeting, stayed on in Kuwait as OAPEC (the Arab oil organization formed by Saudi Arabia, Kuwait, and Libya in 1968) to formulate an oil policy for the war. The same divisions persisted that had characterized the meeting in early September. Saudi Arabia was still urging a cautious policy that did not include an embargo. Faisal had reportedly sent a message to Nixon via his Foreign Minister that "if the United

States becomes too obvious a resupply agent for Israel in the present war, it would be almost impossible for him to withstand pressure to halt oil shipments."[18]   At the Kuwait meeting on October 17, Saudi Arabia successfully resisted attempts to set an embargo against the United States, whose resupply effort to Israel had just begun. Instead it was decided to implement a general 5 percent cutback in production plus a 5 percent reduction per month until occupied lands were liberated and Palestinian rights restored. The reaction in industry and government circles was one of relief, sensing that the oil producers could not have done much less and still done something. The belief that "the effects thus far appear to be more psychological than actual" summed up initial reactions.[19]   The next day, following increasing evidence of massive U.S. military support for Israel, Saudi Arabia announced that its production would be cut 10 percent immediately. Abu Dhabi announced an embargo on all shipments to the United States. One day later, following Nixon's request for congressional authorization of $2.2 billion in military aid to Israel, Saudi Arabia also declared an embargo against the United States. By October 22 all the producers had embargoed shipments to the United States, some had added Holland to the list as well, and Iraq had nationalized Dutch holdings (through Shell) in the Basra Petroleum Company. Kuwait declared an overall production cut of 25 percent, including amounts embargoed to the United States. Saudi Arabia added its embargoed amounts to the 10 percent cutback, bringing the total cutback as a percentage of planned output to more than 25 percent. In addition, each country had its own list of "friendly," "neutral," and "aiding the enemy" states.

In order to standardize the embargo and the production cuts, another meeting of oil ministers was held November 4, where the overall cutback was set at 25 percent. For the North African states, this meant further reduction. For Saudi

Arabia it actually allowed for a slight increase in production. On balance this meeting did not sharply further reduce the amount of oil available. It did, however, take measures to prevent oil from reaching embargoed destinations indirectly. This applied mainly to European and Caribbean refineries which ship products to the United States. A special supervisory committee composed of the oil ministers of Saudi Arabia, Algeria, Kuwait, and Libya was set up to enforce the embargo, and a meeting of foreign ministers was scheduled to coordinate the division of consuming countries into "friendly, neutral, or aiding the enemy" categories. Ministers Yamani of Saudi Arabia and Abdessalam of Algeria were designated as emissaries who would tour the industrialized countries explaining the details and purpose of the cutbacks and embargo and pressing for diplomatic endorsement of Arab aims vis-à-vis Israel.[20] Iraq followed its own policy of expropriation of Western oil interests and embargoed shipments to the United States, Holland, and eventually Portugal, South Africa, and Rhodesia but disassociated itself from the OAPEC production cutbacks. In a formal statement in December, Sadam Hussain, vice-president of the Iraqi Revolutionary Command Council, condemned the cutbacks as having been devised by "reactionary ruling circles well-known for their links with America."[21] The embargo is sound, he said, but not enough. Nationalization is necessary. As for the policy of production cutbacks, they "generally harmed other countries more than America" and "led to results which run counter to its stated purpose." It is a "serious political mistake," he warned, to implement policies which tend to hurt allies and potential allies (Europe and Japan) more than avowed enemies (United States). "The occurrence of a suffocating economic crisis in Western Europe and Japan may drive them to issue relatively good statements now, but these countries will find themselves forced in the next phase to abandon their independent policies and rely more and more on America." Hussain also cited the recent price hikes as being "conducted in a hysterical manner" and increasing the profits of the international companies as much as anyone else, making them stronger than the independent European and Japanese companies. Unmentioned by Hussain was the fact that Iraq—just now embarking on production expansion after a dozen years of stagnation and being more dependent on those revenues than any other country—had the most to lose under the OAPEC production cutback policy. When criticism of the Iraqi policy came, it was not from the reactionaries but from Algeria, which pointed out that the effectiveness of the embargo depended on the tightness of oil supplies in general, which in turn depended on the degree of compliance with the OAPEC cutbacks.

The production cutbacks initially had the desired effect. In November both Japan and the European Economic Community (EEC) issued formal statements referring to Israeli occupation of 1967 territory as illegal. Individually the European countries, mainly Britain and France, moved quickly to set up special deals with Saudi Arabia, Libya, Iraq, and other producers, exchanging crude oil for arms or capital goods. From Japan something more was wanted: a total break in diplomatic and economic relations with Israel. Europe was partially rewarded when OAPEC announced that the projected 5 percent reduction for December would not be applied to European supplies. At the end of November an Arab summit, boycotted by Iraq and Libya, endorsed the Saudi-led oil program, saying that reductions would continue until the cut in oil revenues equaled 25 percent of 1972 revenues. Taking the price hikes into consideration, this would allow production cuts of up to 75 percent.

At a meeting on December 9, the Arab oil ministers, again Iraq excepted, decreed a further 5 percent cutback for January, which would amount to a loss of some 750,000 b/d and a total reduction, on paper at least, of more than 5 million b/d. Although the oil moves by the Arab states did evoke some favorable diplomatic statements by the consuming countries, there were real questions being raised about the actual shortage effect of the cutback and embargo. Forecasts of worldwide economic recession were revised as a distinct lack of shortages showed up. In late December tanker unloadings in European ports were higher than ever. Rotterdam, for example, a boycotted port, took in 5.6 million tons of crude one week in mid-December, just under the normal figure of 6 million tons. After a meeting of the oil committee of the Organization for Economic Cooperation and Development (OECD), officials said that while oil would not be scarce in the coming weeks and months, consumers would have to pay "dearly" for it.[22] In Washington in December, Yamani met with Secretary Kissinger and indicated that the embargo and cutbacks would be modified when Israel began its withdrawal. This represented a significant shift from the original demand for total and unconditional withdrawal. Yamani also indicated how far he was prepared to look beyond the embargo by reiterating earlier predictions that Saudi production could reach 20 million b/d, with much of that geared to U.S. markets. Yamani also took the time to meet with oil company executives in New York; one later predicted that even if the embargo were kept for its "symbolic" value, production and shipments of Arab oil to the United States would increase by February.[23]

The Arab stand on the cutbacks had visibly softened by the end of December. At a meeting in Kuwait on December 25, OAPEC announced a 10 percent hike in output in place of the scheduled 5 percent cut, bringing the total cut from the September production level to 15, rather than 25, percent. "We only intended to attract world attention to the injustice that befell the Arabs," said Yamani.[24] The early success of Kissinger in negotiating a cease-fire agreement between Egypt and Israel on terms that left Egyptian forces on the east side of the Suez Canal was considered to be a major factor, and an OAPEC communiqué "noted with satisfaction the gradual change in the trends of American public opinion which has begun to show a tangible understanding of the Arab problem and the expansionist policy of Israel."[25] There were reports that King Faisal was seeking a formal American declaration supporting the principle of Israeli withdrawal before lifting the embargo. The Americans, meanwhile, were counting on Sadat to persuade Faisal and the other Gulf oil producers to lift it on the basis of the "revolution" in U.S. policy in the Middle East as demonstrated by Kissinger's shuttle diplomacy.

The world of oil was being affected from another direction in these months, related to but separate from the embargo. The spiraling market prices that led to the October 16 price hike by OPEC continued to mount, accelerated by the cutback-induced shortages, and, for the United States, the price it had to pay for premium nonembargoed oil from non-Arab sources. European, Japanese, and American independents bid up the price on all available royalty and participation crude. European and Japanese governments made particular efforts to line up sizable exchanges of crude for arms and industrial goods. A Nigerian auction of crude in November reportedly brought high offers of more than $16 a barrel. At the end of November Iran announced an auction of 109 million barrels, and the *Wall Street Journal* commented that bidding was likely to be spirited. Spirited it certainly was: Iranian officials announced later that bids of more than $17 a barrel were received, and that virtually all of the high bids had come from American independents. The usual "villains," the Japanese and Europeans, had dropped out once quotations left the $10 range. OPEC, meanwhile, had asked the majors to submit proposals for setting a new price schedule based on the market but indicated after a meeting in late November that none of the company proposals had been constructive. A meeting to decide on prices was then set for December 22 in Teheran. The auction in Iran ten days earlier had tested the water.

The OPEC decision in Teheran was to increase the price of crude oil by 128 percent, raising the posted price to $11.65 and the government take to $7. The Shah announced the price hike with a declaration: "The industrial world will have to realize that the end of the era of their terrific progress and even more terrific income and wealth based on cheap oil is finished." A British official commented, "The last chicken of colonialism is coming home to roost."[26] The new price, scheduled to take effect January 1, was 470 percent higher than the price a year earlier. OPEC officials, however, described the new price as moderate based on comparison with the costs of alternative energy sources and with the actual auction prices received by Iran and other countries. The Shah and others had reportedly pressed for a higher price, which was closer to that received at auction. Saudi resistance was reported to be the main factor in arriving at the lower compromise figure.

At the beginning of the new year, then, this was the situation: the whopping price-hike of late December left government officials in a state of shock as they tried to plot out the implications of the price rise for balance of payments and the transfer of financial wealth and power to the oil producers. Although the spokesmen for the major companies dutifully expressed surprise for the record, it was quickly apparent that the companies stood to gain as much as the producing countries by the new price plateau. The companies' enthusiasm for the price hikes was not matched by that of consumer country government officials, but the United States, at least, could get some satisfaction that its economy would be hurt least. Companies weren't paying $17 a barrel in Iran to sell it for any less in New York—where prices as high as $27 were reported. The average profit per barrel for the major companies was nowhere near $10, but it was well over $1 a barrel, or more than triple the rate of the previous few years. Oil stocks, at an all-time high, were revalued upwards, as were domestic reserves in the ground, for vast paper profits on inventory. Domestic prices were hiked to more than $5 a barrel for "old" oil and $10 for "new" oil, even though actual production costs had not risen.[27]   In the Middle East, President Sadat had already brought pressure on the oil producers to relax the cutback, and an end to the embargo was in sight with the completion of the Egyptian-Israeli disengagement accord.

The Oil Weapon in the United States

It was clear at the beginning of 1974 that, barring another outbreak of war between Israel and the Arab states, the worst effects of the oil weapon had already been felt. Europe had suffered little, partly because of a mild winter and partly because the actual reductions were not as great as claimed. The same could be said in general for the United States, although in this case the continuing embargo did restrict imports below demand levels. A study of tanker movements by the International Longshoremen's Association calculated that in December, the month of the deepest cutbacks, oil shipments from the Persian Gulf were only 7.4 percent below the September, or prewar, level.[28]   Japan, which had not yet been accorded the status of a "friendly" country by OAPEC, seemed to be the country worst hit by the oil measures, but even there the crisis was caused more by the sharp increase in prices fueling an already severe rate of inflation than shortages per se. The ILA study calculated that production increases by non-Arab producers like Indonesia, Nigeria, and

Iran had kept the overall global shortfall to around 5 percent in December. With the nominal production cutback now reduced to 15 percent, observers expected the situation to ease into a rough equilibrium of supply and demand, albeit at sharply higher prices for the consumers.

In the United States the continuing embargo restricted imports below demand levels. Semiofficial (American Petroleum Institute) figures showed a declining rate of imports in the last weeks of December and the first weeks of January, coinciding with the arrivals of the last tankers loaded before the embargo. Gasoline stocks were down but heating oil stocks were up from levels of a year earlier. Attempts to check and verify this information with actual refinery statistics on runs of crude oil met a familiar refusal of the companies to divulge such information and led the *New York Times* to write: "The energy crisis is a dramatic paradox: crude oil flows in huge quantities, but information about it has been cut to a murky trickle."[29] Rather than harm the economies of the United States and Europe, the Arab oil embargo provided a short-term crisis atmosphere during which long-term policies could be set that shared the common feature of higher prices for energy resources. As noted by the industry-financed Conference Board Energy Information Center report, these effects hit hardest low-income persons and groups domestically and the non-oil-producing countries of Asia, Africa, and Latin America internationally.[30]

Most of the published figures purporting to demonstrate the effectiveness of the oil embargo are contradictory and self-serving. Before the embargo, when the Nixon administration was interested in minimizing the potential threat in the face of the decision to resupply Israel, it was pointed out that the Arab producers supplied only 6 percent of U.S. crude imports. A month later, when the companies wanted to emphasize the effectiveness of the embargo in order to deflect criticism for shortages and dislocations, the percentage was upped by including imports of refined products from Caribbean and European refineries dependent on Arab crude. Between early October and late November various statements by Interior Department and State Department spokesmen assessed U.S. imports at from 1.2 million b/d to 3 million b/d. These wild and ever growing estimates certainly helped create the panic buying and price-hikes in the weeks that followed.[31]   Reliable statistics include the following: U.S. crude and refined imports for 1973 were averaging 6.2 million b/d just before the embargo, of which 1.6 million b/d were from Arab countries, and half of that from Saudi Arabia. Arab supplies accounted for 26 percent of total imports and 9 percent of total consumption, then running under 17 million b/d.[32]   In addition, the embargo affected U.S. military purchases, resulting in a 325,000 b/d drain on domestic reserves to supply the U.S. armed forces and the client regimes in South Vietnam and Cambodia.[33]   Assuming a totally effective embargo, then, U.S. supplies would be reduced by about 10 percent.[34]   The embargo, of course, was far from totally effective.

The effects of the embargo and cutback were partially offset by increased production in non-Arab countries. United States imports from Iran increased by some 200,000 b/d, and from Indonesia by at least 100,000 b/d. According to American Petroleum Institute figures, the 6.2 million b/d import average dropped to 5.5 million b/d in December and to a low of 4.5 million b/d in mid-February. Interestingly, the API figures show a decline in domestic production, from 9.3 million b/d to 9.1 million b/d, coinciding with the embargo period—an unexplainable phenomenon in light of skyrocketing prices designed to increase supply. At the worst point, in mid-February, combined import and domestic figures totaled less than 14 million b/d.[35]   A warm winter and conservation measures reduced demand to around 16 million b/d. No more than 1 million b/d of the shortfall could be attributed to the embargo.

When the Shah of Iran suggested in a CBS interview in late February that more oil than ever was getting into the United States, energy czar William Simon accused him of "inexcusable and reckless remarks."[36] Actually, the Shah had said much the same thing almost two months earlier, while relaxing in St. Moritz. Then Simon had the Commerce Department quietly classify all information regarding the origin of crude imports,[37] and asserted that the embargo was fully effective. After this first tiff in memory between a shah and a czar, the American Petroleum Institute told the *Journal of Commerce* that the "Shah's remarks may have had some validity a few weeks ago, when there was substantial leakage. Lately, though, we've seen a dramatic drop in our imports which indicates that the embargo has become very tight."[38]

According to a Sunoco executive, leakage at that time was running as high as two million b/d.[39] While that estimate seems grossly exaggerated, it is clear that the "dramatic drop in our imports" was related to a policy shared by the major companies of not importing crude oil that they would have to share with crude-short (mainly small independent) refiners under the allocation system mandated by Congress. While Simon was proclaiming the effectiveness of the embargo, the companies were openly suggesting that as much as 1 million b/d could be made available if the allocation rules were revised.[40] Crude oil imports over the embargo period were up 11 percent from the previous winter, and on March 1, U.S. stocks of crude and refined products were nearly 8 percent higher than the previous year.[41] Gasoline shortages suddenly evaporated in March, although the embargo was still officially on. Perhaps some explanation lies in the fact that companies like Chevron (Standard of California) were raking in an extra half-million dollars *per day* thanks to price-hikes of 30 to 55 percent.[42]

More exact figures on "leakage" became available in April when the Commerce Department declassified the crude import data for the embargo period. Libya, which was blamed in the media for most of the leakage, did export 4.8 million barrels to the United States in November and 1.2 million barrels in December, but none after that. Imports from Saudi Arabia, however, were much higher, amounting to 18 million barrels in November and 7 million barrels in December. Moreover, Saudi oil continued to arrive, albeit at a sharply reduced rate, in January (257,187 barrels) and February (552,212 barrels). Tunisia, Algeria, Kuwait, and the United Arab Emirates were also responsible for small shipments during the embargo. Although the figures do not specify, it is fair to presume that most of the oil imported from Europe during that period was actually Arab oil.[43]

It is not too surprising that the giant companies like Exxon and Chevron continued to assert in the face of these facts that they were not responsible for any leakage, and that the embargo was responsible for the shortages of products during this period. More astounding is the virtuoso performance by Simon, now promoted to Treasury Secretary, in assuring the public that "not one drop" of crude was leaked from Saudi Arabia, even though imports from that country averaged one-half their normal rate during the embargo.[44] Using the conjurer's tricks he learned while becoming a multimillionaire Wall Street investment banker, Simon has tried to convince us that it took up to four months for Saudi oil to reach U.S. ports, a trip that normally takes four to six weeks. Small wonder that Simon's assistant Eric Zausner, when pressed by the *Journal of Commerce* for a "clarification," admitted, "I've had a rough time sorting out these numbers myself."[45]

The political and economic chicanery of the major oil companies and the Nixon administration in relation to the

Arab oil embargo make it virtually impossible to measure with precision the actual impact of the embargo on the American economy. Higher domestic prices tended to make other projects like oil shale and coal gasification economical, a factor which was strengthened by the vast subsidies promised under Nixon's "Project Independence." Alternative fuels like coal more than doubled in price and electric utility rates reflected these costs. The Federal Power Commission has used the "crisis" to grant large price increases to natural gas producers, and the industry is pressing on with its bid for complete deregulation of prices.[46] Simon, as Treasury Secretary, pushed for greater tax "incentives," such as very rapid depreciation write-offs for the energy industry, while opposing individual tax reductions as "inflationary."[47] As for congressional ability to effect even moderate reforms in the industry's operation and regulation, the first half of 1974 certainly bears out the prediction of top industry lobbyist Charles Walker last January: "When it all shakes out, I think we'll see a lot of rhetoric and some action but not that much in terms of radical change affecting the industry."[48] Oil company profits were described in the financial press as "embarrassing," and by the Federal Energy Office as "whoppers," and even these reported increases have been criticized as understatements.[49] Walter Heller, former chairman of the Council of Economic Advisors, calculated that the "siphoning off of consumers' dollars away from other goods and services" represents "an effective reduction in real income" of some $20 billion in 1974.[50] This rise in sales *proceeds* will be realized on a lower sales *volume*, representing a saving for the companies of about $4 billion. Of the $24 billion differential, $8 billion will go to foreign governments, while $3 billion will go to the U.S. government *if* a windfall profits tax is enacted. This leaves a minimum of $13 billion as increased earnings, and Heller assumes that "petroleum accountants will likely be inventive enough to keep a considerable part of this windfall from showing on the bottom line." The overall effect of the oil weapon was to enhance, at least temporarily, the capacity of the giant oil and energy companies to gouge the consuming public and consolidate their monopoly position not only in the energy sector but within the economy as a whole.

Political and Strategic Dimensions of the Oil Weapon

The political and strategic implications of the oil weapon must be evaluated from two directions: (1) the role of the oil weapon in the struggle for political hegemony among various national and class forces in the Middle East; and (2) the role of the oil weapon in the competitive struggles among the industrialized capitalist countries of Europe, Japan, and the United States. In both of these areas we will focus on the strategy and diplomatic tactics of the United States under the direction of Henry Kissinger.

Kissinger's fundamental approach has been to reestablish strong American political and economic influence by encouraging and promoting the alliance between Faisal and Sadat. It is doubtful that Kissinger (unlike many U.S. policy-makers) ever believed that American interests in the Middle East could be assured by a militarily dominant Israel.[51]   During the visits of Israeli and Arab leaders to Washington in February and March 1973, unnamed Middle East experts stressed that "there seems to be no cause for urgency and that a strong case could be made for preserving the status quo" (i.e., continued Israeli occupation of Sinai, Golan, and the West Bank).[52]   In Kissinger's view, stability in the area had to rest on the viability of moderate regimes with firm political bases. One condition for that was demonstrated by the continued political turbulence in Egypt following the

1967 war. The Arab-Israeli impasse had to be solved on terms that gave those moderate regimes a stake, both economically and politically, and on terms that provided defense against more radical elements internally and elsewhere in the Arab world. The key moderate in this scenario is Sadat, who moved after 1971 to transform Egypt into a private enterprise economy allied with the United States rather than the Soviet Union, which was open to Saudi and other capital investments. The October War provided the crisis that could be used to justify an appropriate shift in U.S. strategy. For Kissinger it was a "golden opportunity."[53]

During the war Kissinger used all available means—mainly Israeli dependence on the United States for rearmament—to help bring about a military cease-fire which left neither party in a clearly superior position. The United States could then provide the indispensable mediation wanted by both sides. A shift in the U.S. stance vis-à-vis Arab war aims was hinted at in Kissinger's October 25 remark that "the conditions that produced this war were clearly intolerable to the Arab nations and that in a process of negotiations it will be necessary to make substantial concessions. . . ."[54] Subsequent reports from Israel that U.S. pressures had forced Israel to resupply the Egyptian Third Army trapped on the east bank of the Suez Canal was additional evidence of some shift. The first achievement of this new policy was the quick restoration of diplomatic ties with Egypt in early November after Kissinger's first trip to the Middle East.

Sadat was seen as one key to renewed U.S. influence in the Arab world. Faisal was the other. Faisal controlled the largest single source of crude oil; in the context of the Arab confrontation with Israel, questions of how to use oil resources as a lever and a weapon were unavoidable. Before the war, it was sufficient for Faisal merely to talk about not raising production. Once the war broke out, Faisal had to act.

At the first meeting of the Arab oil-producing states, held more than ten days after the fighting broke out and then only as a result of public pressure from Sadat, Faisal initially blocked strong proposals for an embargo of the United States, and of course refused to follow the Iraqi example of nationalization. Only after President Nixon proposed $2.2 billion in arms aid to Israel did Faisal consent to the embargo. Faisal certainly wanted to pressure the United States into taking a more "pro-Arab" stand, but he was also compelled to participate in the embargo or face a loss of influence in the Arab world, and perhaps even his throne. At the end of October, Yamani told a visiting delegation of U.S. congressmen that "King Faisal has done his best in the last two weeks to represent American interests. . . . We did not want the embargo. We hope that we can do something, but there must be something that we can show as change. . . ."[55]

Secretary Kissinger evidently thought that the Egyptian-Israeli cease-fire and POW exchange might be change enough. He asserted at a press conference on November 21 that "economic pressures" against the United States, while understandable during the war, were "inappropriate" during peace negotiations; that U.S. policy would not be influenced by such pressures; and that unspecified "counter-measures" might be necessary.[56] The result was a temporary escalation of rhetoric as Sheikh Yamani responded from Europe that countermeasures would be met by an 80 percent production cut and that oil facilities would be sabotaged in the event of any military moves.[57] A week later Defense Secretary Schlesinger announced that U.S. naval forces in the Indian Ocean would be beefed up because of U.S. "enhanced interest" in the Persian Gulf area growing out of the oil embargo.[58] Kissinger, meanwhile, was busy lobbying among members of Congress that the October War and the embargo precluded any settlement resembling the prewar balance of forces, that Israel's international isolation was increasing and even public support at home was contingent on Israeli flexibility, and that this might be the Israel's last chance to secure a peace not totally imposed on it.[59]

This pattern of tough talk from Schlesinger and conciliatory diplomacy from Kissinger continued over the next two months, during which time the Israeli-Egyptian disengagement was worked out and the production cuts were eased for Europe and Japan. Kissinger initiated an American campaign to take the lead in developing new relationships between oil-producing and consuming countries, and numerous hints were dropped from Arab and American sources that the days of the embargo were numbered. On January 22 Kissinger suggested that he had gotten pledges from certain Arab leaders that the Egyptian-Israeli disengagement warranted an early end to the embargo:

> We have had every reason to believe that success in the negotiations would mark a major step toward ending the oil embargo. We would therefore think that failure to end the embargo in a reasonable time would be highly inappropriate and would raise serious questions of confidence in our minds with respect to the Arab nations with whom we have dealt on this issue.[60]

At that moment, President Sadat was in the midst of a six-day trip to other Arab capitals, including the oil producers, to "explain" the disengagement accord and argue for an end to the embargo.[61] Faisal was understood to be disposed to lifting the embargo, but there was sharp opposition from Syria and Algeria, as well as Libya and Iraq. Syria was openly critical of Sadat's willingness to disengage the Suez front without any negotiations on the Golan, permitting the Israelis to concentrate forces there. Any proposal to weaken or end the embargo would thus have had small chance of acceptance by a good number of Arab states without Syrian concurrence or at least visible progress on the

Golan disengagement. It is possible that Saudi Arabia told the United States that is would try to have the embargo softened at an upcoming meeting of OAPEC scheduled for February 14 in Tripoli. This possibility was immediately sabotaged by none other than President Nixon when he announced, in his State of the Union address on January 30, that on the basis of his contact with unnamed "friendly Arab leaders" he could announce that "an urgent meeting will be called in the immediate future to discuss the lifting of the oil embargo."[62]

One does not have to possess the diplomatic acumen of a Metternich to realize that besides being a misrepresentation of the facts (the meeting had long been scheduled) Nixon's statement amounted to a virtual challenge to the "friendly Arab leaders" to produce an end to the embargo, a challenge certain to fail in the face of a majority consensus among the Arab leaders that their demands had not nearly been met and that progress on the Syrian front was essential. Any slim chance that Sadat and Faisal could have carried the February 14 meeting was thus eliminated. The question is why.

It is possible that the Nixon statement was simply another attempt to build up his tattered political image by projecting himself as in control of the "energy crisis" and in fact indispensable to its solution. His characterization of the meeting as a new development of which only he and the "friendly Arab leaders" had knowledge would support this interpretation. On the other hand, Kissinger told a congressional committee the next day that Faisal and others had indeed indicated they would recommend lifting the embargo at the February 14 meeting. Following the reports of Nixon's statement, Syria's President Assad flew to Riyadh and received "firm pledges" from Faisal and Kuwait's ruler, Sheikh al-Sabah, that the embargo would be maintained until a disengagement accord for Golan had been worked out on Syrian terms, meaning "part of a plan for a total Israeli withdrawal" and "an assurance of Palestinian rights." On

February 6, Kissinger repeated his expectation that the embargo would be lifted and said that failure to do so would be "a form of blackmail." The result was predictable: the Arab oil ministers' meeting was "postponed indefinitely at the request of Saudi Arabia and Egypt," reportedly because they needed more time to persuade other leaders to lift the embargo.[63] Instead there was a meeting of Presidents Sadat, Assad, Boumedienne, and King Faisal in Algiers that produced no public decisions but was followed by an announcement in Washington that Kissinger would soon be back on the diplomatic beat between Tel Aviv and Damascus.

It seems likely that Saudi Arabia and Egypt called off the Tripoli meeting at the last minute out of fear that a majority would vote to maintain the embargo, perhaps with specific conditions regarding a Syrian settlement that would have delayed any end for months. Given the fact that this whole episode is precisely contrary to Kissinger's style of secret diplomacy, the events suggest that this small confrontation may have been staged to bolster Faisal's image and influence in the Arab world and give him a chance to stand up to U.S. blandishments. This larger role for Faisal was in the U.S. interest as well as in Saudi Arabia's. In any case, the confrontation between Kissinger and Faisal came and went. A short while later, at the beginning of March, Saudi officials in Washington said that an end to the embargo would be announced shortly after Kissinger returned from his current visit to the Middle East. Egypt then proposed that the Arab oil ministers meet on March 10 in Cairo, indicating that enough apparent progress had been made in the Syrian-Israeli disengagement talks to justify renewed consideration of the embargo. This time U.S. officials refused to comment on the probability that the embargo would, in fact, be lifted.[64]

As the date for the Cairo meeting approached, the splits within the Arab world over the embargo question became sharper and more public. Saudi Arabia made an official statement supporting the end of the embargo. Syria, Algeria, and Libya were opposed and insisted that the meeting be held in Tripoli as originally scheduled. Sadat evidently hoped that the more pro-American environment of Cairo would help carry the swing votes, like Kuwait, to end the embargo. The Cairo meeting failed to get off the ground when Syrian, Libyan, and Algerian representatives refused to attend. Algeria was to have chaired the meeting, and in the view of Yamani and others, Algerian presence at the meeting was necessary to neutralize radical nationalist and leftist opposition to ending the embargo. The meeting was rescheduled for March 13 in Tripoli. In the interim the Egyptians and Saudis held strategy meetings in Cairo with the other Gulf producers and announced "full agreement on the oil situation" the day before the meeting. At the Tripoli meeting the Arab producers reportedly agreed to lift the embargo, but postponed any official announcement until an OPEC meeting in Vienna the following Sunday, presumably so as not to antagonize Libya, which continued to oppose the decision.[65]

When the meeting opened on Sunday, March 17, Syria made a formal proposal to lift the embargo provisionally until June, when a formal decision would have to be made in order to keep the embargo lifted. Egypt and Saudi Arabia rejected this proposal, insisting that the embargo be lifted without conditions. This resolution carried. Algeria's last-minute addendum that the decision be subject to review was allowed by Egypt and Saudi Arabia with the clear understanding that they would prevent any reimposition of the embargo. On this basis Kissinger quickly noted, "We do not believe it is probable that the embargo would be reimposed,"[66] and indicated that he based that assumption on Saudi and Egyptian assurances to stand fast against such a move.

The official announcement finally came on March 18 in

Vienna, with unofficial predictions that production levels would be restored to prewar levels. Saudi Arabia pledged an increase in production of 1 million b/d for the U.S. market. Syria and Libya maintained their opposition to the decision. The embargo on Holland, Portugal, South Africa, and Rhodesia was not lifted.

This embargo meeting coincided with an OPEC meeting on prices, in which Algeria, Indonesia, and Iran reportedly proposed a further 15 percent hike in posted prices, a move favored by a clear majority of OPEC countries. However, Saudi Arabia's Yamani reiterated his views that prices were already too high and that they should be reduced so as not to jeopardize the economic and political stability of the industrialized capitalist countries. In what was described as a "stormy session," Kuwait and Abu Dhabi were said to have placed in the conference record a censure of Saudi Arabia, accusing it of discouraging buyers from paying high auction prices. Yamani threatened to break from OPEC and set lower prices if the proposed increase was voted. According to the *New York Times* account of the Vienna meeting:

> On the major issues of the embargo and the oil prices decided here, Saudi Arabia virtually imposed conditions that were closely in line with American desires, with considerable risk to the unity of the Arab countries and the world's major oil-producing nations.[67]

Meanwhile, reports from different international oil brokers revealed that Aramco parent companies were offering large volumes of crude in Europe at prices well below current market levels.

Since the embargo was lifted, Saudi and American interests have been further cemented by official agreements to expand economic, political, and military ties "in ways that will enhance the stability of the Middle East."[68] The initial focus of the agreements was on internal security: Prince Fahd,

Faisal's brother and heir-apparent, headed a delegation to Washington in June. He is Minister of Interior and in charge of internal security, in addition to heading the Saudi Economic Affairs Committee. One agreement involves a U.S. commitment to re-equip and train the National Guard, responsible for the security of Faisal and the Royal House and also stationed around the oil-production facilities. At least 5,000 guardsmen have been guarding Aramco installations since 1967. It has been disclosed that two large fires at Aramco's Ras Tanura refinery in 1973 were caused by Palestinian sabotage. Only by strengthening his security forces will Faisal feel confident of refusing to participate in any future embargo. This agreement, which expands the National Guard by at least two battalions, came on the heels of earlier agreements under which the United States is to provide $750 million in naval equipment and training, and the Saudis are to purchase some 200 jet fighters. United States military personnel number more than 200, in addition to an undetermined number of former military men working on contract to U.S. arms firms like Northrop, Raytheon, Bendix, and Lockheed. The head of a British consulting firm which was recently awarded a contract to recruit 400 key personnel for the Saudi air force described his job as "setting up a quasi-military operation."[69]

The situation in the year after the embargo's end finds United States-Saudi ties stronger than ever, with special efforts being made to involve American corporations in Saudi economic development and to channel Saudi oil revenues into U.S. securities and markets. Both countries seem to be gearing up for the time when the contradictions will prevent Faisal from playing both sides of the fence. The dimensions of these developments have aroused more than a little suspicion about the motives of both parties. Following the general accord on economic and military cooperation, some Saudi and Arab circles concerned are now asking themselves whether this move towards strengthening economic and defense cooperation with Saudi Arabia is evidence of a sincere long term intention on the part of the U.S. to assist in the industrialization and economic development of Saudi Arabia . . . or whether, on the contrary, it represents no more than a transitory tactical maneuver designed to outflank and torpedo European and Japanese efforts to secure bilateral oil deals of their own, with a view to promoting what may be the fundamental U.S. objective of securing a global approach to energy questions under American hegemony.[70]

During the last decade the relationship between the United States and its erstwhile allies (in the industrialized capitalist world) has been one of increasing tension and competition in the economic and political sphere. This was nowhere more pronounced than in the politics of oil and the Middle East. The attempts of some European states, notably France and Italy, to achieve some measure of autonomy with regard to energy supplies and to reduce their dependence on the major international companies had led to some of the most profound changes in the industry to date, beginning with the famous ENI joint venture with Iran back in 1957. After the June War of 1967, France took a markedly detached position toward Israel and renewed deteriorated relations with a number of Arab countries. In Iraq, this stance gained France special treatment in the phase of oil nationalization. Japan, too, during this period, strove mightily to develop some independence vis-à-vis the major oil companies. This effort was concentrated in Indonesia and, to a significant extent, in the Persian Gulf. The vehicle for achieving this independence generally took the form of a national oil company, or strong state financial backing of private national firms. The growing role of national oil companies in the producing states, through nationalization, participation, joint ventures, and marketing of royalty crude, made an increasing number of commercial and barter deals possible, with the result that a small but growing share of world oil trade was moving outside the channels dominated by the major companies. The move on the part of the OPEC countries to focus on industrialization as the cornerstone of their oil policies is bound to accelerate the trend. ENI of Italy articulated from a European perspective the strategy behind these moves in its 1971 annual review, calling for a "basic European policy" in which national oil companies would "assume positions of increasing commitment and responsibility with regard to energy imports." According to ENI Vice-President Francesco Forte, writing in the London *Times:*

> These agreements should be concluded between the national company of each supplying country and national companies of the EEC countries acting jointly. . . . They would be agreements at the national company level to ensure flexibility and economic efficiency, but they would be fitted into national government plans and the terms of EEC directives.
>
> The agreements have to cover the supply of large quantities of crude oil in predetermined terms, exploration financed by the companies of the purchasing countries under ancillary contracts, and the provision by the purchasing country of goods and services in exchange for oil purchased, contributing to local economic development as part of general collaboration with the EEC.
>
> These agreements should be ample in scope and number. Clearly they could not cover all needs, but they would replace the existing machinery, under which the supplying countries and the great international groups confront one another, while the European countries, which are the really interested parties, take no part in the decisions but have to put up with the consequences.[72]

While some of the specifics of this European analysis would be debated by the oil producers, the trend was much the same on their side. The Eighth Arab Petroleum Congress, in Algiers in June 1972, resolved that "the oil-producing

Arab countries should practice direct and effective control over their oil industry," "should establish direct contact with markets importing Arab oil," and "should conclude agreements for the sale of oil through their national companies and the importing companies in these markets."[73] In a presentation to the Thirty-second (Extraordinary) Conference of OPEC in March 1973, Iraqi Oil Minister Hamadi discussed the future direction of oil relationships between producers and consumers. Industrial projects, particularly those like refining and petrochemicals with guaranteed export markets in industrial countries, should be part of all future oil deals, which will presumably be new types of arrangements like service contracts and joint ventures. As for marketing of crude oil,

> national oil companies . . . have at their disposal increasing volumes of crude oil to sell in the world market. . . . Instead of being marketed in conventional ways, these volumes of crude oil could be disposed of in package deals of five years' duration specifying not only the terms of normal free market sales like price, payment, etc., but also the implementation of a number of development projects in the seller country on terms to be agreed upon. In this way, the marketing of crude oil would evolve into a relationship between national oil companies and groupings of project implementing firms from the oil consuming countries over periods of time longer than one year and involving more than one spot deal—thereby bringing about a more stable relationship.[74]

One reflection of this evolving relationship between the oil producers and the consuming countries was a growing unhappiness on the part of European countries with U.S. policy in the Middle East. They viewed the United States-Israeli alliance as myopic, perhaps because the United States was not as threatened as they by the potential cutoff of Arab oil.[75] The extent of this divergence became immediately manifest with the outbreak of fighting in October 1973. In the NATO alliance only Portugal allowed the United States to use its territory for resupplying jet fighters and other arms to Israel. West Germany demanded that U.S. arms stocked at NATO bases there not be sent to the war zone. In a series of secret NATO meetings during the was, U.S. attempts to have their NATO allies "take steps to chill their trade and political relations with the Soviet Union" were rebuffed. The allies' failure to wholeheartedly support the United States in the "alert" confrontation on October 24 caused Kissinger to remark that he was "disgusted" with NATO.[76] A joint statement by the European Economic Community on November 5, supporting Arab demands for total Israeli withdrawal, and a later statement by Japan to the same effect, strained relations even further. Britain and France in particular initiated a series of high-level contracts with individual oil producers (including Iran) in order to secure crude supplies. The United States regarded this as irresponsible and divisive, and argued that a joint response through a consuming nations organization or through the OECD (Organization for Economic Cooperation and Development) was the only way to deal with the oil producers on supply or price questions. Kissinger opened a public campaign in this direction in a speech in London in December in which he proposed that

> the nations of Europe, North America and Japan establish an energy action group of senior and prestigious individuals, with a mandate to develop within three months an initial action program for collaboration in all areas of the energy problem.[77]

"The energy crisis of 1973," he continued, "should become the economic equivalent of the Sputnik challenge of 1957."

What was forthcoming was neither a group as broad as the one Kissinger proposed nor even a joint European position. Although OECD and EEC had their energy committees busy drawing up plans and forecasts, the individual countries went much their own way in practice. France was in the forefront of those wanting to establish a special relationship between Europe and the Arab world, while West Germany and Holland pressed for a common European stand that would share supplies and shortages even in the face of Arab opposition. Bilateral dealings, the threat of competitive devaluations, and export subsidies to cope with the higher costs of oil led the United States to propose a meeting of major oil-consuming nations in Washington on February 11. The American delegation would be headed by Kissinger, Simon, and Treasury Secretary George Schultz. The Common Market countries, plus Canada, Japan, and Norway, were invited. The reservations held by the French and others toward U.S. leadership in formulating joint energy policies made the meeting's prospects very limited. The oil producers warned against the formation of any kind of counter cartel. In an interview in *Le Monde*, Algerian President Boumedienne called the Washington conference a "plan designed to prevent contacts between the oil-producing and consuming countries":

> The Europeans at the present moment have the possibility of laying the basis for a long-term cooperation which would guarantee their oil supplies for 25 years in exchange for their participation in the development of an area in which they are vitally interested. The Washington Conference aims to slow down the new policy of direct links between producers and consumers. This is the policy of the future and may well mark the beginning of the end of the system imposed by the cartel, which made everyone quake but which has been broken by the oil weapon. In addition, the real policy of the United States is not to lower prices but to control the sources of energy and thus ensure its political power. This is the truth. If the Europeans yield before the American "big stick" they will once again return to the sidelines of history. It is a question of choice.[78]

The conference, which the French continued to regard as simply "an exchange of views," produced nothing immediately notable. The final communiqué, "dotted with French demurrers," spoke of "a comprehensive action program to deal with all facets of the world energy situation," including emergency allocations, research, financial, and monetary measures to avoid competitive devaluations.[79] Special working committees, as well as the existing OECD apparatus, were to prepare detailed plans along these lines for future meetings. In early March the EEC announced plans for a high-level conference on long-term economic cooperation between Europe and the Arab world, following suggestions to that end from Arab representatives at an earlier EEC meeting in December. Kissinger interpreted this as a French-inspired move that would wreck any chance for his goal of coordination among all the industrial countries. The State Department officially objected to the EEC plan, complaining that there had been no consultations with the United States, a line rejected even by the Germans. Agence France Presse quoted a "very high American official" as saying "off the record":

> We want to avoid a confrontation between the Europeans and ourselves, but this confrontation has become inevitable. The Europeans cannot compete with the United States in the Middle East, and if we fight them there we will eventually win. The Arab countries need the United States more than they need Europe.[80]

A month later came the joint announcement from Washington and Riyadh that the two governments would hold negotiations "to expand and give more concrete expression" to their relationship.[81] When the full series of bilateral agreements was signed later in early June, the State Department had to go through some verbal gymnastics to show how they were different from the European and Japanese deals they had been criticizing.[82] This agreement, plus those signed with Egypt during President Nixon's quick trip in June and Treasury Secretary Simon's in July, reflects renewed U.S. determination to be the paramount power in the Middle East, and if possible, in the whole international energy sector.

OPEC Strengthens the Crude Oil Price Floor

The summer of 1974 was marked by two OPEC conferences: one in Quito, Ecuador, in late June and the other in Vienna in mid-September. Both showed evidence of a general resolve on the part of the oil-exporting states to consolidate crude oil prices at the high level set in the aftermath of the October War. The June meeting was characterized by most observers as a rather rancorous affair, with only Yamani arguing that the existing level of crude prices was ruining the global capitalist economy and, among other things, giving unfair advantage to the socialist bloc, with its relative self-sufficiency in energy resources. Yamani's analysis, which sounds remarkably like that of Kissinger, led him to disassociate Saudi Arabia from the OPEC decision to raise slightly the royalty on the crude oil still belonging to the companies (equity crude, or about 40 percent of current production). Following the meeting, Algeria's Belaid Abdessalam defended the existing price level, saying that the October War "at the most played the role of a catalyst in taking a decision which was already well prepared and well justified on the economic level."[83] Abdessalam warned that attempts to use Saudi Arabia's capacity for increased production to drive down prices would amount to "a dangerous game" which would lead to a coordinated reduction of output by the other OPEC producers. Average crude prices at the beginning of the summer were about $9 a barrel in the Gulf and $13 a barrel at Mediterranean ports.

Following the Quito meeting, Yamani rushed to New York to open talks with Aramco owners regarding a Saudi takeover of the giant oil-producing operation, of which it now owns 60 percent. He then headed back to Saudi Arabia in time to greet Treasury Secretary Simon during his tour of the Middle East in mid-July. Prior to their meeting Yamani declared that despite his disagreement with the current OPEC price level, "we are not prepared to go outside this framework by undertaking separate negotiations on prices with consumer governments." He attempted to get around this commitment by announcing at Simon's departure that Saudi Arabia would hold an auction for crude in early August with no floor on acceptable bids, a move that Simon hailed as guaranteed to create a new, lower price floor.[84] The auction, though, was "postponed" indefinitely without explanation following a meeting of Saudi Arabia's Higher Consultative Council on Oil and Minerals, chaired by King Faisal's brother Fahd and including his son Saud. This apparent reversal may be linked to the Saudi negotiating position in the ongoing talks with Aramco. A more immediate factor, however, may have been the ruling family's unwillingness to risk political isolation among the other Arab and Middle Eastern producers.

Simultaneous with these developments, Kuwait succeeded in getting Gulf and British Petroleum to "buy back" Kuwait's 60 percent share of production for about $11 a barrel, 2 percent more than the going "buy-back" rate. The State Department issued a statement "deploring" this effective increase in crude price level. For Gulf and BP, of course, it was relatively painless: any increases are passed on to the consumer. Competing companies or governments will have to pay the same $11 per barrel, but without the 40 percent of production that the big companies get at the much lower rate of just over $7 a barrel. The *average* price for the giants remains around $9; all others pay $11. This enhances the ability of the majors to consolidate markets and profits against challenges from "independents." The majors got a further boost when Kuwait and other countries cut back production from the state-owned 60 percent rather than the company-owned 40 percent, thus further lowering the average price per barrel, to well under $9, and at the same time limiting the amount of crude available to other buyers. During the month of August, production was down in Saudi Arabia as well as in the other countries.

The September OPEC meeting featured an instant replay of the June meeting. All the producers except Saudi Arabia levied an increase in royalty and income tax rates to net them an average increased take of 33 cents per barrel, which was calculated to be equivalent to the rise in world inflation over the preceding three months. This represented a compromise in itself, with some producers arguing that the inflation rate should be calculated at the annual rate, closer to 14 percent. The Saudis once again dissociated themselves from this decision, which raises the cost of company-owned crude (40 percent), but quietly increases the price of "buy-back" crude (60 percent) in line with the Kuwait rate. The average increase in the cost of Saudi crude is thus 13 cents per barrel.[85]

Behind these complicated and confusing maneuvers lies a continuing struggle for power and influence in the world of Middle East oil between the United States, allied with Saudi Arabia and the other Gulf states, and Europe, allied with Algeria, Libya, and Iraq. Of course, the lines are not quite that neatly drawn, but one can see a pattern in these recent developments of an attempt by the more progressive oil-producing states, led by Algeria, to offset the post-October War alliance between the United States and Saudi Arabia and Egypt by promoting trade-and-aid deals with European countries and Japan. In this light, the present moves by OPEC to trim both the profits of the major companies and their access to crude oil seem designed to benefit the smaller independent and state-owned companies and develop new channels for the flow of resources not controlled by the seven giants.

The October War and the oil weapon seen this close at hand have had the paradoxical result of disrupting the political and economic hierarchies that had prevailed for decades—but in a way that at least temporarily strengthened the political role of the United States. The initiative of bourgeois nationalist forces in Egypt and elsewhere in the Arab world to undertake a limited military drive against Israeli occupation forces was matched by the readiness of the United States, under Kissinger's direction, to mediate a settlement that would "liberate" most of the Arab territory seized in 1967 and, in so doing, consolidate the state power of Egypt in the hands of the "moderate" forces led by Anwar Sadat and his clique of landowners, tycoons, and technocrats. One hundred years after Khedive Ismail led Egypt to bankruptcy at the hands of British and French financiers, leading to direct British colonial rule, Sadat is banking on the United States, in conjunction with Arab oil money, to transform Egypt into a "big Lebanon." Similar trends, although much less pronounced, are evident in Syria.

The oil weapon has enhanced the political role of Faisal and other reactionary rulers in the Gulf, and the steep rise in government revenues from higher oil prices has provided them with enormous economic and political clout in the Middle East and beyond. This continues a political trend set in motion by the defeat in 1967 of the radical nationalist Arab states, Egypt and Syria. On the edge of the Arab world, Iran's predominant military position has been strengthened further and its tacit strategic alliance with Israel is complemented by economic involvement in Egypt as well.

While some of the contradictions which led to the war and the oil weapon have been partially defused or deflected—most importantly the question of the occupied Egyptian and Syrian territories—the overall material conditions have not been much affected. The liberalization of the Egyptian economy coupled with the continued suppression of leftist political and trade union elements could give rise to serious conflicts in the near future. Sadat's consolidation of power within the existing political structures may not be enough to contain these tensions. In other Arab states as well, the stability (so desired by the United States) that seems to prevail is relatively recent, and there is no way to measure its comparative strength or fragility. It is important to not underestimate the strength of international and class forces behind the current political alignment but at the same time to understand its limits. One weakness of the existing balance was pointed out by the Iraqi leader, Sadam Hussein, in an interview in *Le Monde Diplomatique* in July 1973:

> Another truth is that no leader in this area or anywhere else is immortal. What the imperialists believe to be a guarantee of their interests is only temporarily so. Who can predict what the future holds? Who can tell us what the future of these states' interests will be if the situation in Saudi Arabia or Iran is modified? We leave it to the imperialist states concerned to answer this question which we believe is not favorable to them. And we believe that one day they will be obliged to answer it.[86]

These recent developments around the October War have also had the effect of accelerating the uncoordinated movement of consuming countries to line up secure supplies of oil in direct bilateral deals with producing states, thereby increasing the amounts of oil moving outside integrated channels of the major companies. At the same time, the inability of the European countries and Japan to coordinate their oil policies on a political or economic level left the overall management of the allocation and distribution crisis to the major companies, and political intervention and mediation to the United States. The vulnerability of those countries to the oil weapon was of course greater than U.S.

vulnerability. The subsequent changes of government in the major European countries—Britain, France, and Germany—though not directly related to the energy crisis have delayed further approaches toward a European-Arab "special relationship." It is doubtful that the crisis surrounding the October War has reversed the trend toward economic competition and political divisiveness among the industrialized capitalist countries—a trend that radical nationalist Arab states like Iraq, Libya, and Algeria have exploited to their own advantage. The transfer of real wealth to a few countries previously outside the temples of power will exacerbate international and internal conflicts for economic advantage. This will heighten the sense of "almost medieval turbulence" that has characterized global class and national relations in the year of the oil weapon.[87]

10
The Nature
of the Crisis

The oil embargo and the price explosion are concrete and specific developments in the global struggle for control of the earth's natural wealth of resources. The energy crisis is on one level a crisis of finite and limited supplies of resources. Recognition of this fact is bound to force a transformation of traditional Western capitalist attitudes toward the exploitation of the earth. In its immediacy, however, the energy crisis does not hinge on an impending physical shortage of resources. More than one bourgeois economist has pointed out in recent months that availability of exhaustible resources is a function of costs of extraction and market prices, whether under idealized competitive conditions or the conditions of monopoly and oligopoly that in fact characterize the extractive industries.[1] In this sense, one can find no better definition of the character of the energy crisis than the analysis offered twenty years ago by the Presidential Commission on Materials Policy (Paley Commission):

> The essence of all aspects of the materials problem is costs. The quantity of materials we can have in the future will be determined in great measure by what we can afford to pay for them, not simply in money but even more importantly in human effort, capital outlay, and other productive energies. The real costs of materials lie in the hours of human work and the amounts of capital required to bring a pound of industrial

material or a unit of energy into useful form. These real costs have for some years been declining, and this decline has helped our living standards to rise. In this Commission's view, today's threat in the materials problem is that this downward trend in real costs may be stopped or reversed tomorrow. . . . The problem is not that we will suddenly wake up to find the last barrel of oil exhausted or the last pound of lead gone, and that economic activity has suddenly collapsed. We face instead the threat of having to devote constantly increasing efforts to win each pound of materials from resources which are dwindling both in quantity and quality. . . .

As the best and most accessible resources are used up, it becomes necessary to work harder and harder to produce more supplies from less accessible and lower quality resources. . . .

The cost problem has always been with us. What concerns us about it today is that we have reached a point in the relationship of our demands to our resources where the upward cost pressures are likely to be far stronger and more difficult to overcome than in the past. . . . Rising costs mean the diversion of more and more manpower and capital from other productive efforts to extract required materials, and total national output of goods and services will be smaller by the amount that this diverted manpower and capital might otherwise have produced. Declining or even lagging productivity in the resource industries will rob the gains made elsewhere in the economy and thereby impair the very dynamic which has given the United States its growth. This is not the sort of ailment that gives dramatic warning of its onset; it creeps upon its victim with insidious slowness.[2]

The materials shortages stemming from the Korean War, along with the threat of revolutionary nationalism in resource-rich countries like Iran and Indonesia, formed the setting of the Paley Commission report (headed by CBS chairman William S. Paley). The suppression of revolutionary threats, coupled with new raw materials production in response to temporarily higher prices, effectively postponed the emergence of the energy crisis until the 1970s. While one cannot take seriously the Commission's implication that the

U.S. economy operates in a manner that minimizes waste and produces goods and services actually needed by its citizens, its fundamental observation is more meaningful today than when published twenty years ago.

Historically, the control of natural resources has been the prerogative of the private corporations of the Western industrialized countries. The coinciding dependence of contemporary industrial societies on petroleum as a key energy commodity and the relative concentration of easily accessible reserves of crude oil in the OPEC countries and particularly in the Persian Gulf-Arabian Peninsula area have facilitated a challenge to Western corporate prerogative. This challenge of the oil-producing states has forced the giant oil companies to shift their accumulation of capital (in the form of earnings) to the markets of the industrialized countries and to manipulate supplies in order to raise prices in those markets. The shortages and price inflation that have characterized the plight of workers and citizens of Western countries are consequences of what is at bottom a political crisis over the question of resource control: who is to decide the terms of extraction, the allocation of capital, the distribution of resources, the division of the rewards, and costs of exploitation.

This book has focused on developments over the quarter century since World War II, a period characterized by unprecedented political and economic unity among the industrialized capitalist countries under the hegemony of the United States. During this period, the most significant threats to this order have revolved around attempts by former colonies and neocolonies in Asia, Africa, and Latin America to establish political and economic self-determination, and aggressive responses to these challenges by the United States and its Western allies. The experience of Vietnam as the most bloody and most devastating of these encounters needs no elaboration here. These conflicts have, in turn, profoundly affected internal political developments in the United States.

The energy crisis, with its roots in the oil producers' challenges to continued Western control, can be understood in the context of the Third World struggle for national liberation and, in some cases, social revolution. Over these postwar decades, competition between the industrial capitalist countries and class conflict within their respective societies has been relatively subdued. It may be of utmost significance that this energy crisis coincides with the weakening of postwar American political and economic hegemony among its industrial allies and the resurgence of class conflict in the industrial countries as the highly integrated capitalist economies approach the end of their Keynesian rope.

Oil Prices and the International Capitalist Economy

Even before the embargo and the spectacular price hikes of late 1973, the industry and the media fostered the notion that even if "they" do not cut off "our" oil, the increasing amounts of money accruing to the "Arabs" (as all the oil-producing countries are invidiously lumped together) will result in the subversion and destruction of our international monetary system. Typical of the racist distortions put forth as analysis is this offering from oil industry guru Walter Levy:

> Not the least of the dangers posed by this extreme concentration of oil power and "unearned" money power is the pervasive and corruptive influence which this will inevitably have on political, economic and commercial actions in both the relatively primitive and unsophisticated societies of the producing countries and the advanced societies of the dependent industrialized nations.[3]

Levy's remarks fly in the face of the evidence. Currency fluctuations and disturbances over the last several years were the work of speculation and hedging by the treasurers of U.S.

multinational corporations and banks—all of whom are white, male, and presumably sophisticated and incorruptible. By the end of 1971, according to the U.S. Tariff Commission, U.S. multinational corporations and banks held $268 billion in short-term liquid assets.[4] Arab and other oil country revenues were at that time barely adequate to meet import needs, with few exceptions. Moreover, most oil country revenues are simply paid into country accounts at a few large banks in New York and London.[5] Oil country revenues, in any event, had little effect on the instability of the monetary system and the decline of the dollar as the basic reserve currency. The global inflation, fueled by chronic U.S. balance-of-payments deficits over the 1950s and 1960s, had already forced the devaluation of the dollar and the suspension of convertibility in August 1971. A second official devaluation occurred in February 1973. This process of disarray and disintegration was pronounced and well developed when the energy crisis emerged in its currently recognizable form.

There is no doubt that the price-hikes after October 1973 have substantially accelerated this process, but speculative and self-serving calculations of the surpluses and deficits have escalated even more. A "confidential" World Bank study predicts that OPEC currency reserves may rise to more than $1.2 *trillion* by 1985.[6] Estimates are more commonly half that amount but without much more substantiation.[7] A more concrete assessment is presented by the Organization for Economic Cooperation and Development (OECD), an official grouping of the Western industrial capitalist countries and Japan. In its *Economic Outlook* for the first half of 1974, OECD economists calculate that member countries will spend some $55 billion more on oil imports in 1974.[8] The oil bill for underdeveloped countries is expected to rise by some $10 billion. Calculations by the London-based *Middle East Economic Digest* estimate Middle East oil earnings for 1974 will amount to about $58 billion, with cumulative earnings for the 1973-1983 decade coming to around $635 billion. [9] Of course such projections are highly tenuous and include assumptions about price and consumption levels that cannot be pinpointed at this time. In this respect it is useful to consider more closely OECD assumptions that oil prices will remain more or less constant at 1974 levels, that oil country imports from the industrialized countries will increase by some $12 billion this year (and an indeterminate amount in the future), helping to offset the deficit, and that the effect of the higher oil prices will tend to reduce industrial country imports to 1973 levels by 1980. All of these assumptions, OECD admits, are "optimistic," but they are not unrealistic.[10]   Under these conditions the OPEC surplus balance of nearly $60 billion in 1974 could decline to $15 billion in 1980, with the cumulative surplus by that year coming to $300 billion. Less reduction of demand in the industrial countries, or a lower level of imports by OPEC countries, could increase this by as much as another $100 billion. Whatever the exact figures, OECD analysts conclude that "the OPEC current surplus is likely to decline significantly during the second half of this decade."

The oil producers' monetary surplus, even if it is likely to be closer to $300 billion than $1 trillion in the next decade, poses a serious challenge to a capitalist monetary system already ravaged by inflation, uneven distribution, and exploitation by the United States as "an instrument of imperial taxation."[11] This prospective shift of monetary wealth is taking place in the context of the creation over the four years, 1970-1973, of new international liquidity (i.e., new money) of nearly $110 billion, more than doubling the total of world reserves. Most of this was a function of the continuing U.S. balance-of-payments deficit.[12] This has taken place mainly in the international Eurodollar market, not subject to the monetary controls of any single national authority. One of the largest New York investment banks, Morgan Guaranty, which incidentally manages a large portion of the Saudi reserves and even has a representative on the board of the Saudi Arabian Monetary Authority,[13] reports that this Eurodollar market grew from $105 billion in 1972, to $150 billion in 1973, and a further $10 billion in the first three months of 1974.[14] The increase in oil prices is related to these developments, but the forces at work are more substantial and more pervasive, a fact that is recognized by the international bankers even if they join in blaming it on the oil producers. As Otmar Emminger, Deputy Governor of the Deutsche Bundesbank and German representative on the Economic Policy Committee of the OECD, commented in June 1973:

> The structural adjustment between the major economies and currencies of the world to the new realities of the 1970s—a normalization process after the previous period of absolute predominance of American industrial power—has been a once-for-all process of historic dimensions. It has strained the system of fixed parities [i.e., the postwar Bretton Woods monetary system] beyond the breaking point.[15]

One can only ponder what the direction of Dr. Emminger's "normalization process" may be, considering that the period preceding the decades of American hegemony was punctuated by fierce competitive struggles among the capitalist industrial countries, twice leading to conflicts known as "world wars." The solution to recurring crises in recent years, Emminger notes, has been "a further turn in the inflationary spiral in the world economy."

> Thus we have a close parallel to what has evolved in domestic economic policies: here, too, there is a growing tendency to resolve economic or social conflicts of all kinds by papering them over with inflationary settlements. Inflation as a general instrument for pacification, for resolving conflicts in the domestic as well as in the international field![16]

The higher oil bill for the importing industrial countries has increased the pressures on a system already breaking apart under the expansionary strains of postwar "boom" policies. The problem, however, is less rooted in the gross size of the overall deficit than in the specific distribution of surpluses and deficits among the industrial countries. As the OECD study describes it, "For the area as a whole there should be no financing problem, but, until satisfactory recycling arrangements have been worked out, this may seem a somewhat academic point to an individual country with a large current deficit."[17] Of the total OECD area reserves of $137 billion, Germany alone holds $33 billion.[18] Germany, the United States, Canada, and the Benelux countries, with 60 percent of OECD's gross national product, are in positions of surplus or balance. Germany and Benelux have strong export positions in industrial and capital goods. Canada is a net exporter of energy and other resources. The United States position has been shored up by the importing countries' need for U.S. financial services, the profits of United States-based international energy companies, and U.S. dominance of high technology export markets. The remaining OECD countries, chiefly Britain, France, Italy, and Japan, with 40 percent of the total GNP, have to share among themselves the estimated total deficit for 1974 of around $45 billion.[19]

The customary mercantilist solution to this problem of imbalance has been for the country in deficit to reduce internal demand (usually meaning a reduction in workers' incomes) and stimulate exports in order to earn the foreign exchange necessary for achieving a balanced or surplus position. This is the thinking that lay behind the successive dollar devaluations of the Nixon administration and similar moves by other countries. Back in late 1972, Commerce Secretary Peter Peterson (now head of the New York investment banking house of Lehman Brothers) warned of the trade implications of an "energy deficit":

> One result could be that all the major deficit countries would find themselves forced to engage in a wild and cannibalistic scramble for external earnings to pay their bills. This could create an extremely rigorous competition for manufacturing exports and the sorts of export subsidies which would be deleterious to the interest of all parties in the long run.[20]

Intergovernmental organizations like OECD and the International Monetary Fund have stressed the need to work cooperatively to head off any "cannibalistic scramble."[21] In May an OECD ministerial meeting produced a draft agreement under which members pledged for a period of one year "to avoid having recourse to unilateral measures . . . to restrict imports" and "to avoid measures to stimulate exports . . . artificially."[22] But the pledge carries no sanctions and one is inclined to be skeptical of the ability of the respective nations to act on the basis of a moral commitment. Italy, the industrial country in the most dire financial straits, had already initiated restrictive import measures even against its Common Market partners. Italy's trade position for the first half of 1974 indicates a deficit "superior to the pessimistic forecasts" of the beginning of the year.[23] Britain is basing its eventual salvation on North Sea oil production, while France and Japan have intensified efforts to secure long-term bilateral trade agreements with major oil producers. France's $4.5 billion deal with Iran to swap nuclear reactors and other industrial projects for oil and cash, and Japan's similar $1 billion arrangement with Iraq are an indication of a probable future pattern of relations. France has openly embarked on a course designed to make it "the key commercial intermediary between Europe and the developing countries of Africa and the Middle East," based on its "historic ties with former colonies that are now major producers of raw materials, and, more significantly, because its coastlines, harbors and river systems provide easy access to Europe's industrial heartland."[24]   Morgan Guaranty analysts concluded in June 1974 that no major country appeared to be willing to "accept the implications of its own balance-of-payments policies," with France, Italy, and Britain moving to eliminate their deficits, and Germany acting to preserve its surplus.[25]

The approach of the United States has been to consolidate its trading position in the Middle East in particular. Even the sale of nuclear reactors to Egypt and Israel during Nixon's tour was motivated primarily by the fear of losing that market to European competition. In June 1974, according to Wall Street's *Journal of Commerce,* "U.S. businessmen have booked all of the hotel rooms in Cairo, Kuwait and Riyadh for months in advance."[26]   The government has been pushing the business community to move faster and further. The Commerce Department has set up a special Commerce Action Group for the Near East to take advantage of what Secretary Dent describes as a "staggering proliferation of marketing opportunities, for everything from consumer goods to whole industrial systems and massive infrastructure projects."[27] Nixon's tour of the area, according to Dent, "has already opened the door to these markets and has created the most receptive possible climate for American goods." United States exports to the Persian Gulf area have increased about 35 percent each year since 1970, "greater than for any major world geographic area except Eastern Europe."[28]   Competition with other industrial countries, though, is bound to be heated. In 1972 the Persian Gulf countries imported 44 percent of their import requirements from Europe, 13 percent from Japan, and only 19 percent from the United States. The desire to improve this position has led to increased diplomatic presence in the area, including new embassies and commercial officers, and State Department appeals to Congress for funds to "top off" salaries of

American experts who can serve as advisors to countries in the area "in order to encourage development of natural resources of interest to the U.S., to encourage a favorable climate for trade, and to stimulate markets for U.S. exports."[29] The most important component of this drive for new markets is in the military sector. In fiscal 1974 the United States sold $8.5 billion worth of arms, $7 billion of that to the Middle East area, and $4 billion to Iran alone.[30]

Countries facing severe short-term deficit problems are forced to rely on conventional financial markets, notably the international banking system dominated by New York and London banks, or loans from other governments. Italy and Britain have approached oil-producing countries for loans. The flow of monetary reserves from the importing countries to the producers remains for the most part in the system, but is much more concentrated in a few of the very largest banks on short-term deposits. The consumers, needing large and long-term loans, are facing an increasingly constricted capital market. The banks with the funds have grown extremely cautious in their lending policies, resulting in an informal blacklist of most underdeveloped countries and Italy as well. Italian central banker Carli was forced to declare that in the developing situation the United States would be the "lender of last resort."[31] West Germany has since stepped in with credits worth several billion dollars, but this is seen by all parties as a very temporary solution. Following Nixon's decision to lift restrictions on exports of capital, the largest component of capital export from the United States was not direct investment but loans by U.S. banks to foreign borrowers, including governments.[32] The United States is competing for oil country funds, notably with its pressure on Saudi Arabia to purchase several billion dollars' worth of special nonnegotiable treasury notes.

One aspect of the crisis facing the international monetary system lies in the lack of any national political authority that could make comprehensive decisions at an international level comparable to the role of central banks (in the United States, the Federal Reserve) in individual countries. According to top private bankers like David Rockefeller of Chase Manhattan, the international banking system cannot extend further loans to cover national deficits. Commenting on developments at the Conference of International Bankers in June 1974, the *New York Times* financial analyst wrote: "In the coming tug-of-war between the prudence of bankers and the desperate need of national governments to borrow money to pay for the higher cost of imported oil, the prudence of bankers is likely to prevail."[33] The closest approximation to such an international financial authority is not the International Monetary Fund, with its nearly global membership, but the Bank of International Settlements (BIS) whose directors are the central bankers of the dominant industrial powers. BIS, which conducts its business under a shroud of secrecy, has been meeting more frequently of late.

The crisis atmosphere in international financial circles has been curiously lacking among some important U.S. policy-makers in the latter half of 1974. At the International Monetary Fund and World Bank meeting in Washington in the first week of October, public and private discussion focused on the need to recycle oil revenues from the producers to the consumers. The general consensus on the need for the expansion of a special lending facility under the auspices of the IMF was met coolly by Treasury Secretary Simon, who said that private financial markets were handling the problem "quite adequately." This view was repeated by Treasury Under-Secretary for Monetary Affairs Jack Bennett at a news conference following the IMF meeting. Bennett, a former Exxon executive, took the view that "there is no proof yet" of the need for more official or multilateral assistance to oil importers. West Germany and the United States, the two industrialized countries with balance-of-payments surpluses, declined to respond to the IMF invitation to contribute to the special oil facility. [34]

This approach, seemingly out of step with the doomsday warnings of Ford and Kissinger, is consistent with the fact that the "private financial markets" now performing the recycling tasks are the large New York banks and their branches in London. This *laissez-faire* approach, reflecting the profits that the banker friends of Simon and Bennett are making, has been endorsed on more than one occasion by the *Wall Street Journal*. An October 3 editorial called "IMF Petrojitters" noted that "a year ago, before the increase in oil prices, officials were flitting about the world worrying about how to recycle $100 billion in petrodollars by 1980. Since then we have already recycled $100 billion in a year." The editorial concluded that "so long as the world's governments don't panic and throw up a bunch of restrictions, the petrodollars will be handled by the most efficient recyclers around, which are the capital markets in London and New York."

A week later, on October 10, the *Journal* ran an article by William Cates, former Deputy Assistant Secretary of the Treasury for Industrial Nations Finance. Arguing that the U.S. balance of trade deficit owing to crude oil prices has been more than compensated for by the inflow of capital from the oil producers, Cates urged the United States to take advantage of the structural shift now taking place in the world economy. "Willy-nilly we are moving from trader to banker, and we must learn to understand and accept our enhanced role. . . . either we step up, when and if required, to the novel role of super banker, with all its current and future problems, or we risk collapse of the present system, with all the misery and upheaval that such a collapse would entail."

This sanguine approach of the Treasury Department and the *Wall Street Journal* might be appropriate if the current economic crisis were in fact rooted in the OPEC price hikes.

Most observers, especially top private and official bankers in the Western countries, recognize that any attempt to deal with the underlying inflationary crisis faces a perhaps unprecedented problem in which virtually the whole OECD area—which is to say all the capitalist industrial countries—is in the middle of "a substantial reduction in aggregate demand in relation to supply capacity, both for internationally traded commodities and within countries."[35]   This classic dilemma of capitalism, forestalled for several decades by inflationary panaceas on domestic and international levels, is being brought to a head by the oil price increases. In the words of the OECD analysts,

> as a corollary to the large deterioration on current account for most countries the financial savings of the various domestic sectors—households, enterprises and governments—must, in total, change by the same amount. . . . Equilibrium occurs when the deterioration in financial positions is willingly undertaken by households, companies or the public sector. . . . The major doubt is whether wage earners will acquiesce to such a situation in 1974.[36]

Elsewhere they predict that "inflation based on oil and commodity prices may ease but there is a danger, however, that high rates of inflation will be kept going by a wage/price spiral, as different groups within the community struggle to offset the large changes in relative prices that have occurred and to maintain their real incomes."[37]

This analysis is shared by the Joint Economic Committee of the U.S. Congress. In their interim report prepared for President Ford's economic "summit," they observed that "large wage increases certainly threaten to become an inflationary factor in the months ahead as workers struggle to recover lost ground." The inflation already experienced, according to the Committee report, can be substantially attributed to the monopoly structure of the American economy. High world prices for oil and other raw materials do not explain why wholesale industrial prices have risen at an annual rate of 35 percent in the past 3 months; why iron and steel prices are up 44 percent in the past year, nonferrous metals 45 percent, industrial chemicals 62 percent; why at a time of reduced demand and production cutbacks, automobile prices are up by $700-800 and more within a single year. . . .

Increasingly, a significant part of the current inflation can be understood only in the context of administered prices in concentrated industries which typically increase despite falling demand . . . [and are] unexplainable except in terms of the ability of concentrated industries to resist competitive forces and to achieve a target return on investment in good times and bad.[38]

It should be helpful at this point to consider briefly the causes and consequences of the general commodity price inflation that has characterized the world economy over the last few years and is responsible in part for the relatively strong balance-of-payments position of the United States at present. The primary beneficiaries of the commodity price increases have been the developed primary producers (United States, Australia, Canada, South Africa) rather than the Third World countries. Commodity prices for the products of the developed primary producers increased 85 percent by the first quarter of 1973 over 1968-1970 averages, while the index for the products of the less developed countries increased only 30 percent. Only after the first quarter of 1973 was this situation reversed, with the percentage increases being 28 percent and 77 percent respectively.[39] Strong commodity prices have boosted reserves of underdeveloped countries by some $15 billion in the last two years, but, as in the industrial countries, the economic benefits have not been shared equally. The Indian subcontinent, Sahelian Africa, and the Caribbean have been economically devastated by oil and other price increases. Moreover, for the Third World exporters as a whole, the OECD analysts have bad news. After noting that the underdeveloped countries will have to *improve* their trade balances with the industrial countries in order to pay their oil bills, they predict that "the terms of trade of primary products relative to manufactures have probably reached a cyclical peak, and may well decline considerably. The cooling off period in the OECD area will, therefore, adversely affect both the price and volume of developing countries' exports."[40] It took until 1970 for most commodity prices to reach their peaks of 1951, during the Korean War boom. Reserves of most minerals are plentiful, and present shortages are due to limited production capacity and sudden increases in demand, based on the uncertainty and disarray within the world capitalist economy.

The OECD analysis may be off the mark in one respect: while the inflationary demand for primary commodities may decline with the deceleration of the advanced capitalist economies and the deflationary impact of the huge commodity inventories that have been built up, present high prices for oil (at least) and for other materials (to some degree) will likely be consolidated at something close to the present level, partly through increased wages and other basic costs in the producing countries, and through the concerted political actions of OPEC and similar organizations. The disarray among the leading industrial countries, coinciding with a relatively high degree of unity among the raw materials producers, may lead to an outcome different from that of the 1950s. The chances of another OPEC-like curtailment of production and price hikes in other commodities are not very great but a steady if uneven push on various commodity prices by producers acting singly and together seems more likely. As before, the impetus may not always come from the Third World. In August 1974 leading Australian mining companies, backed by their government, initiated demands for higher prices from Japanese steel companies that depend on Australia for close to one-half their imported ore. One Japanese executive commented, "We have been concerned for some time now about the obvious rise of nationalism in those countries supplying resources to the Japanese economy."[41] Initial talks were held in "utmost secrecy" in order to "keep news of developments from reaching suppliers in India, Brazil and South Africa. But this is clearly an impossibility and accounts for the gloom in Japanese steel circles." This uncoordinated but unrelenting pressure on mineral resource prices accounts for the concern of U.S. strategists like Fred Bergsten that prices would rise "with sufficient subtlety that tough U.S. responses would be difficult to mobilize."[42]

Another aspect of the commodity price boom lies in the extent to which, in the words of the OECD analysts, the boom was largely speculative, based on general uncertainty in the economic outlook and particularly on the expectations of inflation. "The boom was unsupported by the underlying level of real final demand."[43] In the United States, perhaps more than anywhere else, the recurring economic and political crises have been reflected in the increasing inability of corporations to secure the capital necessary for continued expansion. The concentration of capital in the equity (stock) markets in the hands of the "institutional investors"—the banks, insurance companies and other financial corporations—combined with the collapse of high-flying merger manipulations of the "go-go years" has led to the decline of the stock markets as a source of new capital.[44] Inflationary pressures have driven up the cost of loan capital. The result has been intense concentration of speculative funds in the commodity markets.

> In the United States, for example, equity prices deflated by the wholesale price index reached, early this year [1974], their lowest level since 1962. This was accompanied by a massive inflow of funds into commodity markets, where forward contracts have become an important instrument for hedging against inflation and for pure speculation.[45]

While the weakness of the stock market has driven investors to the commodity market, the inability to raise capital through new stock issues has led to heavy bank borrowing by corporations in order to build up inventory. As the *New York Times* put it: "As long as the rate of inflation has been higher than the prime interest rate, it has worked to a corporation's advantage to borrow for inventory accumulation, which in turn requires more money to finance." [46] Now that commodity speculation has been pushed toward its probable upper limit and corporations find it increasingly hard to finance existing inventories, deflationary pressures may well take over as inventories are reduced, leading to an increase in the already record number of bankruptcies and a possible collapse of the debt structure. [47] It is important to appreciate the variety and the dimensions of the pressures operating on the system and see how those pressures are related to the energy crisis.

It remains to point out that the relatively strong U.S. position vis-à-vis the other capitalist countries is based on several factors which are probably not permanent. One is the large flow of petrodollars, or oil country earnings, into U.S. capital markets, aiding the balance of payments and strengthening the dollar against other currencies. The other is the surge in U.S. commodity export prices which, as has been pointed out, have probably reached a peak. Industrial exports like steel have done well in the last few years, but competition from Japan, Germany, and other countries is beginning to cut into markets in this country as well as internationally. Even the higher prices that those competitors must pay for raw materials will not offset this trend.

It seems likely that while the roots of the energy crisis in the demand of oil-producing countries for a greater share in the world product (as inadequately reflected in money payments) have produced fundamental changes in the structure and direction of the world capitalist economy, previous trends will not be entirely displaced. These certainly must include the declining competitiveness of U.S. exports due to the slower growth of U.S. productivity, to the point where American advantage is restricted to high-technology goods like computers, nuclear reactors, and agricultural commodities.[48] Similarly, the resurgent strength of the dollar has been due precisely to the need of consuming countries to pay for oil imports in dollars and, to a lesser extent, in sterling. Thus the reserves previously held by France and Japan, for example, are being transferred to the accounts of the oil producers. There is little reason to believe that either the new or the old holders will be indefinitely content with monetary relationships based on previous privilege rather than a materially more productive economy.

The Transfer of Wealth to the Oil-Producing Countries

An examination of the features of the energy crisis must focus on the financial ramifications of the new world of oil. In the industrial countries, analysis has often been supplanted by an almost hysterical outpouring of slogans and accusations that tend to exaggerate the projected overall monetary imbalance and to assume that monetary transfers, real or imagined, correspond to actual transfers of wealth. One effect of this tendency is to insinuate that the Western economic crisis is primarily the result of the unilateral price-hikes by the oil producers when in fact, according to the OECD study, "energy induced price increases have directly added about 1½ percentage points to consumer prices, or *one-fifth of the overall increase.*"[49] The total revenues of the largest producer, Saudi Arabia, are estimated at $17 billion for 1974. Even if this is an underestimate, it is well under the earnings of the largest company, Exxon, for just the *first half* of the year: $21.3 billion. Total Middle East oil earnings for 1974 will be in the vicinity of $58 billion. While this is far in excess of previous earnings, it is still only about two-thirds of the Pentagon budget for fiscal year 1975 of $86.2 billion.[50] Another perspective on the new oil prices is reflected in the OECD breakdown of the mid-1974 composite selling price of oil products in Europe of $24 per barrel: 40 percent goes to the producing countries; 36 percent to the consuming country in the form of various taxes; and 24 percent to the companies for costs and profits. The breakdown prior to October 1973 was 17 percent, 54 percent and 29 percent respectively.[51]

The producing countries not surprisingly take a view that differs considerably from that of the companies and the Western governments. President Boumedienne of Algeria, in an interview in *Le Monde* just prior to the Washington conference of the big industrial consuming countries organized by Henry Kissinger in February 1974, said:

> We do not find oil too expensive. For us it is machinery, technicians, the cost of knowledge, studies and money which are too expensive. The man who goes hungry, who rides a donkey, who wants to learn to read, does not have the same preoccupations as the one who goes for a drive on Sunday, and for him the price of wheat is more important than the price of asphalt. The price of wheat has quintupled. For a long time the price of iron has not moved, but by how much has the price of a tractor increased? The problem facing the world is much larger than oil alone or even raw materials: it concerns the relations between the developed countries and the others in every field. This is the heart of the question.[52]

The question of the transfer of wealth is more than a question of political and economic rectitude. It is necessary to ask whether the dollars and sterling flowing into the bank accounts of the oil-producing countries represent any more than "mere alterations in book-entries in banks in Zurich, London and New York—for lack of adequate structures for the absorption, and so for the consumption, of commodities and services that could be imported by the oil producing countries."[53]   Because of the fact that the international capitalist monetary system is still based on the dollar, for the countries that pay for oil not with their own national currency but with dollars

> these countries do indeed pay for their oil, though not to the Arab countries but to the United States. In the end, if the Arabs keep their holdings in dollars, a gigantic triangular "fool's fair" will be the result. In exchange for oil, the West European countries will supply the United States with equivalent real values [i.e., actual goods and services], and the USA will receive as a gift—in exchange for unrepayable paper—not only the oil that it purchases itself, but also the value equivalent to that which the European countries purchase.

This projection correctly suggests that monetary reserves and industrial establishments in and of themselves should not be confused with economic development, but at best indicate some potential for development: that is, the material improvement of people's lives through the provision of necessary goods and services. However, most of the oil-producing states will be able to *materialize* a large part of their revenues. The primary exceptions are Saudi Arabia, Kuwait, and the Gulf emirates, and even those states will materialize their revenues to the amount of several tens of billions of dollars.[54] The end result will be something more than a "fool's fair." To the extent that oil earnings are materialized, the real cost of energy for the industrialized countries, including the United States, will have risen: real cost being the amount of labor and capital needed to produce or purchase a given unit of energy.[55]

On one level this transfer takes place with the importation of goods and services for immediate consumption. U.S. food exports to Iran and Saudi Arabia, for example, have doubled and tripled over the last year.[56] The most significant transfers, however, are expected to be on the level of capital goods and infrastructure designed to build productive national economies that will no longer depend on oil revenues as the primary or sole source of national income. Every oil-producing state is committed at least verbally to this approach, and there is little reason to doubt that significant industrialization will take place in the area over the next decade. The problem remains to understand the character of the industrialization and development process in each country and in the region as a whole: its relationship to developments in the international capitalist economy and contradictions in the global process of capital accumulation.

In addition to making an examination of specific developments in individual countries, it is important to remember that developments on the national level cannot be isolated, any more than they can be ignored. In the radical nationalist states—Algeria, Iraq, and Libya—the stress appears to be on development of an industrial and agricultural base that will facilitate the growth of a genuinely national and independent economy, with most or all of its major components being state-owned. Initial emphasis is on those industries able to thrive on the abundance of energy and hydrocarbon resources, like petrochemicals and metal manufacturing. Algeria is the most advanced along these lines. To what extent this strategy can be successful remains to be seen. In the oil industry itself, these countries have gone the furthest in the process of nationalization and the creation of national companies to supplant the various functions previously monopolized by the integrated oil giants. The oil companies have largely been relegated to the role of purchasers on long- or short-term contract. (In Libya this process is still in a more embryonic stage.) Western firms provide capital goods and services such as plant or infrastructure construction on a contract basis. None of these countries has been investing revenues in the United States or other Western countries, buying resorts in the Bahamas, or loaning money to industrial countries. It is to this group that one should look, as in the past, for significant alterations in the existing web of relationships between industrialized and underdeveloped oil-producing countries and for possible support, direct or indirect, of revolutionary movements elsewhere in the Third World. It is these countries, too, which have the closest ties to the socialist camp.

Among the other Middle East oil-producing countries, Iran is very significant in several respects: it has the population and resource base to become a regional power; it has the longest tradition of modern revolutionary activity, even though that activity has been sharply suppressed over the last two decades. It has the most experience in developing national industries and infrastructure, although primarily under the aegis of the American multinational corporations. Iranian strategy under the Shah has been to bolster rather than undermine the existing international economic system dominated by the United States and Europe in a way that enhances Iran's role in that economy and commensurately increases its rewards. This is reflected in the favorable treatment accorded to the largest oil companies and in the enormous expansion of the defense and police establishments. It is reflected in the loan of more than a billion dollars to Britain and the purchase of 25 percent interest in Germany's giant Krupp steel works. It is reflected in the apish concentration of income and wealth in the hands of a small, dominant elite and a system of corruption unrivaled in the Middle East, where it faces stiff competition for this dubious honor.[57] United States and other multinational corporations have come in as "partners" in joint ventures in petrochemicals, auto manufacturing, copper mining, and even Pancake House franchises. Of the opportunities awaiting enterprising international capitalists, *Business Week* writes:

> Foreign investors looking for partners with experience in the local market and knowhow in dealing with the Iranian bureaucracy link up with established local businessmen and even, in some cases, with members of the royal family. Prince Abdul Reza, the Shah's half-brother, for example, formerly owned a share in Deere and Co.'s tractor assembly plant. In setting up the operation, recalls Deere Vice-President Robert Hanson, the Prince "had no trouble getting the right answers from the proper segments of the Iranian government." Other companies invite the Pahlevi Foundation, the Shah's personal investment vehicle, to buy a stake in their operations. The Shah originally set up the foundation with proceeds from the sale of royal holdings in the land reform of the 1950's. Among other things, the foundation owns Iran Air, the national airlines, the Teheran Sheraton hotel, a controlling interest in Bank Omran, and shares in such companies as General Motors of Iran. Through the foundation, observers say, the Shah is in fact "the first capitalist of the country." In the U.S., where it is represented by the law firm of former Secretary of State William Rogers, the foundation recently bought a property on New York's Fifth Avenue, where it will put up a 32-story building to house a variety of Iranian activities.[58]

Similar scenarios of "development" are emerging on the Arab side of the Gulf, especially among the emirates. An American bank executive describes the area to the *Wall Street Journal* as "a banking paradise, with business all over the place."[59] Similar sentiments flourish among businessmen in other fields as well. Among all the Arab Gulf countries, Kuwait has the most experience in managing its surplus oil revenues through investment in Western countries and in development projects elsewhere in the Arab world. There, too, the existence of a politically conscious sector operating through a limited but functioning parliamentary system has tended to limit however slightly the role of foreign financial interests. In other words, Kuwait has gone further than any of the Arabian Peninsula countries in developing a national bourgeoisie which plays a role in setting national policy.

Saudi Arabia is larger in terms of population and resources than any other state on the peninsula, and of course has the lion's share of proven oil reserves. It is the country with the largest potential revenue surplus, and its anachronistic social order and reactionary political system have sharply limited the distribution of that wealth in the past to the royal family, its tribal allies, and an expanding government bureaucracy and private merchant sector. Like Iran, the Saudi military and internal security budgets have expanded proportionately with rising revenues and rising political threats. Unlike Iran, Saudi experience in the public and private sector to utilize these sums to undertake development projects is barely ten years old—dating from Faisal's accession to power in 1964.[60] One result is that the Saudis are even more dependent than the Iranians on the role of foreign corporations and advisors. Economic planning has been contracted into the worthy hands of the Stanford Research Institute. An information bulletin from the U.S. embassy in Jiddah prepared in April 1973 designed to alert American firms to the advantages in Saudi Arabia notes that the prevalent "turnkey" deals in which a foreign company undertakes the construction of a project for a set fee and then withdraws would have to be modified in the Saudi case to "turnkey-plus": "This would mean a follow-on contract for operation and maintenance of the facility" with the training of Saudis to take over as quickly as possible. In every area, port facilities, communications systems, transportation, mineral development, hotels and housing, the embassy bulletin points out that "the interface between industrial development, urban infrastructure, and development of a viable and self-sufficient agricultural and food industry" remains the "most promising opportunity for systems-oriented firms from the United States."[61]

The degree to which the "development" of the oil-rich

countries like Saudi Arabia is in the hands of Western firms and advisors should incline us to assess very cautiously the extent to which a fundamental transformation of relationships is likely. The traditional concerns of the industrialized countries, to maintain access to supplies of raw materials and to new markets for consumer and capital goods, still represent the basis for their participation in these "development" projects. To the need for raw materials and markets the new situation adds a degree of dependence on the oil-producing countries for investment capital. The producing countries are without doubt in a better bargaining position with the industrialized countries and their multinational corporations than most underdeveloped countries. As long as the oil countries remain unable or unwilling to undertake (individually or collectively) tasks of independent development in which production and capital accumulation are controlled by their peoples, that control is destined to remain in the hands of local elites who identify themselves with the Western countries.

An example of this kind of control, actual and potential, in the new world of oil is indicated in the planned role of the Chase Manhattan Bank of New York in the management of Saudi monetary reserves. In an internal memorandum dated November 20, 1973, it is noted that "the Chase Manhattan Bank and its top management have had a close, continuous professional relationship with the Saudi Arabian Government, and key Saudi officials and citizens," including the management of more than 20 percent of Saudi liquid monetary reserves totaling over $5 billion and "management support" for a proposed Saudi Arabian Industrial Finance Corporation. The Chase memo notes the need for

a relatively small group of international private sector "chosen instruments" organized to act as conduits for establishing strong bilateral relationships between Saudi Arabia and their respective countries through reciprocal large scale investment and development efforts.

According to the Chase report, efforts will be concentrated on developing a cadre of competent Saudi managers to run these projects and supervise domestic and international investments in a manner that reflects unspecified "shared objectives"—shared that is by King Faisal and his royal entourage on the one hand and David Rockefeller and his entourage on the other. The report specifically refers to an offer made by David Rockefeller to Faisal to initiate "a series of small interchanges or conferences . . . between key United States, international, and Saudi leaders as part of a broad effort to promote understanding of and a dialogue on the political and economic concerns of Saudi Arabia and its neighbors." The application of this understanding so assiduously promoted by the Chase strategist is reflected in the role of Saudi Oil Minister Yamani within OPEC. At the OPEC conference in Ecuador in June 1974 Yamani reportedly argued:

> Whether they liked it or not, . . . the oil-producing countries are themselves part and parcel of the Western economic system and therefore any lasting damage to that system also constitutes a threat to their own long-term interests, particularly when the present level of oil prices gives such a significant competitive advantage to the Communist bloc which is by and large self-sufficient in energy resources.[62]

The course of events during the months following the oil embargo suggests that future developments are not dependent solely on the predilections of the various regimes and their agents. Despite Yamani's public campaign for lower oil prices even Saudi Arabia has not undertaken actions that would set it irrevocably against the course of the other OPEC producers. Production cuts by Kuwait and Venezuela to maintain high prices in the face of oversupply have been matched by similar cutbacks by Aramco in Saudi Arabia. Production limits and higher prices are also favored by the largest oil companies as well. In focusing on basic trends, in treating the events of the last year in the context of the previous twenty years, the ability of the international capitalist system to extend its domain by broadening and "internationalizing" the ranks of its managers and even ruling class must be taken seriously and its limitations realistically evaluated. This represents the "Merrill, Lynch & Yamani" solution to the energy crisis.

In Conclusion

This attempt to conclude an historical analysis of events so close at hand, to definitively summarize trends that are still emerging, is as frustrating as it is necessary. We are in the midst of a revolutionary period in which it seems probable that the oil embargo and the OPEC-sponsored price explosion will be landmark events, although we are too close to them now to conclusively evaluate their significance. The annual *Strategic Survey* published by London's prestigious International Institute for Strategic Studies begins its review of the events of 1973 with a description of the oil weapon as producing

> the greatest shock, the most potent sense of a new era, of any event of recent years . . . change, drastic not only by recent standards but even in some respects by those of the two centuries since the Industrial Revolution . . . this was the first time that major industrial states had to bow to pressure from pre-industrial ones. . . . [The Arab] victory upset the hierarchies of power long enjoyed, or resented, according to one's station, and opened up prospects of quite new political balances. By the same token, it was by far the biggest extension of the world's effective political arena since the Chinese Revolution.[63]

We have repeatedly referred to the political and economic ambiguity surrounding these events. The oil embargo helped Saudi Arabia consolidate and strengthen its role in the Arab world. The price hikes and the shortages created by the embargo facilitated an enormous and unprecedented extraction of wealth from consumers by the large oil companies. Nevertheless, it is our assessment that the long-term benefits of these actions will not accrue to the countries and the regimes, but to the peoples of the Middle East and of the rest of the Third World and the industrialized countries as well, in part by helping to precipitate the crises necessary to bring about a redistribution of resources and wealth.

This is not by any means an inevitable or foreordained process. Just as the success of OPEC and the oil producers cannot be artificially isolated from the upsurge of economic nationalism and movements for political and social liberation in the Middle East and elsewhere over the last decades, so it will remain necessary for those movements of liberation and revolution to grow. We can speak abstractly on paper of "economic and social forces," of "contradictions" and "dialectics," but it is when people act together in organized fashion to concretely affect the conditions of their material existence that these contradictions are heightened and new forces attain power. For this we must look behind the structure of governments, organizations, and corporations whose agreements, alliances, conflicts, and collaborations reflect, at bottom, the strengths and concrete interests of different classes of people: of the masses in the slums and the peasants subsisting in an impoverished countryside; of the rulers and elites in palaces and corporate board rooms. In the case of the oil-producing countries of the Middle East and the region as a whole, the political and organizational strength of different classes and sectors has been instrumental in forcing regimes to adopt united and militant policies to secure higher income (even though concentrated in relatively few hands) and the potential for industrial and social development. Wealth is concentrated in those countries whose populations and needs are inversely related to oil revenues. Development is structured in such a way as to maintain the same unequal distribution of wealth and power. While we have focused on the need for social revolution in the most reactionary of those states, this should not indicate any satisfaction that the regimes generally regarded as progressive are immune to popular demands for genuinely revolutionary and socialist alternatives. With the increase of wealth and the creation and expansion of new social forces, namely a working class, an urban proletariat, new tensions and contradictions are emerging that cannot be bought off with the new wealth or suppressed with the modern police state apparatus most of these countries are acquiring. The seeds of these revolutionary struggles of the future are being sown throughout the area. In Arabia, the struggle led by the Popular Front for the Liberation of Oman in Dhofar has brought about the combined intervention of local client states like Jordan and Pakistan, self-appointed guardians like Iran, and of course the traditional imperial powers, Britain and the United States.

Just as the struggle for higher oil prices represents a partial and inadequate reflection of the more basic struggle in the Middle East for a more equitable and just social order, so too that basic struggle is going on across the globe. In the space of just a few months in 1974, the dictatorship in Portugal has collapsed, bringing closer the end of white minority rule and, to some degree, Western corporate control of the people and resources of southern Africa. In Greece and Ethiopia similar events have unfolded. Replacement by progressive, much less revolutionary, regimes cannot be assumed, but the break up of the old order is evident in many places. In South Asia the acute contradictions stemming from the backward state of productive forces under local bourgeois regimes in India, Pakistan, and Sri Lanka are mounting toward greater and greater explosive potential. The comparatively secure economic positions of the socialist countries, especially

China, have been enhanced by the rising prices of energy and other resources. The energy crisis and the concurrent food crisis of the capitalist world will undoubtedly focus attention on the productive possibilities of a social order organized around meeting human needs rather than accumulating wealth and profit in a few hands.

What is perhaps the most important feature of the new state of affairs developing out of the crises of the capitalist world is the extension of global contradictions between abundance and scarcity, wealth and destitution, to the countries of the industrialized West: the United States, Europe, and Japan. In addition to the directly higher costs for energy and fuel-based services, the general inflationary surge of prices combined with the need of the U.S. economy to improve its competitive position has put the emphasis on increased productivity for American workers, which can be translated into speed-ups and other forms of oppression and exploitation. As the authors of the *Strategic Survey* put it: "The interaction between the energy crisis and social tensions in the advanced industrial economies could well prove the most powerful of all shaping forces for the future."[64]

It is these broader possibilities of revolutionary change that are the most significant consequences of the changes in the world oil industry. The major companies have strong control over the transportation, refining, and marketing of oil and its products and competing forms of energy in all the major consuming areas. The companies' access to Middle East crude is still substantial, although this access will be progressively reduced by increasing state control in the producing countries and by the intervention of consuming states like Japan, France, and others to secure crude supplies outside the channels dominated by the major companies. The companies will continue to exist and function within the industrial countries in a commanding fashion, until the point at which the existing system is modified or done away with.

The system of relationships that facilitates the flow of resources from the producing countries to the consumers has been left intact to date. The oil companies, after all, exist to provide for the essential needs of the capitalist economies, and not the other way around. Only the terms under which that transfer of resources is made have changed. The present state of affairs is only an immediate point in a long and difficult struggle for change and revolution; it is not any kind of arrival.

No one disputes that resources, including energy resources, are there for the future if they are used wisely and well. Who is to control them, and who is to profit from that control, will be the focus of struggle in all countries and regions in the coming decades. One alternative is that posed in San Francisco in 1973 at the International Industrial Conference by the U.S. government spokesman, William Ruckleshaus, to an audience that included the heads of the major global corporations and Saudi Oil Minister Yamani. Ruckleshaus responded to Yamani's call for a UN-sponsored conference on World Resource Management by noting, in the words of the *Journal of Commerce,* that "it is important that a complete planetary resources management mechanism be in being by the year 2000 when shortages are expected to peak." For those people of the world who do not want to be dependent on a system operated by and for the Yamanis and Rockefellers and shahs and multinational corporations, the alternative is to struggle for a world where resources are shared by the people who produce them and the people who need them—the workers and citizens of all regions who are being played off against one another, now under the name of the energy crisis, for the profit of a few.

Notes

Chapter 1: Middle East Oil: The Beginnings

1. For further details on the political economy of the Middle East before oil, the following are recommended: Ahmad el Kodsy, "Nationalism and Class Struggles in the Arab World," in Ahmad el Kodsy and Eli Lobel, *The Arab World and Israel* (New York, 1970); Vladimir Lutsky, *Modern History of the Arab Countries* (Moscow, 1969); Charles Issawi, ed., *The Economic History of the Middle East* (Chicago, 1966); Bernard Lewis, *The Emergence of Modern Turkey*, 2nd ed. (London, 1968); Doreen Warriner, *Land and Poverty in the Middle East* (London, 1948); and Maxime Rodinson, *Islam and Capitalism* (New York, 1974). On the Arabian peninsula in particular, the best treatment is: Fred Halliday, *Arabia Without Sultans* (London, 1974).

2. The formation and location of crude oil deposits is an imperfectly understood phenomenon for which several conflicting scientific explanations exist. Oil is a type of hydrocarbon formed by a combination of bacterial and mineral catalyzing processes that break up organic matter under certain temperature and pressure conditions. Following this conversion, another combination of conditions is necessary to form collectors, or pockets, inside the earth to retain the hydrocarbon compounds. A combination of geostatic and dynamic pressures and capillary and molecular forces displace the hydrocarbon from the oil-bearing strata to the collector strata. Only hydrocarbon compounds which have been subjected to this immensely complex (and here oversimplified) set of processes become oil deposits. Groups of deposits created at the same time under the same conditions are known as oil fields. Obviously, the geophysical variation in type and size of deposit and field is enor-

mous. Some unique combination of these forces and structural conditions has produced the concentration of large and easily recoverable deposits in the Middle East, and especially in the Persian Gulf region. One useful summary of the processes, and the scientific debates about them, is A. G. Alexin, "Some Problems of Petroleum Geology," in *Techniques of Petroleum Development*, Proceedings of the United Nations Interregional Seminar, New York, January 23 to February 21, 1962.

3. Herbert Feis, *Seen From E.A.: Three International Episodes* (New York, 1947), p. 94.

4. George W. Stocking, *Middle East Oil* (Vanderbilt, 1970), p. 10.

5. Ibid., p. 11.

6. Ibid., p. 15.

7. Ibid., p. 18.

8. Ibid., p. 21.

9. Ibid., p. 22.

10. See "Iran: Ten Years After the 'White Revolution,' " *MERIP Reports*, no. 18.

11. Arnold Klieman, *Foundations of British Policy in the Arab World* (Baltimore, 1972), pp. 77, 87.

12. Stocking, p. 51.

13. Ibid., p. 52

14. John DeNovo, "The Movement for an Aggressive American Oil Policy Abroad, 1918-1920," *American Historical Review* (July 1956), p. 858.

15. Gerald D. Nash, *United States Oil Policy, 1890-1964* (Pittsburgh, 1968), p. 49.

16. Ibid., p. 26.

17. For a good treatment of the development of government-industry ties in an earlier period, see Gabriel Kolko, *The Triumph of Conservatism* (Chicago, 1967). Nash's book on U.S. oil policy is very sympathetic to the industry, but the careful reader can get a useful survey of the World War I period. Ellis Hawley's *The New Deal and the Problem of Monopoly* (Princeton, 1966) carries the story through the 1930s.

18. DeNovo, pp. 854-76.

19. John DeNovo, *American Interests in the Middle East, 1900-1939* (Minneapolis, 1963), p. 186.

20. DeNovo, *American Interests*, p. 189.

21. Ibid.

22. Walter Measday, "A History of Federal Cooperation with the Petroleum Industry," in U.S. Senate, Hearings before the Committee on the Judiciary, Subcommittee on Antitrust and Monopoly, *Governmental Intervention in the Market Mechanism: The Petroleum Industry*, 11 March-2 April 1969 (Washington, 1969), p. 579.

23. Ibid.

24. Ibid., p. 581.

25. Stocking, p. 113; quoting a State Department petroleum advisor.

26. Ibid., pp. 114-16.

Chapter 2: World War II and the Consolidation of American Oil Interests

1. Herbert Feis, *Seen From E.A.: Three International Episodes*, p. 93.

2. Ibid., p. 94. The reserves figure comes from Raymond Mikesell and Hollis Chenery, *Arabian Oil* (Chapel Hill, 1949), p. 177. This is before the American companies got a foothold in Iran and is based on overly conservative estimates of Saudi reserves. The actual share under American control as early as 1946 was probably closer to 50 percent.

3. We put the term "pro-Nazi" in quotes not because the insurgents did not express sympathies for or seek assistance from the Axis powers. What is at question is whether they would have represented a more reactionary and repressive order than already existed under British tutelage.

4. Caltex was the wholly owned foreign marketing subsidiary of Socal and Texaco, while the Saudi joint-producing subsidiary was the California Arabian Standard Oil Company (Casoc). In early 1944 Casoc became Arabian American Oil Company.

5. The details of this high-level influence peddling came to light after the war when Moffett sued the Caltex parent Socal for $6 million, asserting that he had never been recompensed for the services rendered. Socal at first denied that Moffett had played any role in setting off the course of events described above, and later retreated by claiming that

he had undertaken his mission voluntarily, as chairman of Caltex and Bahrein Petroleum Company, and thus an interested party in his own right. The jury awarded Moffett $1.7 million but this was set aside on appeal on the grounds that influence peddling was contrary to public policy. This virtuous stand saved Socal $1.7 million.

Moffett had taken Socal to court earlier for $100,000 for out of pocket expenses incurred while on leave as vice-president to serve as Federal Housing Administrator in 1934-1935. In a letter to the company he claimed he "was really doing more work and was in a much more helpful position for the Standard Oil Co. than if I had remained in the office at 30 Rockefeller Plaza." He was apparently alluding to his role in convincing the Attorney General to drop antitrust proceedings against Socal at that time. This dispute was settled out of court for $25,000.

New York Times, 24 January 1947, 22 May 1948, 2 February 1949, 4 February 1949; Stocking, pp. 92-94; Harvey O'Connor, *The Empire of Oil* (New York, 1955), pp. 281-83.

6. *Foreign Relations of the United States* (hereafter cited as *FR*), 1944, V, pp. 734-36. On the navy overcharge question see the references on the Moffett suit against Socal, above, and Mikesell and Chenery, pp. 132-37. The total aid figure is from the *New York Times*, 2 February 1949, and the Lend Lease figure is from Raymond Mikesell, "Monetary Problems of Saudi Arabia," *Middle East Journal* (April 1947). Roosevelt's message and the company warning are found in Stocking, p. 97. The Interdepartmental Committee is discussed in detail in Feis, pp. 110-22.

7. *FR*, p. 736 (emphasis added).

8. Stocking, p. 98.

9. Ibid., pp. 99, 101.

10. Draft memorandum to President Truman, in *FR*, 1945, VIII, p. 45. Details on company-government interlocks are in the *New York Times*, 6 March 1947 and 25 January 1948. According to Moffett, a large number of military officers in charge of procurement found lucrative positions after the war with Aramco and Caltex.

11. *FR*, 1943, IV, p. 942.

12. Text of memorandum in *FR*, 1944, V, pp. 27-33. The reference to "unilateral political intervention" was spelled out in a memo a year

later by State's Acting Chief of the Petroleum Division, John Loftus: "We want a cessation of British political interventionism in the process of obtaining petroleum concessions in areas directly or indirectly dependent upon British sovereignty. This political interventionism in the past has taken the form of interposing the innumerable and ingenious obstructions of administrative procedure in the path of efforts by United States nationals to obtain concessions in areas within the British sphere of political influence." Loftus goes on to cite Gulf's efforts in securing the Kuwait concession as "the most extreme example, but numerous others could be adduced." *FR*, 1945, VIII, pp. 51-52.

13. *FR*, 1944, V, p. 756.

14. Feis, p. 135.

15. *New York Times*, 3 March 1946.

16. *FR*, 1946, VII, p. 41.

17. *FR*, 1946, VII, p. 44.

18. *FR*, 1947, V, p. 648.

19. Quoted in Richard Barnet, *The Roots of War* (New York, 1972), p. 162.

20. *FR*, 1947, V, p. 504.

21. *FR*, 1946, VII, p. 10.

22. *FR*, 1947, V, p. 516. The opening address by Acting Secretary of State Dean Acheson contained these cautionary lines: "In view of the false significance which undoubtedly would be attributed by certain elements to these discussions if it should become known that they are taking place, it is extremely important that every practical measure be taken to keep the fact that we are holding them from becoming public" (ibid., p. 564). The parallels of interest and propaganda of the post-World War II era with the corresponding period after World War I are too striking to pass without comment. Once again, the "energy crisis" was of prominent concern in the form of a potentially crippling "oil shortage." Once again, Britain as well as Bolshevism was seen as a potential threat to American oil interests in the Middle East. This time, of course, the overwhelming economic superiority of the United States relegated this British threat to the realm of fantasy. What was at stake was not British rivalry but whether Britain would completely acquiesce to U.S. pressures or only partially do so. The "Pentagon Talks," where questions of policy and presence in the Middle East were hammered out

diplomatically if not frankly, were comparable to the Washington Naval Talks of 1923, which hashed out British-American relations for that period.

23. Richard A. Freeland, *The Truman Doctrine and the Origins of McCarthyism* (New York, 1973), p. 89.

24. *FR*, 1946, VII, p. 530.

25. *FR*, 1946, VII, p. 526; Harry S. Truman, *Years of Trial and Hope* (New York, 1956), p. 95.

26. *New York Times*, 15, 16, 17, 18, 22, 25 July 1946; 3, 10 August 1946. The Americans in Saudi Arabia faced no such problems from the "natives." According to a contemporary account in the *New York Times* (28 November 1946), "No labor unions exist among Aramco workers, there being no such institutions in the country and strikes being unpopular with King Ibn Saud." It is worth noting that in Palestine during the Arab-Jewish civil war one of the few instances of Arab-Jewish solidarity occurred in the context of a strike against the Iraq Petroleum Company refinery and port facilities. This occurred despite the fact that the Jewish workers (less than 1 percent) were in the Histadrut while the Arab workers operated through a company union (*New York Times*, 17 March 1947).

27. *FR*, 1946, VII, p. 632.

28. It is possible that the publication of the long-overdue 1948 volume of the *Foreign Relations of the United States* series for the Middle East will provide some more clues.

29. Former CIA agent Miles Copeland indicates that the United States had a role in bringing Zaim to power. See Copeland's *The Game of Nations* (New York, 1970). In Kuwait, any potential anti-American moves were neutralized by the timely gift by the companies to the Sheikh of a 125-foot gold-trimmed yacht, refitted for more than $1 million (*New York Times*, 10 January 1949).

30. Testimony of Ambassador George McGhee to the Senate Foreign Relations Subcommittee on Multinational Corporations, 28 January 1974, released 24 February 1974.

31. Lloyd Gardner, *Economic Aspects of New Deal Diplomacy* (Boston, 1971), p. 229.

32. Telegram from Secretary of State Byrnes to Ambassador Murray in Teheran, 8 April 1946. Murray responded by reporting that Prime

Minister Qavam "understood and recalled that we had already made clear our position in this regard. He remarked that there was plenty of time." *FR*, 1946, VII, p. 413.

33. Charles Issawi and Mohammed Yeganeh, *The Economics of Middle Eastern Oil* (New York, 1962), pp. 121-22, 183; Stocking, p. 154.

34. See Europa Publications, *The Middle East and North Africa, 1973-1974* (London, 1973), p. 294.

35. *New York Times*, 18 March 1951.

36. John A. Loftus, "Petroleum in International Relations," *Department of State Bulletin* (August 1945), p. 174.

37. Krock, *Memoirs* (New York, 1968), p. 262.

38. *New York Times*, 17 May 1951 (emphasis added).

39. Statistics in Issawi and Yeganeh, p. 183.

40. Stocking, pp. 153-56; Joyce and Gabriel Kolko, *The Limits of Power* (New York, 1972), pp. 417-20; Michael Tanzer, *The Political Economy of Oil and The Underdeveloped Countries* (Boston, 1969), pp. 321-26. The head of the American police training mission was General Norman Schwartzkopf, formerly of the New Jersey State Police, who attained fame in connection with the Lindbergh baby kidnapping. Prior to going to Iran he entertained millions as the narrator for the popular radio show "Gangbusters" (Barnet, p. 202).

41. DeNovo, *American Interests*, p. 132.

42. Kolko, p. 419.

43. Senate Foreign Relations Subcommittee on Multinational Corporations, "The International Petroleum Cartel, the Iranian Consortium and U.S. National Security," 21 February 1974, p. 68.

44. Secretary of State Acheson wrote in 1952 that prosecution of the suit "will inevitably be interpreted by the peoples of the [Middle East] as a statement that, were it not for such a conspiracy, they would be getting a higher return from their oil resources. This will, of course, strengthen the movement for renegotiation of the present concession agreements and may give some encouragement to those groups urging nationalization" (ibid., p. 5).

45. The text of the consortium agreement is in ibid., pp. 95-116. The two quotes are from P.H. Frankel, British oil expert, quoted in Tanzer, p. 380; and Issawi and Yeganeh, p. 122.

Chapter 3: The Bonanza Years

1. Issawi and Yeganeh, pp. 188-89.
2. Stocking, p. 403.
3. Ibid., pp. 423-24.
4. Ibid., pp. 395-99.
5. Feis, p. 156.
6. Kolko, p. 447; see also, Freeland, *Truman Doctrine*.
7. Kolko, p. 447.
8. Horst Menderhausen, "Dollar Shortage and Oil Surplus in 1949-1950," *Essays in International Finance*, no. 11 (Princeton, 1950), pp. 33-34.
9. Kolko, p. 461; Helmut Frank, *Crude Oil Prices in the Middle East* (New York, 1966), pp. 29-91; M.A. Adelman, *The World Petroleum Market* (Baltimore, 1972), pp. 131-59.
10. Federal Trade Commission, *The International Petroleum Cartel* (Washington, 1952). All references are to the summary of the report reprinted in U.S. Senate, Hearings, *Governmental Intervention* (1969), p. 551.
11. U.S Senate, Hearings before the Committee on Interior and Insular Affairs, *Oil And Gas Import Issues*, Part 3, 10, 11, and 22 January 1973 (Washington, 1973), p. 1003.
12. FTC, p. 553.
13. U.S. Senate, Hearings, *Oil and Gas Import Issues,* p. 1003.
14. FTC, pp. 561-62.
15. FTC, pp. 570-72.
16. Sir Rupert Hay, "The Impact of the Oil Industry on the Persian Gulf Shaykdoms," *Middle East Journal* (Autumn 1955), p. 364.
17. For comparative figures, see Issawi and Yeganeh, p. 53.
18. Ibid., pp. 110-13.

Chapter 4: Oil Politics and Economic Nationalism

1. *New York Times* (editorial), 6 August 1954.
2. While the popular media talked of Suez in familiarly lurid anatomical metaphors ("jugular" is the resounding favorite), oil industry media consistently regarded the actual "emergency" as minor,

calling for no more than a rerouting of tankers. The Foreign Petroleum Supply Committee, a government-sponsored industry group set up after the Iranian nationalization in 1951, had done all the necessary planning and conniving. See, for example, *Petroleum Week*, 11 May 1956. On the political background to Suez, see Erskine Childers, *The Road to Suez* (London, 1962); and on the role of the workers, see George Lenczowski, *Oil and State in the Middle East* (Ithaca, 1960), pp. 253-80; and *Petroleum Week* issues in April 1956.

3.   *New York Times*, 20, 22 November 1956; 6, 14, 21 December 1956.

4.   Aramco's role in building up Saud testifies to the fact that behind all the chatter about "International Communism," the real stake in the Middle East, as before, was control of oil resources. Aramco's role is mentioned by Childers, p. 313, and is supported by the hoopla about Saud in oil industry publications. See, particularly, *Petroleum Week*, 7 December 1956; 15 February and 29 March 1957. The State Department's slave estimate is in *Middle East Journal*, chronology of events, Spring 1957.

5.   The Council on Foreign Relations account is in John Campbell, *Defense of the Middle East* (New York, 1960), p. 130. On Syria, see Patrick Seale, *The Struggle for Syria* (London, 1965) and Tabitha Petran, *Syria* (London, 1973). On Lebanon, see Richard Barnet, *Intervention and Revolution* (Cleveland, 1968). In light of what was to follow in Jordan and Syria, it is surprising that no attention was paid to a paragraph in *Petroleum Week* (16 November 1956) at the height of the Suez conflict referring to secret British plans that were "more than wishful thinking" for a "new map of the Middle East" that would come about with the takeover of Jordan by Iraq and the division of Syria between Turkey and Iraq.

6.   *New York Times*, 12 July 1958.

7.   The U.S.-British decision not to invade Iraq was reported in the *New York Times*, 18 July 1957. The decision to land troops was no doubt facilitated by the fact that at the time there were three reinforced Marine battalions with the Sixth Fleet, rather than the usual one, and the Fleet itself had seventy-four instead of the usual fifty-one ships, including three attack carriers (*New York Times*, 15 July 1958).

8.   Material for the following paragraphs can be found as follows: The assessment of Mattei's role in P.H. Frankel, *Mattei: Oil and*

Power Politics (London, 1966), p. 95. ENI's first joint venture was with Egypt earlier in 1957, but it was the Iran deal which really shook up the industry. For details, see Muhammad Mughraby, *Permanent Sovereignty over Oil Resources* (Beirut, 1966), pp. 62-116. For industry reactions, see Stocking, p. 169; and Moseley, p. 269.

9. Industry concentration ratios are in Federal Trade Commission, "Investigation of the Petroleum Industry," a report submitted to the U.S. Senate, Permanent Subcommittee on Investigations (Washington, 1973), p. 13. This account of the import controls legislation is based on M.A. Adelman, *The World Petroleum Market* (Baltimore, 1972), pp. 148-59.

10. O'Connor, *World Crisis in Oil*, pp. 387-404.

11. For much of the information and insights in this section I am indebted to Steven Duguid, who has been willing to share the results of his doctoral research at Simon Fraser University. Some of the details can be found in Stocking and in Lenczowski, as well as in the industry press of the period. Needless to say, responsibility for accuracy and interpretation on these questions rests with me alone. One point that should be noted is that since this study focuses specifically on developments in Middle East oil, it does not discuss fully the role of Venezuela in the events leading up to the creation of OPEC. For such an appreciation, see Fuad Rouhani, *History of OPEC*, and Zuhayr Mikdashi, *The Community of Oil Exporting Countries.*

Chapter 5: Middle East Oil in the 1960s

1. For a detailed account of the Beirut meetng and company reaction to OPEC, see Moseley, pp. 293-97.

2. Cost and price data are from Adelman, *The World Petroleum Market*, pp. 48, 183, 209. There is a bias in these figures that minimizes company profits, which I will explain below when we calculate gross company profits for selected years in the 1960s.

3. Quoted in Stocking, p. 364.

4. This account largely follows that of Abbas Alnasrawi, "Collective Bargaining Power in OPEC," *Journal of World Trade Law* (March-April 1973), pp. 188-207. Details of the negotiations were presented by the OPEC Secretariat in a paper presented to the Fifth Arab Petroleum

Congress in March 1965, entitled "OPEC and the Principle of Negotiation."

5. "OPEC and the Principle of Negotiation," paper presented by OPEC Secretariat to the Fifth Arab Petroleum Congress, March 1965, pp. 7, 15.

6. In this discussion I was greatly aided by the willingness of Steve Duguid to share with me an early draft of his work on the political role and ideology of the Arab oil technocrats.

7. Stocking, pp. 200-269.

8. Ibid., p. 217.

9. Ibid., p. 229.

10. Ibid., p. 315.

11. OPEC figures in ibid., p. 462.

12. Figures are from Marwan Iskandar, "Development Plans in Oil Exporting Countries," in Zuhayr Mikdashi, ed. *Continuity and Change in the World Oil Industry* (Beirut, 1970). For more detailed discussions of developments in Iraq and Iran in these years, see "The Baathi Revolution in Iraq," *MERIP Reports*, no. 12; and "Iran: Ten Years After the 'White Revolution,' " *MERIP Reports*, no. 18.

13. Arthur Schlesinger, *A Thousand Days* (New York, 1965), p. 566; Michael Brecher, *The Foreign Policy System of Israel* (New Haven, 1972), pp. 44-45.

14. Interestingly, Kennedy people with prime responsibility for managing the Yemen crisis included Robert Komer of the National Security Council staff who later organized the Vietnam pacification program, and Ellsworth Bunker, later trouble-shooter in the Dominican Republic and long-time ambassador to Saigon, and now serving as the U.S. delegate to the Middle East peace talks in Geneva.

15. Quoted in *Middle East Record 1960* (Tel Aviv, 1961), p. 374.

16. *Middle East Record 1967* (Tel Aviv, 1970), p. 52.

17. *Middle East Economic Survey (MEES)*, 1 October 1965.

18. *MEES*, 18 November 1966.

19. *MEES*, 1 April 1966 and 21 April 1967.

20. Eugene Rostow, "The Middle Eastern Crisis in the Perspective of World Affairs," *International Affairs* (London, April 1971), p. 280.

21. *MEES*, 2 June 1967.

22. Ibid., 8, 16, 23, 30 June; 14, 21, 28 July 1967. For the U.S. role in the war, see Abdullah Schliefer, *Fall of Jerusalem* (New York: 1972).

23. *MEES*, 25 August 1967.

24. Ibid.

25. Ibid., 1 September 1967.

26. Ibid.

27. Ibid., 8 September 1967.

28. Ibid., 22 September 1967. See also 15 September 1967, and Alnasrawi, p. 194.

29. Figures from *Petroleum Press Service*, May 1972, p. 167.

30. Production figures are from OPEC statistics as tabled in Stocking, p. 451; Abu Dhabi and Qatar figures are in *Arab Oil and Gas* (Beirut), 1 December 1973, p. 37. Adelman's price calculations are in *World Petroleum Market*, p. 183. About production costs, Adelman writes: "Few if any published cost estimates are explained; fewer are reproducible; nearly all are irrelevant" (p. 45). His own are on p. 76. Average net investment figures in Issawi and Yeganeh, p. 188, consistently average just below 10 percent. Libyan profits jumped to $202 million in 1964. International Monetary Fund, *Balance of Payments Yearbook*, vol. 19 (1968). My thanks to Odeh Aburdeneh for pointing out this source to me. Unfortunately the fact that IMF figures for most other countries are based on posted prices makes it an unreliable guide for these estimates.

31. Production figures are in *Arab Oil and Gas* (Beirut), 1 December 1973, p. 37. Adelman's profit estimate is on p. 209.

32. Commerce Department figures are in *Survey of Current Business* (September 1967), p. 45, Table 6. The 1970 figure is cited in Moseley, p. 419. Production figures from *Arab Oil and Gas*, 1 December 1973, p. 37.

Chapter 6: Monopoly at Home

1. See the table of controlling companies in *United Mine Workers' Journal* (15-31 July 1973) and reproduced in *MERIP Reports*, no. 21 (October 1973), p. 12. Other figures are from U.S. House of Representatives, Select Committee on Small Business, *Concentration by Competing Raw Fuel Industries in the Energy Market and Its Impact on Small Business* (Washington, July 1971), pp. 43, 47.

2. Bruce Netschert, "The Energy Company: Monopoly Trend in

the Energy Markets," *Bulletin of the Atomic Scientists* (October 1971), p. 15; James Ridgeway, *The Last Play* (New York, 1973), p. 195; U.S. House, *Concentration*, p. 427; *Nuclear Industry* (June 1973), p. 16; *Wall Street Journal*, 12 January 1973. More details can be found in the AEC submission to the Justice Department in June 1971, in U.S. House, *Concentration*, pp. 233-46.

3. *Wall Street Journal*, 20 March 1973.

4. Federal Trade Commission, "Investigation," pp. 14, 18, 21.

5. U.S. House, *Concentration*, p. 41; Netschert, p. 17.

6. *Nuclear Industry* (April 1973).

7. *Platt's Oilgram News Service*, 15 September 1972.

8. Charles Wheatley, American Public Gas Association, in U.S. House, *Concentration*, pp. 355-56; also David Schwartz, FPC staff economist, testimony before the Senate Antitrust and Monopoly Subcommittee (unpublished transcript, June 1973), p. 335.

9. Testimony before the Senate Antitrust Monopoly Subcommittee, June 1973, p. 227.

10. James Halverson, director of the FTC Antitrust Division; and Schwartz's testimony, pp. 227, 316.

11. In the testimony to the Senate Antitrust and Monopoly Subcommittee.

12. See Les Aspin, "Big Oil and the Nixon Campaign," *Nation* (16 February 1974).

13. Schwartz's testimony, p. 335.

14. Charles Wheatley in U.S. House of Representatives, Committee on Interior and Insular Affairs, *Fuel and Energy Resources*, 1972, Part 2 (Washington, April 1972), p. 431. See also, the speech by the late Senator Estes Kefauver, *Congressional Record* (1 March 1957), reprinted in Senate Antitrust and Monopoly Subcommittee, *Competitive Aspects of Oil Shale Development* (Washington, D.C., 1967), pp. 463 ff. Hitler's war effort was fueled almost entirely by gas made from coal.

15. Letter of OEP Acting Director Darrell Trent to Senator Jackson, in Senate Committee on Interior and Insular Affairs, *Fuel Shortages* (Washington, 1973), pp. 299-300, 301; and Governor Francis Sargent, p. 220.

16. *Journal of Commerce*, 26 September 1973.

17. *Oil and Gas Journal*, 24 April 1972.

18. FTC, "Investigation," p. 18. Many have been built in Puerto

Rico, the Virgin Islands, and other Caribbean colonies. Much of the oil refined there comes from the Middle East and is exported to the United States where it is classified as Caribbean in origin. See *Oil and Gas Journal*, 24 April 1972; *Platt's Oilgram News Service*, 29 May 1973; "Billions in Projects Wait for Feds Energy Message," *Oil Daily*, 18 April 1973; Senate Committee on Government Operations, *Staff Study of the Oversight and Efficiency of Executive Agencies as It Relates to Recent Fuel Shortages* (July 1973), pp. 20-27.

19. *Platt's Oilgram News Service*, 18 August 1972.

20. FTC, "Investigation," p. 42.

21. William Pelley, vice-president of Bankers Trust, in U.S. Senate, Committee on Interior and Insular Affairs, *Financial Requirements of the Nation's Energy Industries* (Washington, March 1973), p. 26.

22. *Oil and Gas Journal*, 10 July 1972.

23. Wheatley, in *Financial Requirements*, p. 259.

24. Forty percent—including chemicals and public utilities, according to George Budzeika, "Lending to Business by New York City Banks," *Bulletin* of the New York University Graduate School of Business Administration (September 1971), p. 13. See charts and tables in Ridgeway, pp. 411-31, and in the testimony of John Wilson, Economics Division of FPC, pp. 39-42 (mimeo). Institutional investors manage pension and trust funds and other large agglomerations of capital (*Business Week*, 22 September 1973, pp. 42-54).

25. Wheatley, pp. 261-63.

26. *Wall Street Journal*, 8 January 1974; *Washington Post*, 29 November and 4 December 1973.

Chapter 7: The Road to Teheran Is Through Tripoli

1. *Wall Street Journal*, 27 March 1974.

2. Testimony of a Continental Oil Company executive before the Senate Foreign Relations Subcommittee on Multinational Corporations, 27 March 1974, p. 843.

3. *Middle East Economic Survey (MEES)*, 22 March and 23 August 1968.

4. Testimony before the Senate Foreign Relations Subcommittee on Multinational Corporations, pp. 856-57; figures on p. 914.

5. Ibid.

6. On the Saudi-Iranian conflict, see *MEES*, February 1968. On U.S. military assistance, see Hearings before the House Foreign Affairs Subcommittee on Near East and South Asia, *New Perspectives on the Persian Gulf* (Washington, D.C., 1973), p. 16.

7. *MEES*, 2 February 1968.

8. *Fortune* (November 1967), p. 77.

9. The purpose of this restrictive condition is clearly to ensure that all countries admitted into the organization would be equally anxious to maintain a purely economic approach to the development of the oil sector to the exclusion, as far as possible, of dangerous political cross-currents. See *MEES*, 12 January 1968.

10. *MEES*, 6 June 1969.

11. *MEES*, 28 February 1969.

12. *Business Week*, 25 October 1969, p. 94.

13. Stocking, p. 374.

14. Moseley, p. 349.

15. Sam Schurr and Paul Homan, *Middle East Oil and the Western World* (New York, 1971), p. 73.

16. Stocking, p. 375.

17. Moseley, pp. 325-26.

18. Testimony in executive session to the Senate Foreign Relations Subcommittee on Multinational Corporations, October 1973.

19. *Wall Street Journal*, 8 February 1972; *New York Times*, 11 February 1974.

20. Details of the strike are exceedingly hard to come by. The usual sources like *Middle East Economic Survey* and *Arab Report and Record* note its impact on oil exports but say nothing of the internal political details and dynamics.

21. *MEES*, 9 May 1969. Other references include *MEES*, 7 February 1969; *Petroleum Intelligence Weekly*, 14 April and 16 June 1969.

22. See NEPCO testimony before Senate Foreign Relations Subcommittee on Multinational Corporations, 27 November 1973.

23. Production figures from "Chronology of the Libyan Oil Negotiations," Senate Foreign Relations Subcommittee on Multinational Corporations (January 1974), pp. 3, 9. Trade figures from *Arab Oil and Gas* (Beirut, March 1974), p. 37.

24. *MEES*, 5 September 1969. Qaddafi quote is from "The Broadlines of the Third Theory," lecture by Muammar Qaddafi (Libyan Ministry of Information and Culture), May 1973. For a thorough treat-

ment of the political economy of Libya and the circumstances that gave rise to the coup, see Carole Collins, "Imperialism and Revolution in Libya," *MERIP Reports*, no. 27 (April 1974); and Ruth First, *Libya: The Elusive Revolution* (London, 1974).

25. Ibid.

26. *MEES*, 30 January 1974.

27. This account of the Libyan negotiations is based on the "Chronology of the Libyan Oil Negotiations, 1970-1971," prepared by the Foreign Affairs division of the Library of Congress for the Senate Foreign Relations Subcommittee on Multinational Corporations (Washington, January 1974), and on contemporary issues of *Middle East Economic Survey* and *Petroleum Intelligence Weekly*, except where noted.

28. Adelman, pp. 250-51; *Petroleum Intelligence Weekly*, 8, 15, and 22 June 1970.

29. These and the following quotes are taken from *MEES*, 29 May 1970.

30. *MEES*, 17 July 1970.

31. *MEES*, 24 July and 14 August 1970.

32. *MEES*, 21 August 1970.

33. Testimony of James Akins before the Senate Foreign Relations Subcommittee on Multinational Corporations, 11 October 1973, p. 6.

34. The producing country has the option of taking its 12.5 percent royalty in crude oil or in cash at posted price levels. Historically, the countries had no outlets to distribute or refine this crude, even domestically, and always took the cash option. The radical technocrats had long been urging that countries take the royalty in crude and use that to develop marketing outlets and skills that would allow the development of a national oil industry. In the market of the 1960s this would have meant less revenue, since market prices were well below posted prices for the most part (except in underdeveloped countries like India where the international companies charged the full posted prices to their local affiliates). In the late 1960s the national oil companies of Iran and Iraq began to market oil, but this was restricted to barter deals with East European countries where the international majors had no access anyway. This period (1970) marks the first time when some countries, notably Libya and Iraq, took a substantial part of their royalty in crude and marketed it directly.

35. *MEES*, 16 October 1970.

36. *Platt's Oilgram*, 7 October 1970.

37. *MEES*, 1 January 1971.

38. *MEES*, 8 January 1971. The freight differential figures were not available until January 29, 1971.

39. This and the following quotes are from *MEES*, 15 January 1971. Details on the Chase meetings are from the testimony of James Akins before the Senate Foreign Relations Subcommittee on Multinational Corporations, 11 October 1973 (made public January 27, 1974). pp. 11, 12.

40. Ibid.

41. Ibid.

42. Ibid.

43. *New York Times*, 18 January 1971. The chronology and other details of the Washington meetings can be found in the testimony of Akins and Irwin to the Senate Foreign Relations Subcommittee on Multinational Corporations, released on January 27 and February 10, 1974, respectively. McCloy testified before the same committee in early February that as the lawyer for several of the giant oil firms he had made contact with President Kennedy and Attorney General Robert Kennedy in early 1961 to inform them that the companies anticipated the potential need to form a united front in the event that the producing countries, then having just formed OPEC, would make a set of joint demands. This would indicate that the companies took the potential of OPEC more seriously than they let on publicly in the early 1960s. It is also significant that it took a full ten years for the "threat" to become serious enough to warrant these steps by the industry.

44. *New York Times*, 23 January 1971. Irwin's trip is discussed by various parties in testimony before the Senate Subcommittee on Multinational Corporations. In light of the running controversy between State Department's Akins and MIT professor M.A. Adelman over the reality of the threat to cut off oil, further elaboration is worthwhile. Adelman asserts that inadvertently or otherwise the United States "invited" the threat by sending Irwin over to tell the Shah, et al., how disastrous it would be and by publicly admitting after an OECD meeting in Paris on January 20 that no contingency plans for a cutoff were being planned. Akins cited the "concerted and simultaneous action" phrase and asserts that a cutoff threat had been made by Libyan Deputy Prime Minister Jallud to "the oil companies" on January 11. "The oil companies" in question were the two independents, Occident-

al and Bunker Hunt. It is entirely likely that such a threat was made, but it was totally irrelevant to the situation in the Gulf. Akins also says similar warnings were privately conveyed to Washington by two unnamed "friendly rulers." Given his propensity to make his case on the basis of the Libyan encounter and the vague OPEC phrase, one should be cautious about the reliability of this unidentified evidence. It was not until the Shah's press conference of January 24 that mention is even made of a cutoff, and then it was only in response to a question about whether such an event was possible. Only after the special OPEC meeting on February 3, after negotiations had broken down yet again, was there a specific threat of an embargo, and that was against *companies* which refused to comply with the legislated price changes. Thus the evidence tends to support Adelman's contention that the cutoff threat was prompted by the U.S. government and companies. But he is wrong to suggest, as he does, that the threat would not have been credible without all the U.S. expressions of "concern." The point here is that although such a threat was entirely credible, it was never really called except by the companies and then only to the point that they could appear to be "submitting" to the OPEC demands under duress. Adelman's case is made in *The World Petroleum Market*, pp. 254-56. Akins' retort is in "The Oil Crisis: This Time the Wolf Is Here," *Foreign Affairs* (April 1973), pp. 472-74.

45. *MEES*, 29 January 1971.

46. *MEES*, 19 February 1974.

47. Ibid.

48. *MEES*, 26 February 1971.

49. *MEES*, 19 and 26 February; 5, 12, and 19 March; 2 April 1974.

50. "Oil Power," *Foreign Affairs* (July 1971), p. 652.

51. Abbas al-Nasrawi, "Collective Bargaining Power in OPEC," *Journal of World Trade Law* (March/April 1973), p. 206.

52. *MEES*, 1 January 1974.

53. Figures and quotes from *Platt's Oilgram* and *Oil and Gas Journal*, quoted in Adelman, p. 252.

54. *Oil Daily*, 18 December 1972.

55. *The Economist* (Special Survey), 17 July 1973, p. 16.

56. P.H. Frankel, "The Current State of World Oil," *MEES*, 6 September 1968.

57. Edith Penrose, "Government Partnership in the Major Concessions," *MEES*, 30 August 1968.

Chapter 8: Controlling Oil Resources

1. *MEES*, 7 June 1968.
2. Mikdashi, p. 220.
3. *MEES*, 6 November 1970.
4. *Wall Street Journal*, 15 December 1971.
5. *MEES*, 26 February, 5 March, 16 April, 18 June, 2 July 1971; 2 January and 2 May 1972; and *Petroleum Press Service*, February 1972.
6. *MEES*, 28 March 1969.
7. *Africa Confidential*, 1 October 1971; see also William Quandt, "Can We Do Business With Radical Nationalists in Algeria: Yes," *Foreign Policy* (Summer 1972), pp. 108-31. For an analysis of Algerian development policy, see Karen Farsoun's article in *MERIP Reports*, no. 35 (February 1975).
8. Stocking, p. 305.
9. *MEES*, 11 July 1969.
10. *MEES*, 14 April 1972; much of this discussion is based on Stocking, Chapter 14.
11. *MEES*, 19 November 1971; 11 February, 19 May, 2 June 1972.
12. *MEES*, 16 June 1972.
13. *MEES*, 23 June 1972.
14. *MEES*, 2 June 1972.
15. Details of the settlement are in *MEES*, 2 March 1973.
16. *MEES*, 2 July 1971.
17. *New York Times*, 22 January 1972.
18. *MEES*, 21 January 1972.
19. *Washington Post*, 29 January 1972.
20. *MEES*, 18 February 1972.
21. *MEES*, 11 March 1972.
22. *MEES*, 19 May 1972.
23. *MEES*, 14 July 1972.
24. *MEES*, 18 August 1972.
25. Ibid.
26. Yamani's speech is reprinted in *MEES*, 22 September 1972.
27. *Arab Oil and Gas, (AOG)*, 1 November 1972.
28. Sarkis' speech is reprinted in *AOG*, 1 December 1972.
29. *AOG*, 16 November 1972.
30. *MEES*, 6 October 1972.

31. *New York Times*, 12 August 1973.

32. *New York Times*, 18 June and 13 September 1973.

33. *MEES*, 17 August 1973.

34. *MEES*, 24 August 1973.

35. *AOG*, 1 September 1973.

36. *MEES*, 2 September 1973.

37. *MEES*, 7 and 14 September 1973. In the statement Jallud several times refers to the "participation or nationalization agreements," indicating no distinction or preference in terminology.

38. *AOG*, 1 February, 1 and 16 April, and 1 May 1974.

39. *MEES*, 23 March 1974.

40. Ibid.

41. *MEES*, 18 May and 21 September 1973.

42. *MEES*, 23 February 1973.

43. *MEES*, 3 November 1972. Emphasis added.

44. Hearings Before the Senate Foreign Relations Subcommittee, *Energy and Foreign Policy*, 30 and 31 May 1973, pp. 87-88.

45. Chandler Morse, "Potentials and Hazards of Direct International Investment in Raw Materials," in Marion Clawson, ed. *Natural Resources and International Development* (Baltimore, 1964), p. 411.

46. See, for example, reports of the speech of Saudi economic planner Hisham Nazer to the American-Arab Association for Commerce and Industry in New York, *New York Times*, 12 September 1973.

Chapter 9: The Oil Weapon, the October War, and the Price Explosion

1. *MEES*, 3 November 1972.

2. See the testimony of Assistant Secretary of State Joseph Sisco and Deputy Assistant Secretary of Defense James Noyes in the hearings before the House Committee on Foreign Affairs, *New Perspectives on the Persian Gulf* (Washington, D.C., 1973).

3. The most detailed discussion of the Nixon Doctrine and its antecedents is Mike Klare's *War Without End* (New York, 1972).

4. Navy Memo to Senate Armed Services Committee, quoted in Judith Miller, "U.S. Navy Still Pressing for Base in Indian Ocean," *Washington Post*, 19 May 1974.

5. *Congressional Record*, 21 May 1973, p. S9446.

6. The report, in Arabic, is reviewed in *Journal of Palestine Studies*, no. 9 (Autumn 1973), pp. 142-44.

7. *MEES*, 15 June 1974.

8. *Washington Post*, 19 April 1973.

9. *Washington Post*, 17 June 1973.

10. *MEES*, 1 June 1973.

11. *MEES*, 14 September 1973.

12. *MEES*, 17 August 1973.

13. *Washington Post*, 7 August 1973.

14. *Wall Street Journal*, 5 and 6 September 1973.

15. *Platt's Oilgram News Service*, 7 September 1973.

16. The Arab-Israeli plan was divulged by the *Times* (London), 26 September 1973. The Saudi approach is discussed in *Platt's Oilgram News Service*, 27 September 1973. *Platt's* notes that this approach was decided after consideration of military options to secure oil supplies presented in a Joint Chiefs of Staff "security seminar" in early August.

17. *MEES*, 7 September 1973.

18. *Washington Star-News*, 12 October 1973.

19. *New York Times*, 18 October 1973.

20. *MEES*, 2 and 16 November 1973.

21. *MEES*, 28 December 1973.

22. *New York Times*, 22 December 1973.

23. *New York Times*, 19 January 1973.

24. *New York Times*, 26 December 1973.

25. *New York Times*, 27 December 1973.

26. *New York Times*, 26 December 1973.

27. The distinction is based on rate of production in 1972. Any oil produced above the 1972 rate, or from a well not in production in 1972, is considered "new" oil. As old wells are shut down and new ones opened, or as more prolific wells are stepped up, a greater proportion of oil becomes "new" and thus not subject to price control. The present proportion of "new" to "old" oil is between 30 and 40 percent.

28. *New York Times*, 6 February 1973

29. *New York Times*, 7 January 1973; *Wall Street Journal*, 7 January 1973.

30. *Journal of Commerce*, 31 January 1973.

31. See the articles in the *Philadelphia Inquirer*, reprinted in *Congressional Record*, 30 January 1974, pp. S774-88.

32. Legislative Reference Service of the Library of Congress, "The

Arab Oil Embargo and the U.S. Oil Shortages: October 1973 to March 1974," prepared at the request of Rep. Dante Fascell, p. 5.

33. *Washington Post*, 16 November 1973; *New York Times*, 1 and 13 December 1973 and 3 February 1974.

34. Other projections, including one by the Library of Congress (not cited above), assume an exceptionally cold winter, but there is still no indication of where they get their outlandish demand figures from.

35. Library of Congress, "The Arab Oil Embargo," pp. 7,8.

36. *Journal of Commerce*, 26 February 1974.

37. *Platt's Oilgram News Service*, 2 and 7 January 1974.

38. *Journal of Commerce*, 26 February 1974.

39. *New York Times*, 22 February 1974.

40. *New York Times*, 22 February 1974; *Journal of Commerce*, 22 February 1974.

41. *New York Times*, 27 March 1974.

42. *New York Times*, 12 April 1974.

43. *Wall Street Journal*, 9 April 1974.

44. *New York Times*, 13 April 1974.

45. *Journal of Commerce*, 10 April 1974.

46. *New York Times*, 31 May and 22 June 1974.

47. *New York Times*, 10 June 1974; *Journal of Commerce*, 17 June 1974.

48. *Wall Street Journal*, 25 January 1974.

49. *Wall Street Journal*, 21 January and 4 April 1974; *New York Times*, 18 and 27 April 1974.

50. *Wall Street Journal*, 8 January 1974.

51. See above. During the October War Jackson wrote a short piece in the *New York Times* in which he concluded that U.S. interests demanded that Israel win the war "decisively." Events of the last six months demonstrate the limits of this approach. For a view of Kissinger's prewar attitude toward Israel's occupation of Arab territories, see *Jerusalem Post Weekly*, 18 December 1973.

52. *New York Times*, 7 February 1973.

53. *New York Times*, 8 November 1973.

54. *New York Times*, 26 October 1973. Although other administration spokesmen, notably the President, Defense Secretary Schlesinger and former Defense Secretary Laird, made more conventional hardline statements, subsequent events show that Kissinger was in charge. See,

for example, Nixon's statement on October 15 that U.S. policy would be like that in 1958 in Lebanon and in 1970 in Jordan, i.e., real or threatened military intervention.

55. "United States Oil Shortage and the Arab-Israeli Conflict," Report of a Study Mission by the House Foreign Affairs Committee (20 December 1973), p. 51.

56. *New York Times*, 22 November 1973.

57. *New York Times*, 23 November 1973.

58. *New York Times*, 1 December 1973.

59. See James Reston's column in the *New York Times*, 9 December 1973.

60. *New York Times*, 23 January 1974.

61. *New York Times*, 30 January 1974.

62. *New York Times*, 1 February 1974.

63. *New York Times*, 5, 7, 14 February 1974; *Wall Street Journal*, 20 February 1974.

64. *Journal of Commerce*, 1 March 1974; *New York Times*, 4 and 6 March 1974; *Wall Street Journal*, 6 March 1974.

65. *New York Times*, 10-14 March 1974.

66. *Journal of Commerce*, 22 March 1974.

67. *New York Times*, 20 March 1974.

68. *Washington Post*, 6 April 1974.

69. *Middle East Economic Digest* (London), 21 June 1974.

70. *MEES*, 5 April 1974.

71. *MEES*, 7 July 1972.

72. Ibid.

73. *Middle East Economic Digest*, 16 June 1972.

74. *MEES*, 18 May 1973.

75. See the General Accounting Office report to the House Foreign Affairs Subcommittee on the Near East and South Asia, "A Summary of European Views on Dependency of the Free World on Middle East Oil," 29 August 1973.

76. *New York Times*, 31 October 1973.

77. *New York Times*, 13 December 1973.

78. *MEES*, 8 February 1974.

79. *New York Times*, 12-14 February 1974.

80. *MEES*, 8 March 1974.

81. *Washington Post*, 6 April 1974.

82. *Journal of Commerce*, 11 June 1974.

83. *MEES*, 5 July 1974.

84. *MEES*, 26 July 1974.

85. *Wall Street Journal*, 5 and 17 September 1974.

86. *MEES*, 13 July 1973.

87. International Institute for Strategic Studies, *Strategic Survey, 1973* (London, 1974), p. 12.

Chapter 10: The Nature of the Crisis

1. Robert Solow, "The Economics of Resources or the Resources of Economics," *American Economic Review* (May 1974), pp. 1-14; and Bernard Nossiter, "Dwindling Resources: A Myth?," *Washington Post*, 15 July 1974.

2. President's Materials Policy Commission (Paley Commission), *Resources for Freedom* (Washington, D.C., 1954), vol. I, p. 13.

3. *Foreign Policy* (Summer 1973), p. 166.

4. *Business Week*, 17 February 1973.

5. *The Economist*, 5 May 1973.

6. *Washington Post*, 28 July 1974.

7. At the International Bankers Conference in June 1974 David Rockefeller estimated oil country surpluses would amount to $70 billion in 1974 and $200 billion by 1976 (*New York Times*, 7 June 1974).

8. OECD, *Economic Outlook* (July 1974), p. 109.

9. *Middle East Economic Digest*, 15 March 1974.

10. OECD, p. 96.

11. David Calleo and Benjamin Rowland, *America and the World Political Economy* (Bloomington, 1973), p. 116.

12. OECD, p. 57.

13. *The Economist*, 5 May 1973.

14. *Wall Street Journal*, 28 June 1974.

15. Otmar Emminger, "Inflation and the International Monetary System," Proceedings of the Tenth Meeting of the Per Jacobson Foundation at the University of Basle, June 16, 1973, p. 15.

16. Ibid., p. 16.

17. OECD, p. 46.

18. Ibid., p. 94.

19. Ibid., p. 47.

20. *Platt's Oilgram News Service*, 15 November 1972.

21. Ibid.

22. *New York Times*, 30 May 1974.

23. *Journal of Commerce*, 8 August 1974.

24. *Journal of Commerce*, 1 July 1974.

25. *Journal of Commerce*, 20 June 1974.

26. Ibid.

27. *Journal of Commerce*, 17 July 1974.

28. Statement of Francois Dickman, State Department Bureau of Near East and South Asian Affairs, in hearings before the House Foreign Affairs Committee, *New Perspectives on the Persian Gulf* (Washington, D.C., 1973), p. 175.

29. Figures and quote from former Deputy Assistant Secretary of State Roger Davies in ibid., pp. 154, 155. Davies was killed in August 1974 while serving as ambassador to Cyprus.

30. *New York Times*, 10 July 1974. The figures do not include $1.5 billion in arms granted to Israel and several million more granted to Jordan and Lebanon. For a full analysis of the role of arms exports in the U.S. economy, see Chris Paine, "The Political Economy of Arms Transfers to the Middle East," *MERIP Reports*, no. 30 (August 1974).

31. *New York Times*, 27 June 1974. By April Italy, Britain, and France had borrowed $25 billion from the Federal Reserve and other central banks. See *Journal of Commerce*, 22 April 1974.

32. OECD, pp. 103, 104. According to former Federal Reserve governor Arthur Brimmer, such loans were up by more than one-third to $8.5 billion (*New York Times*, 18 July 1974).

33. *New York Times*, 10 June 1974.

34. *New York Times*, 2 and 5 October 1974.

35. OECD, p. 6.

36. Ibid., p. 17.

37. Ibid., p. 5.

38. Interim Report of the Joint Economic Committee, Washington, D.C., 21 September 1974, p. 3.

39. OECD, p. 36.

40. Ibid., p. 39.

41. *Journal of Commerce*, 22 August 1974.

42. Emma Rothschild, "The Next Crisis," *New York Review of Books*, 4 April 1974, p. 31.

43. OECD, p. 33.

44. *Business Week*, 22 September 1973.

45. OECD, p. 35.

46. *New York Times*, 21 July 1974.

47. See the intermittent columns of J. Roger Wallace in the *Journal of Commerce*, and "Keynesian Chickens Come Home to Roost," *Monthly Review* (April 1974).

48. Theodore Geiger, "A Note on U.S. Comparative Advantages, Productivity and Price Competitiveness," in National Planning Association, *U.S. Foreign Economic Policy for the 1970s: A New Approach to New Realities* (Washington, D.C., November 1971), pp. 62, 63.

49. OECD, p. 34. Emphasis added.

50. Saudi and other Middle East country earnings are in *MEED*, 15 March 1974. The Saudi budget for FY 1975 estimates total revenues at $26 billion, with $14 billion in surplus (*New York Times*, 7 September 1974). Exxon earnings are from the *New York Times*, 20 July 1974. The Pentagon budget can be found in the Brookings Institution's *Setting National Priorities: The 1975 Budget* (Washington, D.C., 1974), p. 72.

51. OECD, p. 33.

52. *MEES*, 8 February 1974.

53. Arghiri Emmanuel, "Myths of Development vs. Myths of Underdevelopment," *New Left Review*, no. 85 (May-June 1974), p. 73.

54. Ibid., pp. 74, 82.

55. William Paley reiterated the theme of the Presidential Commission cited earlier in a U.S. Steel-sponsored ad in the *Wall Street Journal*, 23 July 1974.

56. *New York Times*, 9 July 1974.

57. See the statement of Professor Marvin Zonis in Hearings before the House Foreign Affairs Committee, *New Perspectives*, pp. 65-70.

58. *Business Week*, 22 June 1974.

59. *Wall Street Journal*, 15 April 1974.

60. See "Saudi Arabia, Bullish on America," *MERIP Reports*, no. 26 (February 1974).

61. *Foreign Economic Trends*, April 1973, p. 8.

62. *MEES*, 21 June 1974.

63. International Institute for Strategic Studies, *Strategic Survey, 1973* (London, 1974), p. 1.

64. *Strategic Survey, 1973*, p. 2.

Index

59592

Benjamin Britten

His Life and Operas

by the same author

STRAVINSKY: THE COMPOSER AND HIS WORKS
TIPPETT AND HIS OPERAS
A HISTORY OF ENGLISH OPERA

BENJAMIN BRITTEN

His Life and Operas

by
ERIC WALTER WHITE

Second Edition, 1983
Edited by John Evans

70100

OAKTON COMMUNITY COLLEGE
DES PLAINES CAMPUS
1600 EAST GOLF ROAD
DES PLAINES, IL 60016

UNIVERSITY OF CALIFORNIA PRESS
Berkeley and Los Angeles

This edition first published in 1983
by the University of California Press
Berkeley and Los Angeles, California

Printed in Great Britain

© Eric Walter White 1948, 1954, 1970 and 1983

Library of Congress Cataloging in Publication Data

White, Eric Walter, 1905–
  Benjamin Britten, his life and operas.

  "Short bibliography": p.
  Includes index.
    1. Britten, Benjamin, 1913–1976.   2. Composers—
England—Biography.   I. Title.
ML410.B853W4   1982   782.1′092′4[B]   82–10882
ISBN 0–520–04893–8
ISBN 0–520–04894–6 (pbk.)

To Bettina Hürlimann-Kiepenheuer

Contents

CONTENTS

Illustrations

ILLUSTRATIONS

ILLUSTRATIONS

Preface to the Second Edition, 1983

When at the end of the Second World War our contacts with Europe were being renewed, it was pleasant to find that our European friends wanted to find out what had been happening in Great Britain during the same period, and in particular what new manifestations there had been in the world of the arts. Sometime in 1945 my friends Martin Hürlimann, founder of the Atlantis publishing house in Zurich, and his wife Bettina, came over to London from Switzerland, and I took them to Sadler's Wells Theatre because I thought they would be interested in *Peter Grimes*, a new English opera that had recently been written by one of our most promising young composers, Benjamin Britten. I was particularly impressed by their enthusiastic reaction to this novelty because I felt it came from members of an audience that was familiar with the main stream of European culture and could therefore be accepted as a reliable judgement.

Within a few weeks of the first performance of *Peter Grimes* so many music and opera critics had turned up in London from different parts of the world that it was clear that in a comparatively short space of time the opera had earned itself a place in the international operatic repertory and Britten had been accepted as an important composer in that field. I was delighted. As an amateur of opera myself, and of English opera in particular, I welcomed this blossoming of native talent and its world-wide recognition.

On their return to Zurich the Hürlimanns wrote me a letter, confirming their interest in *Peter Grimes* and offering me a contract for a short book on Benjamin Britten to be published (in German) in their series of Atlantis Musikbücher. This invitation I accepted with pleasure as soon as I could be assured of the approval of Britten himself. This was quickly forthcoming; and henceforward whenever I consulted Britten on any of his music I could be certain of receiving a detailed and most considerate reply; but this did not alter the fact that the responsibility for the final text of these writings was mine alone.

Benjamin Britten: eine Skizze von Leben und Werk was published in Zurich in November 1948 in a German translation by Bettina and Martin Hürlimann; and the English edition was published a few days later by

Boosey & Hawkes in London. The text included separate essays on *Peter Grimes*, *The Rape of Lucretia* and *Albert Herring*. This proved so useful that in 1954 it was decided to bring it up to date by including chapters on *Paul Bunyan*, *The Beggar's Opera*, *The Little Sweep*, *Billy Budd* and *Gloriana*. A further revision and enlargement took place in 1970, when the book, now lavishly illustrated, was published by Faber and Faber in association with Boosey & Hawkes. New chapters were added on *The Turn of the Screw*, *Noye's Fludde*, *A Midsummer Night's Dream*, *Curlew River*, *The Burning Fiery Furnace* and *The Prodigal Son*.

I am deeply grateful to John Evans, Research Scholar to The Britten Estate, who has undertaken the editorial responsibility for the present edition. He has incorporated my new chapters on *Owen Wingrave* and *Death in Venice*, has updated the bibliography and list of published works, and has revised the text, where recent scholarship has shed new light on aspects of Britten's career.

Acknowledgements are also due to a number of other individuals who have assisted me in my researches for the earlier editions—W. H. Auden, Prince Ludwig of Hesse and the Rhine, Elizabeth Mayer and William Plomer (all now, sadly, deceased); and Henry Boys, Eric Crozier, Imogen Holst, Iris Lemare, Donald Mitchell, Peter Pears, Myfanwy Piper and Basil Wright. I am grateful to The Britten Estate, Boosey & Hawkes and Faber Music for permission to quote music examples from copyright material.

These little essays on the operas lay no claims to offer musical analyses in depth. Their aim has been simply to try to answer some of the more obvious questions an astute listener is likely to ask.

Those of us who took part in the renaissance of English music which occurred during Britten's lifetime realized that we were lucky enough to have been involved in an important moment of music history. When comes such another Golden Age?

E. W. W.

Part One
LIFE

A page from Britten's manuscript full score of 'L'Enfance' from *Quatre Chansons Françaises* (1928)

I
Early Works and Training

Edward Benjamin Britten was born at Lowestoft, Suffolk, on St Cecilia's Day (22 November) 1913, the youngest of four children. His father was a dental surgeon, and his mother a keen amateur singer, who acted for some years as secretary of the Lowestoft Choral Society. The Britten family lived in a house directly facing the North Sea.

Music was an early love of his. He started to compose at the age of five. He had piano lessons from Miss Ethel Astle, a local teacher, when he was eight, and started viola lessons with Mrs Audrey Alston of Norwich about two years later.

There were other activities too. Some years afterwards when he was accorded the freedom of his home town, he recalled his appearance on the stage of the Sparrow's Nest as 'a very small boy, dressed in skin-coloured tights, with madly curly hair, trying desperately to remember the lines spoken by Tom the water-baby, sitting on the lap of Mrs Do-as-you-would-be-done-by'* played on this occasion by his mother. It seems appropriate that this musician should have shown such an early interest in the stage.

His activity as a juvenile composer was phenomenal. He himself has described these early efforts as follows:† 'I remember the first time I tried, the result looked rather like the Forth Bridge, in other words hundreds of dots all over the page connected by long lines all joined together in beautiful curves. I am afraid it was the pattern on the paper which I was interested in and when I asked my mother to play it, her look of horror upset me considerably. My next efforts were much more conscious of *sound*. I had started playing the piano and wrote elaborate tone poems usually lasting about twenty seconds, inspired by terrific events in my home life such as the departure of my father for London, the appearance in my life of a new girl friend or even a wreck at sea. My later efforts luckily got away from these emotional inspirations and I began to write sonatas and quartets which were not connected in any direct way with life . . . At school I somehow managed

* From Benjamin Britten's speech at Lowestoft, 28 July 1951.
† From *The Composer and the Listener*, a broadcast talk by Benjamin Britten, 7 November 1946.

(*Left*) Benjamin Britten aged four (*Right*) Britten aged about seven

Britten with his family and friends at Lowestoft

to be able to fit in a great deal of writing with the extremely busy life that everyone leads at school . . . I wrote symphony after symphony, song after song, a tone poem called Chaos and Cosmos, although I fear I was not sure what these terms really mean.' By the time he left his preparatory school, South Lodge, at the age of fourteen to enter Gresham's School, Holt, he had already written ten piano sonatas, six string quartets, three suites for piano, an oratorio, and dozens of songs. Some of his piano music and three of the songs were arranged later for string orchestra in the *Simple Symphony* (1934). Other songs (*e.g.* the collection *Tit for Tat*) have been published in the course of time. There is no doubt that these early compositions provide convincing evidence of his musical precocity.

As a child he heard little music outside his home, with the exception of the local choral society concerts, some occasional chamber music concerts, and the Norfolk and Norwich Triennial Festival. At the 1924 Festival, he was present when Frank Bridge conducted his suite *The Sea* and, in his own words,* 'was knocked sideways'. Three years later when Bridge came to the 1927 Festival, his viola teacher, Audrey Alston, took her young pupil to meet the composer. 'We got on splendidly' Britten wrote years later,† 'and I spent the next morning with him going over some of my music . . . From that moment I used to go regularly to him, staying with him in Eastbourne or in London, in the holidays from my prep school. Even though I was barely in my teens, this was immensely serious and professional study; and the lessons were mammoth. I remember one that started at half past ten, and at tea-time Mrs Bridge came in and said, "Really, Frank, you must give the boy a break." Often I used to end these marathons in tears; not that he was beastly to me, but the concentrated strain was too much for me . . . This strictness was the product of nothing but professionalism. Bridge insisted on the absolutely clear relationship of what was in my mind to what was on the paper. I used to get sent to the other side of the room; Bridge would play what I'd written and demand if it was what I'd really meant . . . He taught me to think and feel through the instruments I was writing for.' This discipline was especially salutary to a person of Britten's phenomenal facility and fluency.

He spent two years at Gresham's School, Holt (from September 1928 to July 1930). During that period his interest in music was as intensive as ever—to the surprise of some of the other boys when they caught him reading orchestral scores in bed!‡ He was taking piano lessons from Harold Samuel in London, and his composition lessons with Frank Bridge

* From 'Britten Looking Back' by Benjamin Britten. *Sunday Telegraph*, 17 November 1963.
† Ibid.
‡ *Cf.* Benjamin Britten's speech on receiving an honorary degree at the University of Hull, 1962, reprinted in the *London Magazine*, October 1963.

Britten with Frank and Ethel Bridge, *c.* 1930

continued during the holidays. When he left school at the age of nearly seventeen, the die was cast and he had made up his mind to make music his career. In this he was backed by his parents; but there were a number of people who couldn't take such a decision seriously. There has always been a section of the population in this country that is suspicious of the arts and finds it difficult to accept the view that they can offer a talented young man a respectable livelihood. In later years he recalled how, at a tennis party at Lowestoft in the summer of 1930, he was asked by some people what career he intended to choose. When he told them he meant to be a composer, they were amazed. 'Yes, but what else?' they asked.* Clearly they looked on music as a spare-time occupation.

But if Britten thought that his troubles were over when he entered for an open scholarship in composition at the Royal College of Music, London, he was much mistaken. His examiners on that occasion were John Ireland, S. P. Waddington, and Ralph Vaughan Williams; and, according to Ireland,† the other two adjudicators were at first against making the award at all, one of them going so far as to say 'What is an English public school boy doing writing music of this kind?' But eventually Ireland managed to convince them, and the award was made.

His period at the College was an unhappy and frustrating one. As he himself said,‡ 'when you are immensely full of energy and ideas, you don't want to waste your time being taken through elementary exercises in dictation'. Although he was still composing reams of music, opportunities for performance inside the College were very restricted. When he failed to get two choral psalms of his performed, Frank Bridge tried to intervene, saying it was important for a young composer to hear what he'd written, 'because without aural experience it was difficult to link notes and sounds'. Vaughan Williams claimed that the singers weren't up to scratch, to which Bridge retorted that it was up to the College to have a sufficiently good chorus and he ought to use his influence. Only one of Britten's works was played at the College during the whole of the time he was a student; and that was his *Sinfonietta* (op. 1) for ten instruments, which was included in the programme of a chamber music concert on 16 March 1933, more than two months after it had received its first performance elsewhere. A String Quartet in D major (composed in 1931) was given a 'play through' by the Stratton Quartet, but it had to wait until 1975 for its first performance.

During the three years he remained at the College he worked under John Ireland for composition and Arthur Benjamin for piano. He won the Ernest Farrar Prize for composition and passed the examination for the Associate-ship of the College as a solo pianist in 1933. The importance of this period resided mainly in the fact that he now had better opportunities for hearing

---

* From *British Composers in Interview*, by Murray Schafer. Faber, 1963.
† Ibid.   ‡ Ibid.

music, especially contemporary music, making friends with other musicians, and generally broadening his intellectual horizon. Among the classics, the strongest influences on his work were Mozart and Schubert, rather than Beethoven and Brahms; his love of Purcell was a later growth. His feeling for melody and lyricism led him to find in Mahler a congenial spirit. He was convinced that in compositions like *Des Knaben Wunderhorn* and the *Kindertotenlieder*, Mahler had expressed the idea behind the music with such success as to achieve real perfection of musical form. Among contemporary composers, he particularly admired the works of Stravinsky, Schoenberg and Berg. He tried (in vain) to get the score of *Pierrot Lunaire* bought for the library at the College. He was deeply impressed by the concert performance of *Wozzeck* that Sir Adrian Boult conducted at the Queen's Hall on 14 March 1934; so when he was given a small travelling scholarship on leaving the College, it was natural that he should express a wish to go to Vienna that autumn to study with Berg. But the College was consulted, and difficulties arose. Britten says,* 'I think, but can't be sure, that the Director, Sir Hugh Allen, put a spoke in the wheel. At any rate, when I said at home during the holidays, "I *am* going to study with Berg, aren't I?" the answer was a firm "No, dear." Pressed, my mother said, "He's not a good influence," which I suspected came from Allen. There was at that time an almost moral prejudice against serial music . . . I think also that there was some confusion in my parents' minds—thinking that "not a good influence" meant morally, not musically. They had been disturbed by traits of rebelliousness and unconventionality, which I had shown in my later school days.'

Outside the College, he pursued the possibility of getting his music performed, with varying degrees of success. M. Montagu-Nathan, when acting as secretary of the Camargo Society, met him at a party in 1932 and, finding he had written a ballet score, prevailed on him to submit it. Whether the manuscript was ever read by Edwin Evans, the Society's chairman, or by Constant Lambert, its conductor, is not clear; but in the course of time the score was returned to the composer and nothing more was heard of it.†

More fruitful was his contact with the Macnaghten-Lemare concerts of new music given at the Ballet Club Theatre (later known as the Mercury Theatre). It was here that the public heard his music for the first time. A *Phantasy for String Quintet* in one movement (posthumously published) and Three Two-Part Songs for female voices to words by Walter de la Mare were given at the concert on 12 December 1932; and the performance was reviewed by *The Times*, *Musical Times* and *Music Lover*. The notice by 'C.D.' in *Music Lover* particularly riled the young composer. 'C.D.' praised the three part-songs as being 'good from one who, I believe, is only 19; even

* From 'Britten Looking Back', op. cit.
† 'A Lost Opportunity' by M. Montagu-Nathan. *Radio Times*, 31 July 1953.

though they were reminiscent in a quite peculiar degree of Walton's latest songs which were heard recently elsewhere'. The Walton songs referred to were the Three Songs from *Façade* which had been performed for the first time at the Wigmore Hall on 10 October 1932. Twenty years later, recalling this first contact with his critics, Britten wrote: 'I was about 17 and three part-songs of mine had been given at a London theatre concert. They were written as a student's exercise, with the voice parts in strict canon. The first was amiably grotesque, the second atmospheric in a cool way, the third lumpily "folky". The only written criticism of this performance damned them entirely—as being obvious copies of Walton's three *Façade* Songs. Now anyone who is interested can see for himself that this is silly nonsense. The Walton Songs are brilliant and sophisticated in the extreme—mine could scarcely have been more childlike and naïve, with not a trace of parody throughout. It is easy to imagine the damping effect of this first notice on a young composer. I was furious and dismayed because I could see there was not a word of truth in it. I was also considerably discouraged.'*

On 31 January 1933, came the *Sinfonietta* which had been written the previous summer; and the programme of the concert on 11 December 1933 included Two Part-Songs for Mixed Choir and another unpublished work, entitled *Alla Quartetto Serioso*—'Go play, boy, play'. A programme note stated: 'This Quartet is not yet finished. There will be five movements in all, the three movements that are ready being the first, second, and fifth, viz. 1. Alla Marcia, 2. Alla Valse (the dance), 5. Alla Burlesca (ragging).'

In 1934, the year in which Holst, Delius and Elgar died, Britten came of age. He had left the College the previous year and was starting the difficult business of earning his living by composing. In addition to the works mentioned above, he had already written the *Phantasy Quartet* for oboe, violin, viola and cello, and a set of choral variations for mixed voices unaccompanied entitled *A Boy was Born*. During the year he was to add the *Simple Symphony*, a piano suite entitled *Holiday Diary*, a suite for violin and piano, and also a collection of twelve children's songs for voice and piano, *Friday Afternoons*, written for a boys' school at Prestatyn, where his brother was headmaster. None of these works had long to wait for performance. The *Phantasy Quartet* was first played by Leon Goossens and the International String Quartet in London, and also at the 1934 ISCM Festival at Florence (5 April). *A Boy was Born* was broadcast by the BBC on 23 February 1934; and when it was given its first public performance at the Macnaghten-Lemare concert of 17 December 1934, it received a glowing review from A. H. Fox-Strangways in the *Observer*.† ['The music] has one mark of mastery' he wrote, 'endless invention and facility. [The composer] takes what he wants, and does not trouble about what other people have

* From 'Variations on a Critical Theme'. *Opera*, March 1952.
† Afterwards reprinted in *Music Observed*, Methuen, 1936.

thought well to take. He rivets attention from the first note onwards: without knowing in the least what is coming, one feels instinctively that this is music it behoves one to listen to and each successive moment strengthens that feeling.'

Not all the critics were as enthusiastic as Fox-Strangways. On several occasions 'J.A.W.' complained in the *Daily Telegraph* that 'the solving of technical problems' seemed to 'occupy the composer's mind to the exclusion of musical ideas'; and William McNaught wrote: 'This young spark is good company for as long as his persiflage remains fresh, which is not very long. To do him justice, his *Sinfonietta* closed down in good time.'

Fortunately this was the period when John Grierson had collected some remarkably talented persons to work on documentary films for the GPO Film Unit; and Britten joined the group in 1935. In the five years 1935-9 he produced incidental music for some thirty documentary films by the GPO Film Unit, two documentary films by other units, and one feature film, *Love from a Stranger* (Trafalgar Films), directed by Rowland V. Lee.

He entered into this work with great zest and seems to have enjoyed its special conditions and restrictions. Some years later, when recalling this period of his life, he said:* 'I had to work quickly, to force myself to work when I didn't want to and to get used to working in all kinds of circumstances. The film company I was working for was not a big commercial one, it was a documentary company and had very little money. I had to write scores not for large orchestras but for six or seven instruments, and to make these instruments make all the effects that each film demanded. I also had to be very ingenious and try to imitate, not necessarily by musical instruments, but in the studio, the natural sounds of every-day life. I well remember the mess we made in the studio one day when trying to fit an appropriate sound to shots of a large ship unloading in a dock. We had pails of water which we slopped everywhere, drain pipes with coal slipping down them, model railways, whistles and every kind of paraphernalia we could think of.' In a short time he built up a considerable reputation in this specialized field. In 1936, shortly after the release of *Coal Face*, a pictorial survey of the coal industry in Great Britain, directed by Grierson with music by Britten, Kurt London wrote:† 'It is astonishing to observe how, with the most scanty material, using only a piano and a speaking chorus, he can make us dispense gladly with realistic sounds. This stylization makes a much stronger impression than a normal musical accompaniment.'

As soon as it was realized that this young composer had a flair for occasional and incidental music, commissions for theatre and radio as well as further film work followed. Britten was prepared to oblige with every type of music, light or serious; and the successful undertaking of these commissions

---

* From *The Composer and the Listener*, op. cit.
† *Film Music* by Kurt London, Faber and Faber, 1936.

showed that he was a reliable business man, who could work quickly and to time, and make the best of the limitations of the particular medium he was writing for and of the resources (however modest) that had been placed at his disposal. These virtues always stood him in good stead. He was often able to complete the greater part of the composition of a new work in his mind so that (as was the case with Mozart) the act of committing it to paper became an almost mechanical process which could be carried out at high speed. There was the added advantage that this enabled him to plan his musical life in advance so that he knew with reasonable certainty when he would be free to compose and by what date he could promise delivery of a new work. Reliability, expeditiousness and an unfailing capacity for hard work: these are formidable assets for an artist to have. When they are allied to an instinctive talent of no mean order, the result is bound to be phenomenal.

II
Collaboration with W. H. Auden

Four years before Britten entered Gresham's School, Holt, W. H. Auden, one of the older boys at the same school, published his first poem, *Woods in Rain*,* a set of octosyllabic couplets in the conventional Georgian style. In 1926 he became co-editor of *Oxford Poetry*; and his undergraduate poems, which had wide currency at the University, showed him at that time to be under the influence of masters like Rainer Maria Rilke and T. S. Eliot. By 1930, when his *Poems* were published by Faber and Faber, he had found his distinctive style.

Poems has rightly been hailed as a landmark of modern poetry. It is also a prophetic book, alive with a premonition of the doom that was so soon to bring disaster to the world. These thirty poems are imbued with a sense of struggle: war between classes, between parties, between members—war between life and death. But there is never any doubt what action has to be taken:

> . . . *never serious misgiving*
> *Occurred to anyone,*
> *Since there could be no question of living*
> *If we did not win.*†

In fact, more important still, the danger is seen to be actual not potential, and the poet analyses the new technique of total warfare with uncanny skill.

> *This is the dragon's day, the devourer's;*
> *Orders are given to the enemy for a time*
> *With underground proliferation of mould,*
> *With constant whisper and the casual question,*
> *To haunt the poisoned in his shunned house,*
> *To destroy the efflorescence of the flesh,*
> *The intricate play of the mind, to enforce*

---

* Published in *Public School Verse*, Vol. IV, 1923–4 (Heinemann), where the author's name was misprinted 'W. H. Arden'.
† *Poems* XII.

28

Conformity with the orthodox bone,
*With organized fear, the articulated skeleton.**

The political awareness of Auden and of several of his contemporary poet friends, including C. Day Lewis, Louis MacNeice and Stephen Spender, was directed during the next few years to the struggle against Fascism in its various manifestations; and their attention was naturally focused on the battleground in Spain, for (as Auden wrote)

> *... the time is short, and*
> *History to the defeated*
> *May say Alas but cannot help nor pardon.*†

Auden was not content, however, merely to write and publish poems. He looked towards stage and screen as being possibly more persuasive pulpits than the printed page. Remembering the didactic force of Bertolt Brecht's 'epic' drama, examples of which (like *Die Dreigroschenoper* and *Mahagonny*) he had seen performed in Germany at the beginning of the thirties, he began to experiment with social and political charades of his own, cast in the form of verse masques or plays. In *The Dance of Death*, which was published in 1933 and produced two years later by the Group Theatre at the Westminster Theatre, London, he aimed at presenting 'a picture of the decline of a class, of how its members dream of a new life, but secretly desire for the old, for there is death in them'. In 1935 he wrote his first full-length verse play, *The Dog beneath the Skin, or Where is Francis?* in collaboration with Christopher Isherwood. As its alternative title implies, it is a quest drama—a search for an heir—and it was produced by the Group Theatre at the Westminster Theatre in January 1936.

Meanwhile, his desire to work for the cinema had led him to approach the GPO Film Unit and ask his friend Basil Wright whether there was any way he could be employed in documentary films. Grierson was delighted to enlist his help; and he was forthwith engaged to write scripts for two films, *Coal Face* and *Night Mail*, that the Unit had in production. As Britten had been commissioned to provide the music for both films, it was necessary to arrange a meeting between the two collaborators. This took place on 5 July 1935, when Basil Wright drove Britten down to Colwall, near Malvern, where Auden was working as a master at a boys' preparatory school called 'The Downs'.

'*Coal Face*' (wrote Basil Wright)‡ 'was a pure experiment with the sound track. Its success as a film was not great, but without it the big success of *Night Mail* could not have been achieved. In *Coal Face* (which was devised and made by Cavalcanti), Auden and Britten used for the first time the

---

* *Poems* XVI.
† From *Spain*, 1937, reprinted in *Another Time*, Faber and Faber, 1940.
‡ From an unpublished letter to E. W. White dated 1 April 1948.

spoken voice reciting from official reports of mine disasters and from lists of coal-mining job-names—in rhythm, sometimes unaccompanied, and sometimes with percussion. *Coal Face* also contained the first musical setting by Britten of words by Auden: the poem beginning "*Oh lurcher-loving collier, black as night*" which was specially written for the film and set for female voices.' It was in *Coal Face*, too, that he had the ingenious idea of accompanying a train going through a tunnel and approaching nearer and nearer by reversing the sound track of a recorded cymbal clash. For *Night Mail*, Britten wrote a special instrumental score; and Basil Wright recalled how when 'the closing music of the film turned out to be too long, he made some fantastically ingenious excisions from the sound track itself'. After *Night Mail*, there was some talk of another film with which Auden and Britten were to be jointly concerned—an elaborate experiment about the negro in Western civilization, *Negroes* or *God's Chillun*—but it was never released. The only other films they were engaged on together were *Calendar of the Year* (1936) and *The Way to the Sea* (1937), the latter being a Strand Film Company documentary which described the electrification of the Portsmouth railway line, and for which Auden wrote a special end-commentary.

On 25 November 1935 Britten signed a contract with the firm of Boosey & Hawkes, the music publishers. Shortly afterwards when the miniature score of his *Sinfonietta* was published, he had a copy specially bound and, on the day it came back from the binders, showed it to Auden who, being on the point of leaving for Spain, wrote in pencil on the flyleaf his poem beginning—

*It's farewell to the drawing-room's civilised cry**

This collaboration, so auspiciously begun, soon developed outside the film world. In 1936 Britten was invited to compose a work for the Norfolk and Norwich Triennial Festival and asked Auden to devise a libretto. Auden chose man's relations to animals as his subject, selecting three poems to illustrate animals as pests, pets and prey—the first an anonymous prayer for deliverance from rats, the second an anonymous dirge on the death of a monkey entitled *Messalina*, the third *Hawking for the Partridge* by T. Ravenscroft—and framed them with an original prologue and epilogue of his own. The title of this symphonic cycle for high voice and orchestra was derived from the opening of the epilogue:†

Our hunting fathers told the story
Of the sadness of the creatures,

---

* This poem was first printed in *The Listener*, 7 February 1937, under the title 'Song for the New Year', and reprinted in *Another Time*, 1940.
† Reprinted in *Look, Stranger!* by W. H. Auden, 1936.

Pitied the limits and the lack
Set in their finished features . . .

Auden did not completely succeed in disciplining his material, and Britten cannot have had an easy task to set it. The resulting score was in places satirical, poignant and savage. According to Scott Goddard,* *Our Hunting Fathers* on its first performance at Norwich on 25 September 1936 'amused the sophisticated, scandalized those among the gentry who caught Auden's words, and left musicians dazzled at so much talent, uneasy that it should be expended on so arid a subject, not knowing whether to consider Britten's daring style as the outcome of courage or foolhardiness'. As for Frank Bridge,† he didn't really like the work, but defended it warmly. After the first performance he gave the young composer a long talking to about the scoring, which he thought didn't work, though he approved of the approach to the individual instruments. He was severe on the last movement 'as being too edgy'; and in the end Britten revised it.

In 1937 came the first of Britten's song cycles, *On This Island*. For this he chose five lyrics by Auden, of which four came from *Look, Stranger!* and the fifth (Nocturne) was extracted from *The Dog beneath the Skin*. This selection was to have been the first of two or more collections with the common title *On This Island*: but in the event only the first set was completed. Here Britten probably showed sound judgement, for the specific gravity of these lyrics written by Auden *dans l'an trentième de son age*, with their mixture of romance, neurosis and satire, do not call for treatment on too extended a scale. When it first appeared, *On This Island* had an exciting quality of contemporaneity—it was the product of two young minds thinking along related lines and working to a common purpose—and although it has subsequently been rather overshadowed by the later song-cycles, it is never likely to lose its special attraction. Here, too, Britten benefited from Frank Bridge's criticism. It appears that the first song originally began with a downward *glissando* on the piano. As Britten said,‡ 'Bridge hated that, and said I was trying to make a side-drum or something non-tonal out of the instrument: on the piano, the gesture ought to be a musical one.' So he rewrote it as a downward D major arpeggio.

After *The Dog beneath the Skin*, two further verse plays came from Auden's collaboration with Isherwood. The Group Theatre produced *The Ascent of F6* at the Mercury Theatre, London (26 February 1937) and *On the Frontier* at the Arts Theatre, Cambridge (14 November 1938). To both of these Britten wrote incidental music scored for piano (four hands) and percussion; and in the case of *On the Frontier* he added parts for two trumpets. Other Group Theatre productions at the Westminster Theatre

---

* From *British Music of Our Time*, edited by A. L. Bacharach, Pelican Books, 1946.
† See 'Britten Looking Back', op. cit.   ‡ Ibid.

31

for which Britten provided music were *Timon of Athens* (November 1935), the *Agamemnon* of Aeschylus in Louis MacNeice's translation (November 1936), and MacNeice's verse play *Out of the Picture* (December 1937).

Early in 1938 the collaboration between Auden and Britten was interrupted by a journey Auden and Isherwood made to the Far East to report on the Sino-Japanese War. At a farewell party in London for the two writers, Hedli Anderson sang three of the Cabaret Songs that Britten had composed in 1937 to words by Auden.* The only other work by Britten and Auden belonging to 1937 was *Hadrian's Wall*, a BBC radio feature.

There was a further occasion, after Auden's return from China, when they worked together. A Festival of Music for the People was organized in London 'by musicians of the progressive movement in Britain'; and for its third concert at the Queen's Hall on 5 April 1939 Britten wrote a *Ballad of Heroes* to honour men of the British Battalion, International Brigade, who had fallen in Spain. This consisted of three movements: (1) Funeral March with words by Randall Swingler; (2) Scherzo (Dance of Death) with words by Auden, and (3) Recitative and Choral with words by both Auden and Swingler. The moral of the work was contained in the Epilogue (Funeral March):

> *To you we speak, you numberless Englishmen,*
> *To remind you of the greatness still among you*
> *Created by these men who go from your towns*
> *To fight for peace, for liberty and for you.*

Britten set the text for tenor (or soprano) solo, chorus and orchestra, with three extra trumpets to be played in a gallery and (later) off; and it was conducted on this occasion by Constant Lambert. The work is pervaded by a feeling of deep bitterness, for at that moment the Spanish Civil War was over, and it seemed as if the forces of reaction were triumphing everywhere in Europe.

There is no doubt that this friendship with Auden had a great effect on Britten. It confirmed his liking for occasional quips and quiddities. It introduced him to Auden's favourite death fixation—the obsession that the illness and death of an individual symbolizes the decay and dissolution of a class—and for a period he seemed to have adopted it so wholeheartedly that it became almost a rule for his more extended compositions to include a key movement entitled Dance of Death or Funeral March. Fortunately, however, he outgrew this idea before it had time to shrivel into a fixed and meaningless cliché. But his most valuable and lasting gains were probably a fuller sense of the artist's political responsibility, a deeper appreciation of

---

* Britten composed a fourth Cabaret Song, entitled *Calypso*, in America in 1939; and all four songs were posthumously published by Faber Music in 1980.

the beauties of poetry, and a growing awareness of the aesthetic problems involved in the alliance of words and music. Through Auden he was introduced, not only to the poetry of some of his contemporaries, but also to that of earlier writers like Donne and Rimbaud.

For his part, Auden always found Britten a wonderful person to work with; and in later years his opera librettos for Stravinsky and Henze doubtless benefited from this collaboration.

III

American Visit

Important though Britten's collaboration with Auden undoubtedly was, it would be wrong to imply that it occupied his working life during the last years of the thirties to the exclusion of other activities. During this period a number of other works were written and performed. In London the Lemare Concerts at the Mercury Theatre continued to feature his music. A *Te Deum* written specially for Maurice Vinden and the Choir of St Mark's, North Audley Street, London, was performed at the Mercury Theatre on 27 January 1936, and provoked a rather curious review from Constant Lambert, who wrote in *The Sunday Referee*: 'Mr Britten is, I admit, rather a problem to me. One cannot but admire his extremely mature and economical methods, yet the rather drab and penitential content of his music leaves me quite unmoved. At the same time he is the most outstanding talent of his generation and I would always go to hear any first performance of his.'

1936 was the year in which he went to Barcelona, where he played his Suite for Violin and Piano with Antonio Brosa at the ISCM Festival. On this occasion he met the young composer Lennox Berkeley, who had been living in Paris for some time, and found him to be a most congenial companion. Years later Berkeley recalled how they had explored the city together.*

> Our schedule was very full. Among the various sideshows for those attending the festival was an afternoon display of folk dancing, held in a small park called Mont Juic. Ben Britten and I went there together and were much taken with some of the tunes that accompanied the dances. Ben wrote them down on old envelopes and various pieces of paper which he produced from his pockets. From these jottings we eventually constructed the orchestral Suite which we called *Mont Juic*. Only later did we discover that Mont Juic was better known as the name of the prison, but despite the risk of seeming to suggest that we had been

* 'Views from Mont Juic' by Lennox Berkeley. *Tempo*, September 1973.

lodged there during our stay in Barcelona, we decided to keep to the title.

Neither of them realized that at that moment Spain was on the brink of civil war. When Britten returned to London, his Suite for Violin and Piano was announced for performance at a Lemare Concert on 1 February 1937, but was withdrawn on account of his mother's death the day before. (His father had died a few years previously.)

In 1937 he obtained his first popular success with the *Variations on a Theme of Frank Bridge*. He accepted an invitation in May of that year to write a new work for the Boyd Neel String Orchestra to play at the Salzburg Festival on 27 August; and Boyd Neel recalled* that in ten days' time he appeared 'with the complete work sketched out. In another four weeks it was fully scored for strings as it stands today, but for the addition of one bar'. The theme he chose came from the second of Frank Bridge's *Three Idylls* for string quartet (1911), and he wrote variations on it, entitled March, Romance, Aria Italiana, Bourrée Classique, Wiener Waltzer, Moto Perpetuo, Funeral March, Chant, and Fugue and Finale, which reveal to the full his resource and skill in dealing with a string orchestra. When some years later he was cross-examined about his preference for this medium, he replied:† 'I am attracted by the many features of the strings. For instance the possibilities of elaborate *divisi*—the effect of many voices of the same kind. There is also the infinite variety of colour—the use of mutes, pizzicato, harmonics and so forth. Then again, there is the great dexterity in technique of string players. Generally speaking, I like to think of the smaller combinations of players, and I deplore the tendency of present-day audiences to expect only the luscious "tutti" effect from an orchestra.'

The work caused a sensation at Salzburg and soon gained for its composer an international reputation. Within less than two years it had been played more than fifty times in various parts of Europe and America.

Late in 1937 he acquired a converted windmill in Snape in Suffolk, a few miles inland from Aldeburgh, and one of the first works he composed there was the Piano Concerto.

The Piano Concerto in D major was first performed at a Promenade Concert at the Queen's Hall on 18 August 1938, with the composer as soloist. In its original form, it comprised four movements: I. Toccata, II. Waltz, III. Recitative and Aria, IV. March. Britten himself explained that the work 'was conceived with the idea of exploiting the various important characteristics of the piano, such as its enormous compass, its percussive quality, and its suitability for figuration; so that it is not by any means a

---

* 'The String Orchestra' by Boyd Neel in *Benjamin Britten: a commentary* (edited by Mitchell and Keller). Rockliff, 1952.
† 'Conversation with Benjamin Britten.' *Tempo*, February 1944.

symphony with piano, but rather a *bravura* concerto with orchestral accompaniment.' Eight years later, the third movement, which had given the impression of being a weak and rhetorical setting of an imaginary script, was withdrawn and an Impromptu (or, more accurately, an air with seven variations) substituted for it. For this new movement Britten used only material contemporary with the original work, drawing on incidental music to a BBC play *King Arthur* by T. H. White and using some of the figuration from the earlier movement.

The American composer, Aaron Copland, has described a happy visit to the Old Mill at Snape that summer.* He had with him the proofs of his school opera, *The Second Hurricane*, which he played over for Britten. In return Britten gave him a run-through of his recently completed Piano Concerto. Copland was 'immediately struck by the obvious flair for idiomatic piano writing in the Concerto, but had some reservations as to the substance of the musical materials'.

Meanwhile, the political scene in Europe was becoming increasingly sombre; and this naturally affected Britten. In 1937 he composed a *Pacifist March* to words by Ronald Duncan for the Peace Pledge Union, which had been founded by the Revd. Canon Dick Sheppard in 1934. The following year, he wrote incidental music for *Advance Democracy*, a documentary film directed by Ralph Bond in association with Basil Wright and produced by the Realist Film Unit in 1939. He also composed a part-song for unaccompanied mixed chorus with the same title and with words by Randall Swingler, but this was not used in the documentary film. It was followed by the *Ballad of Heroes* mentioned in the previous chapter. This, together with the incidental music for J. B. Priestley's play *Johnson over Jordan* and the radio adaptation of T. H. White's *The Sword in the Stone*, was the last music Britten wrote before leaving England for America.

This was a period of great depression and unrest. Rather than become helpless victims of a new Fascist or Nazi order with its attendant persecution and misery, many persons were looking for salvation to the New World and considering the possibility of emigration. A lead in this direction was given by Auden. He and Isherwood visited America early in 1939, and by the outbreak of war he had decided that only in the United States could he find the complete anonymity he needed if he was to break away from the European literary family and let his individual genius develop in full independence. Louis MacNeice quotes him as saying that 'an artist ought either to live where he has live roots or where he has no roots at all; that in England today the artist feels essentially lonely, twisted in dying roots, always in opposition to a group'.†

* From 'A Visit to Snape' by Aaron Copland. *Tribute to Benjamin Britten on his Fiftieth Birthday*, Faber, 1963.
† Letter from Louis MacNeice printed in *Horizon*, July 1940.

There were a number of reasons that led Britten also to leave England in May 1939. The darkening political situation was one. Then he was dissatisfied with the reception of his work in this country and had a growing sense of frustration as an artist, which he felt might be dissipated by a change of scene—in his own words,* he was 'muddled, fed-up and looking for work, longing to be used'. But the dominant factor was certainly Auden's personal example and his decision to become a citizen of the United States; and when Britten first reached America, it was his firm intention to do likewise.

Britten was accompanied on this voyage by his friend Peter Pears, who had been a member of the BBC Singers from 1936 to 1938 and had already toured America twice with the New English Singers. They went first to Canada and were then invited to New York to hear the first American performance of the *Variations on a Theme of Frank Bridge* by the New York Philharmonic on 21 August 1939. Eleven days later Auden was sitting in one of the dives on 52nd Street,

> *Uncertain and afraid*
> *As the clever hopes expire*
> *Of a low dishonest decade:*
> *Waves of anger and fear*
> *Circulate over the bright*
> *And darkened lands of the earth,*
> *Obsessing our private lives;*
> *The unmentionable odour of death*
> *Offends the September night.†*

For Britten and Pears there was a certain degree of consolation in the fact that Pears had sympathetic friends who offered them the hospitality of their home on Long Island. There was even a substitute 'home from home' atmosphere, for 'when, driving out to Amityville, Britten read the nostalgic word "Suffolk" on the signpost, he was delighted to be by the sea again and in his native county, although so far from home'.‡ He stayed there for the greater part of the next two and a half years, with the exception of a few months spent with Auden in Brooklyn early in 1940 and a visit to California in 1941.

The Violin Concerto in D minor was one of the first of his works to be written in America—it was finished at St Jovite in the Province of Quebec, Canada, in September 1939 and received its first performance (28 March 1940) by the New York Philharmonic under John Barbirolli, with Antonio

---

* *On Receiving the First Aspen Award*, Faber, 1964.
† From *Another Time*.
‡ From 'Benjamin Britten: Another Purcell', by Phoebe Douglas. *Town and Country*, December 1947.

Brosa as soloist. Although the violin part, which was edited by Brosa himself,* calls for great virtuosity from its executant, the Concerto is by no means an extended violin solo with orchestral accompaniment, but represents a real advance on the Piano Concerto in powers of construction. Another composition belonging to this period is *Canadian Carnival* (*Kermesse Canadienne*), a light-hearted frolic for symphony orchestra.

But more important than either of these was *Les Illuminations*, which was finished at Amityville on 25 October 1939. When this song cycle for high voice and string orchestra received its first complete performance† in London on 30 January 1940, it not only confirmed the uniformly favourable impression made by the *Variations on a Theme of Frank Bridge* two years previously, but also showed Britten to be a song-writer of exceptional range and subtlety. When Rimbaud wrote the prose poems in *Les Illuminations*, he was living in London with Verlaine, spending much of his time in the East End and the docks. Remembering the prophetic Latin tag *Tu Vates eris* that Apollo had inscribed on his brow when he was a schoolboy of fourteen, Rimbaud was arrogantly confident that through the stimulus of debauch and vice he could attain the power of supernatural vision. In poems like *Métropolitain*, *Villes*, and *Parade* (the last two of which are included in Britten's song cycle) he seemed to have discovered all the monstrous significance of a modern industrial capital. Contact with the fierce alchemy of these poems quickened a new nerve in Britten's musical sensibility. The words and music fused in a sudden startling outburst of heat and energy. '*J'ai seul la clef de cette parade sauvage.*' Britten had found the key as well as Rimbaud; and the door opened on to a hitherto unknown surrealist world.

The *Diversions* for piano and orchestra were written in Maine in the summer of 1940 for the one-armed Viennese pianist, Paul Wittgenstein, who reserved the sole rights of performance until 1951. This work consisted of 'eleven straightforward and concise variations on a simple musical scheme'; and in an introductory comment to the concerto, Britten said: 'I was attracted from the start by the problems involved in writing a work for this particular medium, especially as I was well acquainted with and extremely enthusiastic about Mr Wittgenstein's skill in overcoming what appear to be insuperable difficulties. In no place in the work did I attempt to imitate a two-handed piano technique, but concentrated on exploiting and emphasiz-ing the single line approach. I have tried to treat the problem in every aspect, as a glance at the list of movements will show: special features are trills and scales in the Recitative, widespread arpeggios in the Nocturne, agility over the keyboard in the Badinerie and Toccata, and repeated notes in the finale

---

* Britten later revised the solo part himself in 1958.

† 'Being Beauteous' and 'Marine' had already been performed at a Promenade concert in August 1939.

Tarantella.' The theme with its fourths shows a certain affinity with that of *The Turn of the Screw.*

The *Sinfonia da Requiem*, written about the same time, had a particularly strange history.

Having been approached through the British Council some time in 1940 and asked whether he would write a symphony for a special festivity connected with the reigning dynasty of a foreign power, Britten agreed in principle, provided it was understood no form of musical jingoism was called for. On further investigation, it appeared that the country in question was Japan and the festivity the 2,600th anniversary of the foundation of the Mikado's dynasty in 660 B.C. by Jimmu Tenno, and that other composers in France, Germany, Italy and Hungary had received similar commissions. In due course, the outline of a *Sinfonia da Requiem* in three movements— *Lacrymosa*, *Dies Irae*, and *Requiem Aeternam*—was submitted to the Japanese authorities and approved. Britten felt that this work, which is permeated with a sense of the terror and ghastliness of war, would not be inappropriate to the occasion in view of the Sino-Japanese conflict. He was wrong, however; and about six months after the completed score had been handed over, he received a furious protest through the Japanese Embassy, complaining that the Christian dogma and liturgical ceremony that lay at the basis of the work were a calculated insult to the Mikado, and rejecting the *Sinfonia* out of hand. With Auden's help he drafted a suitable reply; but shortly afterwards the Japanese attacked the Americans at Pearl Harbor, and thenceforth all communications were severed. The first performance of the *Sinfonia da Requiem* was given on 30 March 1941, by the New York Philharmonic under Barbirolli. There is no record of a performance in Tokyo, although it seems possible the work may have been put into rehearsal there sometime in 1941.

About this time Britten's interest in oriental music was stimulated by his meeting Colin McPhee; and the two musicians joined together in making gramophone records of some Balinese music which McPhee had transcribed for two pianos—an interesting anticipation of the oriental part of the score of *The Prince of the Pagodas.*

In the list of Britten's published works to which opus numbers were given, two gaps occurred between opus 15 (the Violin Concerto) and opus 18 (*Les Illuminations*). These were reserved for *Young Apollo*, a work for piano, string quartet and string orchestra written for the Canadian Broadcasting Corporation, first performed in Toronto in August 1939, but subsequently withdrawn; and a group of about half a dozen choral settings of poems of Gerard Manley Hopkins, which Britten held back from publication and performance as being not up to standard. A work for which no opus number seems to have been reserved was *Paul Bunyan*, written early in 1941. Thirty-five years later the score and text were revised and the work was given a stage

39

performance at the Maltings Concert Hall, Snape, in the summer of 1976; and when it was finally published, it was given the opus number 17. This choral operetta, in which Britten's partnership with Auden was resumed, was put on for a week's run at the Brander Matthews Hall, Columbia University, New York, in May 1941. Their treatment of this American legend came in for some hard knocks—Britten himself wrote a few years later 'the critics damned it unmercifully, but the public seemed to find something enjoyable in the performance'.*

Britten wrote incidental music for two Columbia Workshop broadcasts while he was in America: a monologue by Auden for Dame May Whitty called *The Dark Valley*, and Auden's adaptation of D. H. Lawrence's *The Rocking Horse Winner*. He also executed various ballet commissions. *Soirées Musicales*, a suite of five movements that he had adapted from Rossini in 1935 for the use of the GPO Film Unit had already been used by Antony Tudor to accompany a ballet called *Soirée Musicale* that was produced by the London Ballet at the Palladium, London, on 26 November 1938. When in 1941 Lincoln Kirstein wanted a new ballet for the South American tour of his American Ballet, Britten composed another suite after Rossini called *Matinées Musicales*, joined this to his *Soirées Musicales* suite, and added the Overture to *La Cenerentola* as a finale. The resulting ballet was called *Divertimento*, and its choreography was composed by Balanchine. About the same time Britten made a new orchestral version of *Les Sylphides* for Ballet Presentations Inc. (Ballet Theater), New York City. There was some talk of his writing a ballet for Eugene Loring and his Dance Players to be called *The Invisible Wife*; but in the end that project came to nothing. Instead, he allowed his *Variations on a Theme of Frank Bridge* to be used by Dance Players as an accompaniment to *Jinx*, a ballet about circus people and their superstitions with choreography by Lew Christensen. This was first performed in 1942 at the National Theatre, New York.

He visited Ethel and Rae Robertson in California in the summer of 1941, and during his stay at Escondido wrote various two-piano compositions for them: the *Introduction and Rondo alla Burlesca*, the *Mazurka Elegiaca* in memory of Paderewski,† and the *Scottish Ballad* for two pianos and orchestra. At the same time he was busy on his String Quartet No. 1, which had been commissioned by that Great American patroness of music, Elizabeth Sprague Coolidge, and was performed for the first time by the Coolidge String Quartet in Los Angeles in September 1941. Later that year it won him the Library of Congress Medal for services to Chamber Music.

But, meanwhile, what had happened to his intention of seeking

---

* *Peter Grimes: Sadler's Wells Opera Books No. 3.*
† This was originally intended for publication in a memorial volume for Paderewski, but it was not included, as it had been mistakenly written for two pianos instead of one as specified.

naturalization as an American citizen? When the Second World War broke out in September 1939, he realized that, had he still been in England, scruples of conscience would have prevented him from becoming a combatant; but he could not help wondering whether, should he decide to return, there might be ways in which his services as a non-combatant might be useful. The progress of the war aggravated his mood of indecision. Louis MacNeice, who had numerous discussions with Auden in the autumn of 1940, wrote: 'For the expatriate there is no Categorical Imperative bidding him return—or stay. Auden, for example, working eight hours a day in New York, is getting somewhere; it might well be "wrong" for him to return. For another artist who felt he was getting nowhere it might be "right" to return.'*

In Britten's case, the mental struggle whether to stay in America or return to England was echoed by a physical illness. He suffered from an acute streptococcal infection during the whole of 1940. (It was perhaps typical of Auden that he should claim that this illness was nothing more than the physical expression of Britten's psychological indecision.) But as he recovered the following year, his course seemed crystal clear. He would not become a naturalized American, but would go back to England. His decision to return was confirmed by a strange incident that was destined to have important repercussions on his future career. During his stay in California he happened to pick up a copy of the BBC's weekly magazine, *The Listener*, for 29 May 1941, and there his eyes were caught by the opening words of an article by E. M. Forster: 'To talk about Crabbe is to talk about England.' Forster was discussing George Crabbe, the East Anglian poet of the late eighteenth and early nineteenth centuries, who had been born at Aldeburgh, not far from Lowestoft. Reading how this bleak little fishing village on the Suffolk coast 'huddles round a flint-towered church and sprawls down to the North Sea—and what a wallop the sea makes as it pounds at the shingle!' Britten realized with a twinge of homesickness that he must not only familiarize himself with Crabbe's poems (which at that time he did not know), but also get back to his native Suffolk as quickly as possible.

But it was not easy to cross the Atlantic at that stage of the war; and he and Peter Pears were kept waiting for nearly six months on the East Coast of the United States before they could obtain a passage in March 1942. The delay had certain compensations, however. It meant he was able to attend a performance of the *Sinfonia da Requiem* under Serge Koussevitzky in Boston. When in the course of conversation the conductor asked him why he had as yet written no full-scale opera, Britten, who had already been turning over in his mind the possibility of quarrying material for an opera out of Crabbe's poem *The Borough*, replied that 'the construction of a scenario,

* From 'Traveller's Return'. *Horizon*, February 1941.

discussions with a librettist, planning the musical architecture, composing preliminary sketches, and writing nearly a thousand pages of orchestral score, demanded a freedom from other work which was an economic impossibility for most young composers'.* Koussevitzky was interested in the Crabbe project and, when they met again some weeks later, announced that he had arranged for the Koussevitzky Music Foundation to put up $1,000 for the opera, which was to be dedicated to the memory of his wife, Natalie, who had recently died.

* *Peter Grimes: Sadler's Wells Opera Books No. 3.*

IV
In Wartime England

In the spring of 1942, the fortunes of Great Britain and her allies seemed to be at their nadir. Their disasters in Russia, the Far East and North Africa were only too apparent, and the pattern of recovery as yet unrevealed. Nevertheless, their faith in ultimate victory was completely unshaken.

Britten returned to a land of black-out, material privations and total mobilization. As a pacifist by conviction, he appeared before a tribunal; but in view of his conscientious objections, he was exempted from active service. He was allowed to continue with his work as a composer, and as a pianist he appeared at the special wartime concerts organized all over the country by the Council for the Encouragement of Music and the Arts (CEMA), which later grew into the Arts Council of Great Britain. Then, as always, he laid great store by his work as a musical executant. In an interview given early in 1944* he said: 'I find it valuable for my activities as a composer to see how listeners react to the music. I also enjoy rehearsals—especially if I am working with sympathetic and intelligent musicians—delving deeper and deeper into the great music of all ages, and learning a lot from it. There are some composers whose music I do not like, but performing it makes me analyse my reasons for the dislike, and so prevents it from becoming just habit or prejudice.'

He had left America on 16 March 1942, sailing in a small Swedish cargo boat, and by the time he reached England after being mewed up for more than a month in a tiny cabin next to the refrigerating plant, he had written two new choral works, the *Hymn to St Cecilia* and *A Ceremony of Carols*. In addition, he brought over from America a setting of *Seven Sonnets of Michelangelo* for tenor and piano, and a number of new works written and performed in the United States, but not yet heard in Great Britain. These were considerable assets. In wartime England, opportunities for hearing new music were all the more welcome because of their infrequency; and Britten's new compositions met with ready acceptance, partly because of their obviously attractive qualities of ease, grace and intelligibility, and

* 'Conversation with Benjamin Britten.' *Tempo*, February 1944.

43

partly also because it was immediately apparent that during his three years' residence in the United States his mind and music had strikingly matured.

Out of Michelangelo's seventy-seven sonnets, Britten chose numbers XVI, XXXI, XXX, LV, XXXVIII, XXXII and XXIV and set them in a fine flowing *bel canto* style. The pure and open vowels of the Renaissance Italian are adequately matched by the extended gesture of the vocal line; and each sonnet stands forth clearly, three-dimensionally, in its setting. The various songs are well contrasted with each other in tempo, mood and key.

These *Sonnets* were written in 1940 specially for Peter Pears and were an immediate success on their first performance by him with the composer at the piano (Wigmore Hall, September, 1942). The two performers had given their first concert together at Oxford in 1937; and they now decided to form a permanent tenor/pianist partnership. Furthermore, most of Britten's subsequent song-cycles, solo canticles, and also certain parts in the operas, were to be composed with Peter Pears's voice in mind, with its masterly style, clear diction, and unique vocal quality.

The *Hymn to St Cecilia* was intended to help restore the old custom of celebrating on 22 November (Britten's birthday) the feast of the patron saint of music—a custom that had been regularly observed in former centuries. Auden's invocation to the saint fully deserves to take its place among the odes of Nicholas Brady and John Dryden as set by Purcell and Handel respectively:

> *Blessed Cecilia, appear in visions*
> *To all musicians, appear and inspire;*
> *Translated Daughter, come down and startle*
> *Composing mortals with immortal fire.*

And listening mortals were certainly startled when on St Cecilia's Day, 1942, the *Hymn* was performed for the first time by the BBC Singers conducted by Leslie Woodgate. The work as a whole made an impression of simplicity, delicacy, sweetness and tranquillity—qualities that were not altogether expected from a composer who before his visit to the United States had acquired a reputation for precocious sophistication. The sweet and languid harmonies of the opening invocation, the brisk scherzo ('*I cannot grow*') with its light broken movement, the candid beauty of the soprano solo, '*O dear white children casual as birds*', and the group of gay cadenzas where the voices flamboyantly imitate violin, drum, flute and trumpet in turn—all these episodes combined to form a work that gave great pleasure and satisfaction to listeners and performers alike.

A Ceremony of Carols was first performed by the Fleet Street Choir, conducted by T. B. Lawrence, in Norwich Castle on 5 December 1942. In some ways it is a pendant to *A Boy was Born*; but whereas the earlier work was in the form of a theme followed by six variations and finale for mixed

44

voices unaccompanied, *A Ceremony of Carols*, which was written for treble voices (generally divided into three parts with occasional solo passages) accompanied by harp, consists of a sequence of nine carols framed by a Procession and Recession with an interlude for harp solo placed about two-thirds of the way through. As with the earlier work, the words are drawn mainly from anonymous medieval carols; but there are also settings of poems by James, John and Robert Wedderburn, Robert Southwell and William Cornish.

Whereas formerly the Church was a powerful patron of the arts, in recent years the divorce between it and the artist has become unfortunately accentuated—to the detriment of both parties. It is always pleasant to find an exception—a church that recognizes the importance of the living artist and is prepared to employ him and display his work; and in the early 1940s a striking example of enlightened patronage in England was provided by the Church of St Matthew, Northampton. The commissions placed through its then incumbent, the Revd. Walter Hussey, included a statue of the Madonna and Child by Henry Moore, a mural painting of the Crucifixion by Graham Sutherland, and a Festival Cantata written by Britten to commemorate the fiftieth anniversary (on 21 September 1943) of the Church's consecration.

For this Cantata, Britten selected a number of passages from *Rejoice in the Lamb*, that strange eighteenth-century poem written by Christopher Smart, while he was in a madhouse. This work is a canticle of general praise, in which not only 'nations and languages', but also 'every creature in which is the breath of life', including the poet's favourite cat, Jeoffry, and the mouse 'a creature of great personal valour', unite in praising God and rejoicing in their service:

> *Hallelujah from the heart of God*
> *And from the hand of the artist inimitable*
> *And from the echo of the heavenly harp*
> *In sweetness magnifical and mighty.*

The music matches the thrilling visionary quality of the words.

Three years later the same church commissioned a work for organ; and Britten composed the *Prelude and Fugue on a Theme of Vittoria*.

Other commissions led to the composition of a small-scale *Te Deum* to commemorate the Centenary Festival of the Church of St Mark, Swindon (24 April 1945) and the *Prelude and Fugue* for 18-part string orchestra, written specially for the tenth anniversary of the Boyd Neel Orchestra and performed at the Wigmore Hall on 23 June 1943.

In 1944 his earlier work for strings, *Simple Symphony*, took on a new lease of life when it was used by Walter Gore (at the suggestion of David Martin) for a light essay in abstract dancing; and it was first performed in this guise by the Ballet Rambert at the Theatre Royal, Bristol (19 November 1944).

45

Of special importance was the *Serenade* for tenor solo, horn and string orchestra (first performed by Peter Pears and Dennis Brain with Walter Goehr and his orchestra at the Wigmore Hall on 15 October 1943). The *Serenade* was written to almost the same scale as *Les Illuminations*; but this time Britten chose an English instead of a French text, and Rimbaud's fragmentary, half-apprehended visions gave way to an exquisitely selected miniature anthology that included the Lyke Wake Dirge, lyrics by Cotton, Tennyson, Blake and Ben Jonson, and as finale a sonnet by Keats. In the words of Edward Sackville-West,* to whom the *Serenade* is dedicated: 'The subject is Night and its prestigia: the lengthening shadow, the distant bugle at sunset, the Baroque panoply of the starry sky, the heavy angels of sleep; but also the cloak of evil—the worm in the heart of the rose, the sense of sin in the heart of man. The whole sequence forms an Elegy or Nocturnal (as Donne would have called it), resuming the thoughts and images suitable to evening.' Just as *A Ceremony of Carols* was framed by a Procession and Recession, the *Serenade* has a horn solo that is played on natural harmonics as Prologue and repeated off-stage as Epilogue.

Simultaneously with his work on the *Serenade*, Britten found time to collaborate with Edward Sackville-West, who was writing a melodrama for broadcasting based on Homer's *Odyssey* and entitled *The Rescue*. In the Preamble to the published text, the author explains that he was experimenting with a form of radio-opera or radio-drama and 'in writing *The Rescue* some of the awkwardnesses incident to radio-drama were automatically removed ... by the operatic nature of the composition, which was deliberately built upon an hypothetical structure of music'.† This meant that the composer was given many opportunities, which included passages of speech-music, instrumental solos associated with various characters, a vocal quartet of gods and goddesses, and brief transitional passages for orchestra between the scenes. Here again Britten's flair for descriptive music did not desert him. Describing the course of their collaboration, Sackville-West paid him this tribute: 'I was continually struck by the unerring instinct with which Britten hit upon the right musical backing for whatever it was I had written, or—alternatively—rose imaginatively to any occasion the script presented for quasi-independent music.' *The Rescue* was first performed by the BBC in two parts on 25 and 26 November 1943.

From this period dated his friendship with Michael Tippett, then musical director at Morley College, Lambeth. Recalling the occasion of their first meeting, Tippett wrote:‡ 'We wanted a tenor soloist for the Gibbons Verse

* 'Music: Some Aspects of the Contemporary Problem' by Edward Sackville-West. *Horizon*, June, July and August 1944.

† *The Rescue*: a melodrama for broadcasting based on Homer's Odyssey, by Edward Sackville-West. Secker & Warburg, 1945.

‡ 'Starting to know Britten' by Michael Tippett. The *London Magazine*, October 1963.

46

Anthem *My Beloved Spake*. Walter Bergmann, who was then chorus-master in Morley, suggested Peter Pears, recently returned from America. When Pears came to rehearsal, Britten came with him.' Tippett was so impressed by Pears and Britten as musical artists that he wrote his vocal cantata, *Boyhood's End*, for them as a duo; and this was performed by them at Morley College in June 1943. A fortnight later Tippett, who was also a conscientious objector, was sentenced to three months imprisonment in Wormwood Scrubs for failing to comply with the conditions of his exemption. While he was serving this sentence, it so happened that Pears and Britten gave a concert in the gaol. The page-turner on this occasion was John Amis, who later described* how the prison authorities were 'bamboozled into thinking that the music to be performed was so complicated that it required the expert services of Michael on the platform to help me to turn over the pages for Britten'.

Shortly after Tippett's release from prison, when Britten asked him what larger works he had written, Tippett mentioned *A Child of Our Time*, an oratorio which he had finished early in the war, but which, on the advice of Walter Goehr, he had put aside in a drawer for the time being. Britten examined the score. In one of the Spirituals he suggested 'the effect could be greatly enhanced by lifting the tenor part suddenly an octave higher†—an improvement that the composer immediately adopted. Britten's enthusiasm did much to help bring about the first performance of *A Child of Our Time* at the Adelphi Theatre, London, on 19 March 1944.

By this time Britten had found a librettist for his opera and was hard at work on the score. While still in the United States, he had thought of asking Christopher Isherwood to collaborate with him; but after his return to England it was clear the choice would have to fall on someone living in England. He already knew Montagu Slater as a poet and dramatist—in fact, he had written incidental music for two of Slater's Left Theatre productions, *Easter 1916* (1935) and *Stay Down Miner* (1936), and for his two one-act verse plays, *The Seven Ages of Man* and *Old Spain*, which were performed in the summer of 1938 as puppet plays at the Mercury Theatre, London, by the Binyon Puppets—and now he asked him to prepare a libretto, based on Crabbe's poem, *The Borough*, which would give special prominence to the character of Peter Grimes. The writing of the text, together with its revisions and corrections, took about eighteen months, and by the end of 1943 all was ready for Britten to begin the work of composition.

Before settling down to this long and arduous job, however, he wrote a special work at the request of an old friend, Lieutenant Richard Wood, who

---

* From 'Wartime Morley' by John Amis in *Michael Tippett*, Faber, 1965.
† From 'Starting to know Britten' by Michael Tippett, op. cit.

was in a prisoner-of-war camp at Eichstätt in Germany. It was to be written for the prisoners' musical festival early in the new year. According to Richard Wood, who organized the festival: '*The Ballad of Little Musgrave and Lady Barnard* for male voices and piano by Benjamin Britten and dedicated to the musicians of OFLAG VII B arrived just in time for us to put it into the programme at the end of the festival (which had started on 18 February 1944). Our resources and capabilities were brilliantly envisioned by the composer and the result was a little work (eight to nine minutes) of great dramatic force. The choir enjoyed singing it enormously in the end, though it was quite foreign to their style. We gave it four times.'

Britten began composing the music of *Peter Grimes* in January 1944; and the score was completed by February of the following year. The question of its production in England was naturally one that exercised his mind. Although opera in Italian had formed a regular part of London's musical diet since the beginning of the eighteenth century, opera in English had never had a real opportunity of taking root. Before the war, it was customary for some special organization or syndicate to arrange an annual eight to ten weeks' season of opera in German, Italian, and sometimes French, at Covent Garden, while it was left to a company like the Carl Rosa (founded in 1875) to tour English versions of the more popular operas through the provinces. During this period a kind of English *Volksoper* was growing up, first at the Old Vic Theatre in the Waterloo Road, London, and after 1931 at Sadler's Wells. This company presented not only operas in English, but also English operas; and before the outbreak of the war it had to its credit productions of Stanford's *The Travelling Companion*, Dame Ethel Smyth's *The Wreckers*, Holst's *Savitri* and Vaughan Williams's *Hugh the Drover* among others. Although after the onset of the German air attacks on London in September 1940 the Sadler's Wells Theatre was closed to the public and used as a rest-centre for evacuees, the opera company just managed to avoid complete disintegration. At first it toured the provinces with a scratch company of twenty-five (including the orchestra) playing simplified versions of operas like *The Marriage of Figaro* and *La Traviata*; but soon, emboldened by the eager response it evoked, it began to build up its singers, chorus and orchestra and to enlarge its repertory. At the time Britten and Pears returned to England from America, the Sadler's Wells Opera Company under the direction of Joan Cross had made the New Theatre, London, its headquarters; and in 1943 Peter Pears joined it and was soon singing leading parts in *Così fan tutte*, *La Traviata*, *The Bartered Bride* and other operas.

In February 1944, in the course of an interview, Britten said:* 'I am passionately interested in seeing a successful permanent national opera in

---

* 'Conversation with Benjamin Britten.' *Tempo*, February 1944.

existence—successful both artistically and materially. And it must be vital and contemporary, too, and depend less on imported "stars" than on a first-rate, young and fresh, permanent company. Sadler's Wells have made a good beginning.' A few months later, thanks largely to the enthusiastic support of the Opera Director, Joan Cross, it was agreed that *Peter Grimes* when completed should be given its first performance by the Sadler's Wells Company and that this production should mark the Company's return to its own theatre, which had now been derequisitioned. The certainty of production—and at an early date—was a fresh incentive to Britten (if one were needed) to finish the score; and, as he later admitted,* 'the qualities of the Opera Company considerably influenced both the shape and the characterization of the opera'.

Rehearsals began on tour—according to Eric Crozier† 'in a Methodist Hall in Sheffield, in a Birmingham gymnasium and in the Civic Hall at Wolverhampton. They were as thorough as circumstances could allow, with a company exhausted by much travel and busy with eight performances each week of other operas'. But, despite all difficulties, the production was ready on time; and so it came about that on 7 June 1945, a month after the capitulation of Germany, Sadler's Wells Theatre reopened with a world first performance of a new English opera. Peter Pears sang the title role; Joan Cross herself appeared as Ellen Orford; the parts of Auntie, Balstrode, Mrs Sedley and Swallow were taken by Edith Coates, Roderick Jones, Valetta Iacopi and Owen Brannigan respectively. The scenery and costumes by Kenneth Green were in an attractively realistic style; Eric Crozier produced; and the orchestra was conducted by Reginald Goodall.‡

Any apprehensions that may have been felt beforehand by the composer, by those concerned in the production, by executants or audience, were all swept away as the actual performance proceeded. The orchestra might be too small to do full justice to the interludes, the stage space too cramped for the full sweep of the action, the idiom of the music unfamiliar, the principals and chorus under-rehearsed, yet the impact of the work was so powerful that when the final chorus reached its climax and the curtain began to fall slowly, signifying not only the end of the opera but also the beginning of another day in the life of the Borough, all who were present realized that *Peter Grimes*, as well as being a masterpiece of its kind, marked the beginning of an operatic career of great promise and perhaps also the dawn of a new period when English opera would flourish in its own right.

---

* *Peter Grimes: Sadler's Wells Opera Books No. 3.*
† *The Rape of Lucretia* (commemorative volume) ed. Eric Crozier. The Bodley Head, 1948.
‡ See Appendix B.

V

First Operatic Successes

The success of *Peter Grimes* was immediate and decisive. The London production was followed during the next three years by others in Stockholm, Basle, Antwerp, Zurich, Tanglewood, Milan, Hamburg, Mannheim, Berlin, Brno, Graz, Copenhagen, Budapest, New York, Stanford and Oldenburg, so that within a comparatively short time Britten's fame as an opera-composer was world-wide, and it became a matter of general interest to know what sort of an opera he was going to write next.

With the end of the war, however, the unstable operatic picture in Great Britain continued to shift and change. Perhaps the most important new factor was the avowed intention of a number of public-spirited persons to reopen the Royal Opera House, Covent Garden, which had been used as a dance-hall during the war years, and to run it as a national lyric theatre for opera and ballet. Messrs Boosey & Hawkes, the music publishers, took a lease of the building from the ground landlords and sub-let the theatre to a special Trust that was set up with Lord Keynes as its first Chairman. Financial backing from the State, which had never before shown any real interest in opera, was secured through the Arts Council; and immediate steps were taken to obtain resident opera and ballet companies. The Sadler's Wells Ballet, which had been created by Ninette de Valois and built up carefully during the last fifteen years from quite humble beginnings, accepted an invitation from the Covent Garden Trust to become its resident ballet company and gave a magnificent performance of Tchaikovsky's *Sleeping Beauty* at the gala reopening on 20 February 1946: but as no existing English opera company seemed suitable for transfer to Covent Garden, the Trustees decided to form their own company from scratch. Clearly it would take some time to assemble and train the singers; and, in actual fact, the new company's first performance was not given until January 1947.

Meanwhile, the general outlook was uncertain. Divided counsels in the management at Sadler's Wells led in March 1946 to a change of policy, 'the withdrawal of *Grimes* from the Wells' repertory at the composer's request, and the resignation of those who believed in the Wells as a progressive

50

centre for British opera.'* In Britten's own words:† 'To some of the singers, writers and musicians involved in *Peter Grimes* this appeared to be the moment to start a group dedicated to the creation of new works, performed with the least possible expense and capable of attracting new audiences by being toured all over the country.' Accordingly, a new company was planned with the object of providing opportunities for the composition and performance of works that would forego the apparatus of large orchestras and choruses. As was explained in a preliminary manifesto:‡ 'The practical aim behind the formation of the Glyndebourne English Opera Company is towards providing a method by which singers of the first rank can devote five months of each year between June and October—slack months in the concert world—entirely to the rehearsal and performance of opera.' The scale of the new venture was to be kept as small as possible—at least to begin with—since it was only thus that 'the principles of high quality in singing, musicianship and preparation can be reconciled with the regular performance of new works'. To speed this venture, Britten agreed to write a new opera for eight singers and twelve musicians, to be produced by the Company at Glyndebourne for a limited run in the summer of 1946 and then to be taken on tour to the provinces and to London.

It happened that a friend of his, Ronald Duncan, had just written a verse play, *This Way to the Tomb*, in the form of a masque with anti-masque—the first part showing how in the fourteenth century Father Antony, Abbot of St Farrara on the island of Zante, withdrew to a mountain height with the firm resolve of fasting unto death, but was there assailed by various temptations, including the deadly sin of pride, and the second part depicting a television relay from the saint's tomb some five centuries later and the unacceptable miracle of the saint's return to life. For its production at the Mercury Theatre on 11 October 1945, Britten wrote incidental music—liturgical chants for a four-part choir using the Latin words of Psalm 69 and a Franciscan hymn, two songs with piano accompaniment for one of the novitiates, and various bits of jazz for piano (four-handed) and percussion in the television scene. These numbers were so well contrived that it was legitimate to claim that the play owed a considerable part of its success to Britten's music; and it was not unexpected, therefore, when this collaboration was carried a stage further by Duncan being invited to write the libretto for Britten's next opera, *The Rape of Lucretia*, the subject of which had already been suggested by Eric Crozier.

* From 'Peter Grimes' by Eric Crozier, originally written for *Music and Letters* in 1946, but rejected by the editor, Eric Blom, as being too controversial, and finally printed in *Opera*, June 1965.

† Special programme note by Benjamin Britten for the Salzburg Festival performance of *The Rape of Lucretia*, 1950.

‡ 'Benjamin Britten's Second Opera' by Eric Crozier. *Tempo*, March 1946.

Meanwhile, Britten had celebrated the completion of the score of *Peter Grimes* by plunging into the composition of various new works. At the beginning of August 1945, just after his return from a tour of Belsen and other German concentration camps which he had undertaken as Yehudi Menuhin's accompanist, he set nine of the *Holy Sonnets of John Donne* for high voice and piano and followed this up by finishing his second String Quartet on 14 October, both works being written as an act of homage to commemorate the 250th anniversary of the death of Henry Purcell which fell on 21 November of that year.

The Holy Sonnets of John Donne stand nearer to the *Seven Sonnets of Michelangelo* than to the Keats Sonnet in the *Serenade*; and over and above their Baroque display of rhetoric, they reveal a facility for musical conceits that not only closely matches the sombre metaphysical imagery of Donne's poetry with its agony of repentance in the shadow of death, but also recalls the fanciful and unorthodox side of Purcell's genius as well. Clearly, Britten's work as concert pianist and accompanist since his return from America had helped to familiarize him with the music of Purcell; and it might be said of him that, like Purcell, he never failed to 'find for words a music that exists in its own right as music'. This tribute is particularly true of the last Sonnet in the sequence, '*Death be not proud*' which, constructed as a *passacaglia* with a firm muscular ground bass, successfully embraces extremes, being both simple and fanciful, sensuous and austere, a masterpiece of feeling and form.

The second String Quartet was Britten's most ambitious attempt since the Piano and Violin Concertos to write a work in which form would be dictated, not by extraneous ideas, but by inner musical necessity. The result was a most impressive essay in sonata form expressed in terms of contemporary idiom.

Britten's interest in Purcell was by no means transitory. Shortly after the commemoration concert of 21 November 1945, he planned jointly with Peter Pears a performing edition of Purcell's works, for which he realized the figured bass with characteristic ingenuity and invention. The first volumes of this edition started to appear in 1946 and included the Golden Sonata and selections from *Orpheus Britannicus*, *Odes and Elegies* and Playford's *Harmonia Sacra*. He also chose a theme of Purcell's for the air on which he constructed a set of variations to accompany a Ministry of Education film on the instruments of the orchestra; and a new performing edition of *Dido and Aeneas* was planned and advertised in the 1946 Glyndebourne programme for production there the following year, though in fact the project did not materialize until 1951.

The sound track for the educational film entitled *Instruments of the Orchestra* was designed to show the orchestra as a whole, its four departments and the individual instruments; but it is written in such a way

Britten with his Rolls-Royce, outside the Old Mill, Snape, 1946

that it could be played as a concert piece as well, with or without spoken commentary. It was planned in the form of variations and a fugue on a dance tune from Purcell's incidental music to *Abdelazar, or The Moor's Revenge*. Every point in this lucid musical exposition was made with such directness and precision that the work became a kind of standard vade-mecum for the young listener.

Before settling down to the task of composing *The Rape of Lucretia*, he allowed himself the relaxation of writing incidental music to *The Dark Tower*, a radio parable play by Louis MacNeice inspired by Robert Browning's poem, '*Childe Roland to the Dark Tower Came*'. This was first broadcast in the BBC Home Service on 21 January 1946, with Britten's music played by a string orchestra of twenty-six players, together with percussion and one trumpet. The author's attitude to the composer's score was enthusiastic. 'Benjamin Britten', he wrote, 'provided this programme with music which is, I think, the best I have heard in a radio play. Without his music *The Dark Tower* lacks a dimension.'

The greater part of the winter and spring of 1946—apart from a concert tour with Peter Pears to Holland and Belgium—was taken up with the composition of *The Rape*; and during this period Britten collaborated closely, not only with his librettist, but also with Eric Crozier and John Piper, who had been chosen as producer and designer respectively. Rehearsals started at Glyndebourne in June; and the first performance was given on 12 July with Ernest Ansermet as conductor. A first-rate double cast had been assembled, the parts of Lucretia, Tarquinius, and the Male and Female Chorus being taken on alternate nights by Kathleen Ferrier and Nancy Evans, Otakar Kraus and Frank Rogier, Peter Pears and Aksel Schiotz, Joan Cross and Flora Nielsen. Reginald Goodall alternated as conductor with Ansermet.*

In this production, special praise should be given to the work of John Piper as designer. The double arcade setting used for the second scenes of Acts I and II was particularly memorable, though, as he himself has admitted,† its parallel arrangement was 'in essence pictorial or architectural rather than theatrical, and it threw a heavy burden on the producer if he was to make the scenes played in front of these arcades "work" in an interesting and lively way'. A real triumph was his drop cloth for Act II, the deep glowing and smouldering colours of which, depicting Our Lord in Majesty, recalled the best English stained glass of the thirteenth and fourteenth centuries.

Some of the critics who reviewed *The Rape* seemed to have difficulty in accepting the premises of the work. Looking for the mass effects of a second

* See Appendix B.
† *The Rape of Lucretia* (commemorative volume) ed. Eric Crozier. The Bodley Head, 1948.

Britten with
W. H. Auden at
Tanglewood, Lenox,
Mass., August 1946

Peter Grimes, they found instead an opera built on austere chamber music lines, without a large chorus or orchestra. The Christian commentary of the single figures who act as Male and Female Chorus was objected to as a wilful anachronism, while the didactic tone of some of their historical information about the Romans and Etruscans—references that for the most part were modified or expunged in the revised version of 1947—was resented as an intrusion. Yet there was no doubt that the audience's attention was held throughout; and by the end of the summer the opera had been played about seventy-five times in Great Britain, the 100th performance being celebrated at Covent Garden on 17 October the following year.

When the tour to Manchester, Liverpool, Edinburgh, Glasgow, London (Sadler's Wells), Oxford and Holland was over, Britten and his supporters decided to relaunch the company in 1947 under the title of the English Opera Group. The arrangement with Glyndebourne had not worked altogether satisfactorily; but it was agreed that the new company, though now independent, would perform there as visitors in June 1947. The Group's aims and objects were restated by its promoters in the following terms: 'We believe the time has come when England, which has always

Britten, Audrey Mildmay (Mrs John Christie) and Eric Crozier at Glyndebourne, 1947

depended on a repertory of foreign works, can create its own operas. . . . This Group will give annual seasons of contemporary opera in English and suitable classical works including those of Purcell. It is part of the Group's purpose to encourage young composers to write for the operatic stage, also to encourage poets and playwrights to tackle the problem of writing libretti in collaboration with composers.' It was made clear that concerts as well as opera performances would come within its purview; and the new works commissioned for 1947 were an opera by Britten and a *Stabat Mater* by Lennox Berkeley.

Immediately after *The Rape of Lucretia*'s short initial run at Glyndebourne, Britten was invited to attend the first American production of *Peter Grimes* at Koussevitzky's Berkshire Music Center at Tanglewood, Lenox, Massachusetts (6, 7 and 8 August). On his return, he wrote an Occasional Overture in C (opus 38) for the opening orchestral concert of the new BBC Third Programme at the end of September. Later he decided to withdraw this composition, and it does not figure in his list of published works. About the same time he provided incidental music for Carl Czinner's New York production of Webster's *The Duchess of Malfi*, with Elizabeth Bergner in the title role.

56

In the winter of 1947, Britten undertook another Continental tour. He visited Zurich and there conducted performances of the *Serenade* and *The Young Person's Guide to the Orchestra* and, together with Peter Pears, gave recitals in Switzerland, Holland, Belgium and Scandinavia. On returning to England, he settled down to the composition of a new opera for the English Opera Group. In the past two years a number of possible themes had been suggested by Ronald Duncan, including *The Canterbury Tales*, *Abelard and Heloise*, and an action based on Jane Austen's *Mansfield Park* to be called *Letters to William;** but in the event the choice of librettist fell on Eric Crozier. The problem confronting the composer and librettist was how to devise an opera made to the same measure as *The Rape of Lucretia*, but as different as possible in subject matter and style—in fact, a comedy instead of a tragedy, with a setting in the late nineteenth century instead of B.C. Guy de Maupassant's short story, *Le Rosier de Madame Husson*, gave them the basic idea; but the Gallic original was so freely adapted into East Anglian terms that the resulting lyrical comedy, *Albert Herring*, which was performed for the first time at Glyndebourne on 20 June, had much of the flavour of an original work.

Once again, scenery and costumes were entrusted to John Piper—with excellent results. Particularly admired were the scene in Mrs Herring's greengrocery in Loxford and the drop curtain showing the market-place with the rich fuscous gloom of its buildings enlivened only by the newly painted signboard of its inn glowing like a drop of blood. Frederick Ashton, widely known for his work as choreographer to the Sadler's Wells Ballet, produced. A first-rate cast was assembled, including Peter Pears in the title role and Joan Cross as Lady Billows, an elderly regal autocrat; and Britten himself conducted.†

Before the Glyndebourne season closed, *The Rape of Lucretia* was revived in a revised version; and both operas were included in the English Opera Group's tour abroad to the International Festival at Scheveningen, the Stadsschouwburg in Amsterdam and the Lucerne International Festival. During their visit to Switzerland, a concert was given at Zurich, at which Lennox Berkeley's *Stabat Mater* was performed for the first time; and back in England, they gave ten opera performances at Covent Garden and toured to Newcastle upon Tyne, Bournemouth and Oxford before dispersing for the winter.

Meanwhile, the Covent Garden Opera Company had prepared a new production of *Peter Grimes* under the direction of Tyrone Guthrie. This opened on 6 November, with Peter Pears and Joan Cross as guest artists singing their original parts; and, after a few performances, their roles were taken over by Richard Lewis and Doris Doree. New scenery and costumes

* See *How to Make Enemies* by Ronald Duncan. Rupert Hart-Davis, 1968.
† See Appendix B.

had been designed by Tanya Moiseiwitsch; but whereas Eric Crozier's production and Kenneth Green's designs at Sadler's Wells had been in a realistic vein (as specified in Montagu Slater's libretto), Tyrone Guthrie and Tanya Moiseiwitch had decided on a more abstract interpretation of the drama. The main Borough set, as used in the first scenes of Acts I and II and the whole of Act III, showed a spacious beach and implied an illimitable sea beyond, but left the Moot Hall, the church and *The Boar* entirely to the imagination. This meant that the dance at the beginning of Act III, which, with its four-piece band, is supposed to take place off-stage in the Moot Hall, was given on the beach in full view of the audience, a change which was not for the better. It must also be admitted that on the vast stage of Covent Garden the scene in *The Boar* failed to produce that feeling of claustrophilia which is so characteristic of English pubs. But, apart from these reservations, the production was a fine one, and it was specially admired in Brussels and Paris when the Covent Garden Company played there the following June.

VI
Festival Interlude

In 1947 Britten moved from Snape to Aldeburgh. There he took a house overlooking the sea that Crabbe had described so lovingly in *The Borough*:

Various and vast, sublime in all its forms,
When lull'd by Zephyrs, or when rous'd by Storms,
Its colours changing, when from Clouds and Sun
Shades after shades upon the surface run.

Shortly after the move he was driving by car from Holland to Switzerland with his friends Peter Pears and Eric Crozier. The English Opera Group had just played *The Rape of Lucretia* and *Albert Herring* at the Holland Festival and was due to appear a few days later at the Lucerne Festival. Eric Crozier has described* how the three of them felt 'proud that England was at last making some contribution to the traditions of international opera. And yet—there was something absurd about travelling so far to win success with British operas that Manchester, Edinburgh and London would not support. The cost of transporting forty people and their scenery was enormously high: despite packed houses in Holland, despite financial support from the British Council in Switzerland, it looked as if we should lose at least £3,000 on twelve Continental performances. It was exciting to represent British music at international festivals, but we could not hope to repeat the experiment another year. It was at this point that Peter Pears had an inspiration. "Why not make our own Festival?" he suggested. "A modest Festival with a few concerts given by friends? Why not have an Aldeburgh Festival?" '

During the following autumn and winter the idea was submitted to the local population, who greeted it with enthusiasm; and although Aldeburgh had only about two and a half thousand inhabitants, £1,400 was subscribed in advance guarantees. The programme was carefully planned; the artistic direction entrusted to the English Opera Group; the Arts Council of Great

---

* 'The Origin of the Aldeburgh Festival' by Eric Crozier. The *Aldeburgh Festival Programme Book*, 1948.

59

LIFE

Britain offered a financial subsidy; and the first Aldeburgh Festival, with the Earl of Harewood as President, opened 5 June 1948, and lasted for nine days.

After the war, the idea of creating new festivals had been sparked off by the Edinburgh Festival (established in 1947); but that was by intent an international festival held in a capital city. What chance of success would a local East Anglian festival have, set in a small seaside town? Writing about the first Aldeburgh Festival shortly after it had been held, E. M. Forster said:* 'A festival should be festive. And it must possess something which is distinctive and which could not be so well presented elsewhere. . . . There exists in Aldeburgh the natural basis for a festival. It can offer a particular tradition, a special atmosphere which does not exist elsewhere in these islands: nothing overwhelming, but something that is its own, something of which it can be proud.' These are certainly prerequisites; but however unique the setting, the artistic direction must be first-rate if a festival is going to establish itself on the national—let alone the international—plane. And this, thanks to Benjamin Britten, the Aldeburgh Festival was able to achieve. At first the names of Benjamin Britten, Eric Crozier, and Peter Pears appeared as Founders, while the artistic direction was handled by the English Opera Group; but in 1955 the style was changed, Britten and Pears appearing as Artistic Directors, and the following year the two of them were joined by Imogen Holst. This triumvirate continued until 1968, when they were joined by Philip Ledger; Colin Graham was added to the group in 1969 and Steuart Bedford in 1974. After Britten's death in 1976, Mstislav Rostropovich was made an Artistic Director in 1978.

The 1948 Festival was a success, and the festival operation immediately established itself on an annual basis. It proved to be local in the best sense. Britten's compositions played an important but not preponderant part in the festival programmes. He and Peter Pears frequently appeared as executants; he and Imogen Holst conducted a number of the concerts. The members of the Aldeburgh Festival Singers were drawn from amateur singers in the neighbourhood. Lectures and exhibitions on East Anglian themes supplemented the opera performances, concerts and recitals. To begin with, performances were confined to Aldeburgh; but gradually, as it was seen how limited was the accommodation that Aldeburgh could provide, the Festival started to expand. The first extension was to Thorpeness, where madrigals were sung in the open air on the Meare in 1949, and the Workmen's Hall and the Country Club were used for cabaret, plays and concerts in subsequent years. Then some of the neighbouring country houses were laid under contribution; and a number of neighbouring East

* 'Looking back on the first Aldeburgh Festival', a talk by E. M. Forster, broadcast in June 1948 and reprinted in the *Aldeburgh Festival Programme Book*, 1949.

60

Anglian churches followed—Blythburgh in 1956, Framlingham in 1957, Orford in 1958 and (further afield) Ely Cathedral in 1964.

Britten saw to it that many contemporary composers were made warmly welcome at the Festival. Visitors from abroad included Francis Poulenc, Hans Werner Henze, Aaron Copland, Zoltán Kodály, and Witold Lutoslawski. A full list of British composers who have participated in the Festival since 1948 would run to over fifty names; but a short list—Malcolm Arnold, David Bedford, Arthur Benjamin, Richard Rodney Bennett, Lennox Berkeley, Harrison Birtwistle, Sir Arthur Bliss, Gordon Crosse, Peter Maxwell Davies, Peter Racine Fricker, Roberto Gerhard, Alexander Goehr, Oliver Knussen, Elizabeth Lutyens, Colin Matthews, Nicholas Maw, Priaulx Rainier, Alan Rawsthorne, Edmund Rubbra, Humphrey Searle, Matyas Seiber, Roger Smalley, John Tavener, Sir Michael Tippett, Sir William Walton, and Malcolm Williamson—is sufficient to show how wide and catholic the representation has been.

Many friends of Britten and Pears have been featured as executants; and some like Julian Bream and George Malcolm have returned to the Festival year after year. The great singer Kathleen Ferrier made only one appearance at Aldeburgh; but that was a most poignant one in the summer of 1952, just a year before her death. Britten himself has described her performance of his *Canticle II: Abraham and Isaac.** 'Many people have said they will never forget the occasion: the beautiful church, her beauty and incredible courage, and the wonderful characterization of her performance, including every changing emotion of the boy Isaac—the boyish nonchalance of the walk up to the fatal hill, his bewilderment, his sudden terror, his touching resignation to his fate—the simplicity of the Envoi, but, above all, combining with the other voice, the remote and ethereal sounds as "God speaketh".'

In later years a remarkable feature has been the strength of the participation of artists from the USSR. Mstislav Rostropovich, the cellist, and his wife Galina Vishnevskaya, the soprano, first appeared at the 1961 Festival. In 1964 Sviatoslav Richter, the pianist, agreed at short notice to join Rostropovich in a special additional recital in the Parish Church (20 June). Close and friendly relations with these Russian executants led directly to the composition of Britten's *Cello Sonata in C*, the *Symphony for Cello and Orchestra* and the three *Suites for Cello Solo* for Rostropovich, and indirectly to the Pushkin song cycle, *The Poet's Echo*.

Most of the opera performances and music recitals were given in the Jubilee Hall, which has a minute stage and a capacity of only about 300. In default of a larger building to accommodate its main events, the Festival was

---

* From 'Three Premières' by Benjamin Britten in *Kathleen Ferrier: a memoir*. London, Hamish Hamilton, 1954.

61

Britten with members of Lancing College choir prior to the first
performance of *Saint Nicolas* at Lancing, 1948

bound to remain small and intimate in scale; and this was indeed the case
during its first nineteen years. It might have been thought that with high
prices and limited accommodation, there would be a tendency for it to
become exclusive and even a little snobbish. But no—from the beginning it
was equally welcomed by local inhabitants and visitors. E. M. Forster had a
revealing anecdote to tell* about the first performance of *Albert Herring* in
the Jubilee Hall in 1948. 'During the first interval a man in a pub said: "I
took a ticket for this show because it is local and I felt I had to. I'd have sold
it to anyone for sixpence earlier on. I wouldn't part with it now for ten
pounds." '

Although the price of tickets for opera and the top concerts never reached
the figure of ten pounds, it is true to say that, even after making allowance
for substantial subsidies from the Arts Council and from the large body of
friendly subscribers that the Festival soon built up, prices remained
comparatively high, and at first it was difficult to see how students and other
young people could afford to go to the Festival in any numbers. In 1958
Princess Margaret of Hesse, who was herself a devoted festival-goer,
decided that 'by some means or another it must be made financially possible
for more young people to attend the Festival'. It was not long before she had
enlisted the help of some friends and opened the fund which bears her name.
A modest start was made in 1959 and about half a dozen students were given

* E. M. Forster, op. cit.

a limited number of tickets for different events. Each year the gates were opened wider and by 1962 the numbers had risen to thirty-four. The rules were that applicants must be under twenty-five years of age and that, although they need not necessarily be music students, they were expected to be playing some active part in the musical life of the community in which they lived or worked.'* This infusion of new blood was most welcome and did much to popularize the Festival among the younger generation.

A number of Britten's works have received their first performance at the Festival. The first Festival of all (1948) opened with his cantata *Saint Nicolas* to words by Eric Crozier. This had been written during the winter of 1947–8 specially for the centenary celebrations of Lancing College in July 1948. Its lay-out for tenor, mixed choirs, string orchestra, pianoforte and percussion and the fact that it included two familiar hymns for congregation and choirs made it particularly suitable for performance in the Aldeburgh Parish Church. Movements like The Birth of Nicolas with the innocent and joyful lilt of the sopranos' tune answered at the end of each verse by the boy Nicolas's treble '*God be glorified!*' and the strangely primitive story of Nicolas and the Pickled Boys fully justified E. M. Forster's comment:† 'It was one of those triumphs outside the rules of art which only the great artist can achieve.' The cantata was subsequently revived at the 1949, 1951, 1955 and 1961 Festivals.

Lachrymae, Reflections on a Song of Dowland, a set of ten variations for viola and piano, was first performed by William Primrose and the composer at the 1950 Festival.‡ The following year the *Six Metamorphoses after Ovid* for oboe solo were written for Joy Boughton and played by her at an open-air concert given on the Meare at Thorpeness. These short pieces take their place with Debussy's *Syrinx* and Stravinsky's Three Pieces for Clarinet Solo as being among the most outstanding compositions for an unaccompanied wind instrument that have been written in this century.

For the 1953 Festival a special work for string orchestra was commissioned. Six variations on the tune of *Sellinger's Round, or, The Beginning of the World* were written by Lennox Berkeley, Arthur Oldham, Humphrey Searle, Michael Tippett and William Walton as well as Britten himself. The identity of the actual composer of each movement was concealed at the festival performances and revealed only later. (Subsequently Michael Tippett took his variation and used it as the second movement of his *Divertimento on 'Sellinger's Round'* for chamber orchestra.)

Two of Britten's song cycles have been launched at the Festival—*Songs from the Chinese* for tenor and guitar (performed by Peter Pears and Julian Bream) in 1958 and *Songs and Proverbs of William Blake* for baritone and

---

* 'The Hesse Students' by Arthur Harrison. *Aldeburgh Festival Programme*, 1963.
† E. M. Forster, op. cit.
‡ Britten orchestrated the *Lachrymae* for viola and strings in 1975.

piano (performed by Dietrich Fischer-Dieskau and the composer) in 1965. *Nocturnal after John Dowland* was specially written for the guitarist Julian Bream and played by him at the 1964 Festival. The *Suite in C* for harp solo, which was specially written for the harpist Osian Ellis, was played by him at the 1969 Festival. Of the compositions written for Rostropovich, the *Sonata in C for Cello and Piano* and the first two *Suites for Cello Solo* received their first performances at the 1961, 1965 and 1968 Festivals respectively. The première of the *Symphony in D for Cello and Orchestra* was planned for a concert at Blythburgh Church on 30 June 1963, but had to be cancelled owing to Rostropovich's illness. Its first performance was eventually given in Moscow on 12 March 1964 with Britten conducting; and the first performance in England followed on 18 June 1964, when the work was heard at Blythburgh Church as originally planned. The third *Suite for Cello Solo* was first performed by Rostropovich at Snape Maltings on 21 December 1974.

A most unusual work was the *Gemini Variations* which received its first performance during the 1965 Festival. Describing its origin,* Britten wrote: 'When in Budapest in the spring of 1964 at a Music Club meeting for children, I was very taken by the musical gifts of two twelve-year-old twins. Each played the piano, one the flute and the other the violin: they sang, sight-read, and answered difficult musical questions. It turned out that they were the sons of one of Budapest's most distinguished flute players. At the end of the meeting they approached me and charmingly, if forcefully, asked me to write them a work. Though I claimed that I was too busy, my refusal was brushed aside: however, I insisted on one small bargaining point: I would do it only if they would write me a long letter telling me all about themselves, their work and their play—in *English*. I felt safe. After a week or two, however, the letter arrived, in vivid and idiosyncratic English, and I felt I must honour my promise. *Gemini Variations* is the result.' The theme used was no. 4 of Zoltán Kodály's *Epigrams* (1954). The boys were Zoltán and Gabriel Jeney.

Occasionally some of Britten's juvenile compositions have been given a hearing. Two movements were salvaged from a Christmas Suite called *Thy King's Birthday* written about 1930 or 1931—these were the fourth movement entitled *New Prince, New Pomp*, for soprano solo and chorus, performed at the 1955 Festival, and the second movement entitled *Sweet was the Song* for contralto solo and chorus of female voices at the 1966 Festival. For the 1969 Festival he dug out some early settings of lyrics by Walter de la Mare that he had written between the ages of fourteen and seventeen, and these were published under the title *Tit for Tat*.

If, as was the intention from the beginning, opera was to provide the

* In *Faber Music News*, Autumn 1966.

backbone of the Festival programmes, it was clear that it would have to be on an intimate scale. Fortunately the existence of the English Opera Group went a long way to overcome the inevitable difficulties inherent in presenting opera in the cramped conditions of the Jubilee Hall. The touring of the Group's productions of *The Rape of Lucretia* and *Albert Herring* in 1947 had been the factor that originally led to the idea of an Aldeburgh Festival: so it was appropriate that the opening opera at the 1948 Festival should be *Albert Herring*. Its East Anglian setting and local connotations proved particularly apt, and it was revived at several subsequent Festivals (in 1949, 1951, 1953, 1957 and 1963). The tenth anniversary performance (1957), when Joan Cross, Peter Pears and others returned to their original roles, was particularly impressive. *The Rape of Lucretia* was presented in 1949 and revived in 1954, 1959 and 1960. In 1949 came the first opera of Britten's to be created at the Festival—the extraordinarily popular entertainment for children of all ages called *Let's Make an Opera!*—and this was revived in 1950 and 1956, and also in 1965 with a revised Introduction. The opera in 1950 was Britten's realization of *The Beggar's Opera* with the keys of some of the movements transposed so that the part of Macheath could be sung by a baritone; in 1963 it was revived with Peter Pears singing the original tenor role. In 1951 (the year of the Festival of Britain) Britten's realization of Purcell's *Dido and Aeneas* was brought to Aldeburgh from London where it had been heard for the first time at the Lyric Theatre, Hammersmith. (It was revived at Aldeburgh in 1962.) *The Turn of the Screw* followed in 1955 (with a revival in 1961); and two years later *Noye's Fludde* was given its first performance in Orford Church and revived there in 1961. The boldest venture of all was in 1960, when the English Opera Group presented *A Midsummer Night's Dream* for the first time in the enlarged Jubilee Hall, a very complicated production to mount on the pocket-handkerchief stage. In 1964 came the première of the first of the parables for church performance, *Curlew River* (revived in 1965); in 1966 the second parable *The Burning Fiery Furnace* (revived in 1967); and in 1968 the third parable *The Prodigal Son* (revived in 1969)—all three of them (like *Noye's Fludde*) produced in Orford Church.

If the attractively edited programmes of the Aldeburgh Festival are examined, it will be seen that (apart from Britten himself) the main composers laid under contribution have been Purcell, Mozart, and Schubert. Not far behind in popularity are Dowland and Byrd, Bach and Haydn, Mahler and Berg, Holst and Bridge.

In addition to the general concert programmes, many of the Festivals have featured an outstanding series of programmes of Church Music specially devised by Imogen Holst and usually given in the Parish Church. These have been presented in conjunction with the BBC Transcription Service.

Lectures, poetry readings and recitals, the occasional play, and numerous exhibitions of paintings, drawings, and sculpture—all these have helped to build up a balanced festival programme.

From the beginning, the Festival had difficulties in finding the right sort of accommodation for its needs. It was difficult enough to fit in concerts, exhibitions, lectures, recitals and the occasional play; but opera was the most difficult of all, for the conditions in the Jubilee Hall were cramped and inadequate. It is true that in 1959/60 certain improvements were carried out which made it easier to produce *A Midsummer Night's Dream* in the summer of 1960. At the same time it was decided to purchase a property in the centre of the town and to adapt it as a festival club with a small gallery for exhibitions. To raise the extra money needed, various friends of the Festival—artists and collectors—were asked to help by contributing something (pictures, sculpture, books, or manuscripts) to an auction that was held by Christie's on 23 March 1961 and raised about £11,480. Among the items sold on this occasion were two important manuscripts of Britten's—the autograph full score (in pencil) of *The Young Person's Guide to the Orchestra* and the first sketches (also in pencil) of *Seven Sonnets of Michelangelo*—both of which were bought by James M. Osborn, Curator of the Osborn Collection in the Yale University Library.* The sketches and original manuscript full score of the *Cantata Academica* (*Carmen Basiliense*) had been bought prior to the sale by Dr and Mrs Paul Sacher of Basle.

The new Festival Club proved a great success; but the main problem of a new hall or opera-house still remained to be solved. In the 1957 Festival Programme Book an article by H. T. Cadbury-Brown called *Notes on an Opera House for Aldeburgh*, carried an illustration showing the site of a Festival Theatre as then proposed. This particular plan came to naught; but ten years later, an imaginative adaptation of part of the Maltings at Snape was carried out, one of the Malt Houses being turned into a concert hall seating about 820 with a restaurant and bar for the public as well as full dressing room facilities for the artists. At the same time the auditorium was made suitable for stereophonic recording by Decca, and a recording room was provided for Decca and the BBC.

The Maltings Concert Hall was opened by Her Majesty The Queen, who was accompanied by the Duke of Edinburgh, on 2 June 1967; and after hearing the first two pieces of music in the Inaugural Concert programme— Britten's setting of the National Anthem for chorus and orchestra, and his overture *The Building of the House* (composed specially for the occasion)— the audience realized that the new hall had extraordinarily fine acoustic

---

* It is worth remembering that at least two other important manuscript scores of Britten's are to be found in the United States. Both *Peter Grimes* and the String Quartet No. 1 are in the Library of Congress, Washington D.C.

properties. Other concerts followed, including the first performance by the Vienna Boys' Choir of Britten's vaudeville for boys and piano called *The Golden Vanity*; and the open platform was used for opera when *A Midsummer Night's Dream* was given in a new production.

The existence of this hall made it possible to plan future Festivals on a larger scale than hitherto: so it was a particularly traumatic experience when on 7 June 1969, shortly after the opening concert of the 1969 Festival, the building was destroyed by fire.

But temporary reverses like this could not affect the unique nature of the Festival. In the first place, there is the special attraction of this remarkable little seaside town. As Britten said in his speech on receiving the Freedom of the Borough of Aldeburgh on 22 October 1962: 'Everyone is charmed by its lovely position between sea and river, and the beach with the fishing boats and life-boat, the fine hotels, the golf course, wonderful river for sailing, the birds, the countryside, our magnificent churches, and of course the shops too.' And then it is the sort of place where friends meet, friendships are renewed, new friendships formed, and there is time to converse.

Referring to the Festival on the same occasion Britten said: 'It is a considerable achievement, in this small Borough in England, that we run year after year a first-class Festival of the Arts, and we make a huge success of it. And when I say "we" I mean "we". This Festival couldn't be the work of just one, two or three people, or a board, or a council—it must be the corporate effort of a whole town.' This is true up to a point: but ultimately everyone must recognize that, had it not been for its resident composer, the Aldeburgh Festival would not have come into being and could not have continued. As Rostropovich wrote after his first visit to Aldeburgh in 1961: 'Britten's energy and capacity for work during the Festival were phenomenal. He was the heart and brain of the Festival. He took part in it as pianist and conductor. He was at all the rehearsals and concerts, he looked into literally every trifle.'* He was indeed an exuberant and indefatigable host.

* From 'Dear Ben . . .' by Mstislav Rostropovich in *Tribute to Benjamin Britten on his Fiftieth Birthday*.

VII
Aldeburgh: Living in Crabbe Street

Britten's life was now following a fairly regular annual pattern. There was usually a new opera on the stocks; various other compositions to be squeezed in; a certain number of engagements as conductor and pianist; a recital tour with Peter Pears at home or abroad; and each summer would bring the busy but refreshing interlude of the Aldeburgh Festival.

About the time of his move to Crag House in Crabbe Street, Aldeburgh, he completed two vocal works: *Canticle I* and *A Charm of Lullabies*. The *Canticle*, a setting of a poem by Francis Quarles for tenor and piano, is a more extended composition for voice than any of the previous song-cycles. Peter Pears, writing just before *Canticle II* was composed, considered it to

Britten and Peter Pears buying vegetables from Jonah Baggott in the High Street, Aldeburgh, *c.* 1948

Britten's studio at Crag House, Aldeburgh

be 'Britten's finest piece of vocal music to date'.* He particularly praised the vocal line as being 'free, melismatic yet controlled, and independent throughout, whether as one or two or three parts in counterpoint or a melody with chordal accompaniment'. *A Charm of Lullabies* reverts to the earlier song-cycle pattern. Britten has chosen five contrasting lullabies—*A Cradle Song* (William Blake), *The Highland Balou* (Robert Burns), *Sephestia's Lullaby* (Robert Greene), *A Charm* (Thomas Randolph) and *The Nurse's Song* (John Philip)—and set them for mezzo-soprano and piano.

For the 1948 season of the English Opera Group, he decided to add to the company's repertory a new version, not of *Dido and Aeneas* as had been promised earlier, but of *The Beggar's Opera*. For this purpose, John Gay's text was slightly revised by Tyrone Guthrie, while Britten, ignoring all the numerous versions of the score that had appeared during the last two and a quarter centuries, went back to the sixty-nine airs originally chosen by Dr Pepusch and included all of them in his version except three. Since his return from America he had produced three volumes of folk song arrangements (two for the British Isles and one for France), which showed such sensitivity to the mood and mode of folk tunes and skill in their setting

* 'The Vocal Music' by Peter Pears in *Benjamin Britten: a commentary* (edited by Mitchell & Keller). Rockliff, 1952.

Eric Crozier, Benjamin Britten and Tyrone Guthrie at a rehearsal for *The Beggar's Opera*, 1948

that it was hardly surprising he should now find himself strongly attracted by the traditional tunes of *The Beggar's Opera*.

For the first performance at the Arts Theatre, Cambridge (24 May, 1948), Tyrone Guthrie was producer. Profiting by a hint in the original Prologue that the opera had previously been performed by the Company of Beggars in their 'great Room at St Giles's', he set the whole opera in this 'great room', which for some reason or other he imagined as a laundry. Unfortunately, the atmosphere of this laundry with its piles of clothes, whether dirty or clean, proved stifling, rather than inspiring, and the production never succeeded in fulfilling the producer's laudable aim of restoring the mordancy of the original satire. Britten wrote the part of Macheath for tenor voice; and the role was admirably sung by Peter Pears. When the opera was revived two years later, the part was transposed for baritone; but the result was not altogether satisfactory, some of the buoyancy and sparkle of the score being thereby lost.

After a week's run in Cambridge where it was conducted by Britten himself and Ivan Clayton, the opera was played by the English Opera Group at the Holland Festival, the Cheltenham Festival of Contemporary Music, the Festival du Littoral Belgique, Birmingham, and the Sadler's Wells Theatre and the People's Palace, London; and during the next two or three years it had separate productions in Austria, Switzerland and Germany.

70

Having produced a Young Person's Guide to the Orchestra, Britten seemed bound sooner or later to write a Young Person's Guide to Opera. The episodes in *Albert Herring* where the three children bounce their ball against the door of Mrs Herring's greengrocery shop and where Miss Wordsworth rehearses them in their festive song were among the most successful passages in that comic opera and seemed to show that such a work might well be written so as to provide parts to be played by children themselves. He believed there were many children in the country who were natural musicians and actors and was anxious to help provide an outlet for their artistic talent. In this he was ably aided and abetted by Eric Crozier. *Let's Make an Opera!* was planned in two parts: the first part showing a group of children who help two or three of their elders to plan, write, compose and rehearse an opera; and the second part being the opera itself, *The Little Sweep*. The music of this one-act opera is not continuous, but consists of eighteen musical numbers, and there is a certain amount of spoken dialogue. It was an ingenious stroke on the part of the authors of this entertainment for young people to implicate the audience in the performance of *The Little Sweep* as well as the children and the three or four professional adults that are called for in the cast. *Let's Make an Opera!* was an immediate success on its first production at the second Aldeburgh Festival (14 June, 1949). The conductor, who plays such an important part in knitting together the different musical strands of this entertainment, was Norman Del Mar; Basil Coleman produced; and the attractive setting was by John Lewis.*

To judge by statistics, *The Little Sweep* is easily Britten's most popular opera to date. As well as many professional productions in Europe, America, Asia and Australia, there have been innumerable amateur performances all over the world, particularly in schools. It is difficult to assess the long-term effect of all this; but Eric Crozier's words should be borne in mind: 'Many children write plays for their own performance. Few, I suspect, attempt opera. Perhaps, with the stimulus of an imaginative example before them, they may be prompted to explore the fascinating possibilities of expression and entertainment that it offers.'†

Britten continued to attract much attention abroad. By 1948 monographs on his work as a composer had appeared in France, Italy and Switzerland as well as Great Britain. In its first three years the English Opera Group had toured its productions of *The Rape of Lucretia*, *Albert Herring* and *The Beggar's Opera* to Holland, Belgium, Switzerland, Denmark and Norway. *Peter Grimes* reached the Scala, Milan, and the Opéra, Paris, in 1947, and the Metropolitan Opera House, New York, the following year. *The Rape of*

* See Appendix B.
† 'An Opera for Young People' by Eric Crozier. *Times Educational Supplement*, 19 March 1949.

Lucretia aroused considerable controversy when it was produced in America. In Chicago it was greeted by one of the local newspapers with the banner headline 'Bold, Bawdy and Beautiful'; but its run in New York at the end of 1947 came to an abrupt end after only twenty-three performances. On that occasion Olin Downes, music critic of the *New York Times*, performed a complete *volte face*. After attending the first night, he praised the work warmly; but a return visit led him to change his mind and he decided it was 'as arrant a piece of musico-dramatic twaddle as has been visited upon the public for years'.* Hostile criticism of Britten was also forthcoming from the USSR. The first All-Union Congress of Soviet Composers that met in Moscow from 19–25 April 1948, energetically denounced him, together with Menotti and Messiaen, as being 'impregnated with extreme subjectivism, mysticism and disgusting facetiousness', while it was said that Stravinsky 'actually showed by his idealization of the Middle Ages that he breathed the spirit of fascism'. But this attack of nerves over the 'reactionary formalist trends in the West' calmed down when Stalin disappeared from the scene; and by 1962 Stravinsky was being amicably received by Krushchev in the Kremlin in the course of a triumphal return to his native country nearly half a century after he had left it, and early in 1963 Britten was being warmly fêted in Moscow and Leningrad during a festival of British music.

During the autumn and winter of 1948–9 Britten was preoccupied with the *Spring Symphony*, the first performance of which was given at the Holland Festival, Amsterdam (9 July, 1949) by kind permission of Koussevitzky, who conducted the first American performance at the Berkshire Festival in Tanglewood (13 August, 1949). For two years he had been planning such a work, 'a symphony not only dealing with the Spring itself, but with the progress of Winter to Spring and the reawakening of the earth and life which that means'.† Apparently his original intention had been to use medieval Latin verse for this purpose; but 'a re-reading of much English lyric verse and a particularly lovely Spring day in East Suffolk, the Suffolk of Constable and Gainsborough', led him to change his mind and to substitute for his choice of Latin verse an anthology of English poems.

The *Spring Symphony* is written for three soloists (soprano, alto and tenor), mixed chorus, boys' choir, and large orchestra, including a cow-horn and a vibraphone. 'It is in the traditional four movement shape of a symphony, but with the movements divided into shorter sections bound together by a similar mood or point of view. Thus after an introduction, which is a prayer, in Winter, for Spring to come, the first movements deal with the arrival of Spring, the cuckoo, the birds, the flowers, the sun and "May months' beauty"; the second movements paint the darker side of

* See 'The Rape of Lucretia', *New York Times*, 30 December 1948, and 'Second Thoughts', *New York Times*, 9 January 1949.
† 'A Note on the Spring Symphony' by Benjamin Britten. *Music Survey*, Spring, 1950.

Britten and E. M. Forster on a walk, 1949

Spring—the fading violets, rain and night; the third is a series of dances, the love of young people; the fourth is a May-day Festival, a kind of bank holiday, which ends with the great thirteenth-century traditional song "Sumer is i-cumen in", sung or rather shouted by the boys.'*

The sentiments of this springtime anthology culled from Edmund Spenser, Thomas Nashe, George Peele, John Clare, John Milton, Robert Herrick, Henry Vaughan, Richard Barnefield, and Anon, are without specific indication of time or period, apart from the account of Elizabethan London in May extracted from Beaumont and Fletcher's *The Knight of the Burning Pestle* which forms the finale, and the four stanzas extracted from Auden's poem '*Out on the lawn I lie in bed*'† where the lulled listener is suddenly startled by the prophetic reference to Poland and war. Britten shows great skill in setting all this heterogeneous material in a unified musical idiom and in ordering it so as to give the work the specific gravity of a symphony instead of the running lightness of a suite.

After the *Spring Symphony* the collaboration with Ronald Duncan was resumed. Britten had already written incidental music, not only for Duncan's verse play *This Way to the Tomb*, but also a fanfare for his translation and adaptation of Jean Cocteau's play *The Eagle Has Two Heads*, which was produced by the Company of Four at the Lyric Theatre, Hammersmith, on 4 September, 1946. He now provided music for

* Ibid.   † From *Look, Stranger!*, 1936.

73

Duncan's new verse play, *Stratton*, which started a brief provincial tour at the Theatre Royal, Brighton, on 3 October, 1949, and asked Duncan to provide the words for *A Wedding Anthem* for the wedding of the Earl of Harewood and Miss Marion Stein. This work for soprano and tenor soli, choir and organ was performed at St Mark's Church, North Audley Street, London, on 29 September, 1949, by Joan Cross, Peter Pears and the Choir of St Mark's which had recently sung a number of Britten's choral works, including the early *Te Deum* (in C major) and *A Boy was Born*.

Shortly after the Harewood wedding, which was attended by the King and Queen and many members of the Royal Family, Britten and Pears left for a concert tour of North America. They crossed to the West Coast of the United States; and, while in Hollywood, Britten saw Carl Ebert rehearsing *Albert Herring* for production by the University of Southern California. Back in England by Christmas, he started to think about his next opera. He had already chosen Herman Melville's posthumous story, *Billy Budd, Foretopman*, as the subject, and E. M. Forster and Eric Crozier had agreed to write the libretto jointly. The work of composition was begun in February 1950 and finished in the autumn of 1951.

Britten accepted a commission from the Arts Council of Great Britain for the opera to be produced in connection with the Festival of Britain, 1951; but for some time the exact destination of the new opera was uncertain. At first it was intended that it should be produced by the Sadler's Wells Opera Company at the 1951 Edinburgh Festival, and this was announced in September 1950. Two months later, however, Sadler's Wells decided they would have to abandon the idea of producing it as it was likely to prove beyond their resources; and in December it was announced that *Billy Budd* would definitely be produced at the Royal Opera House, Covent Garden, in the autumn of 1951.

During the twenty months or so that *Billy Budd* was being composed, Britten allowed few extraneous matters to distract him. The only other compositions belonging to this period were the *Lachrymae*, for viola and piano (1950), and the *Six Metamorphoses after Ovid* for solo oboe (1951), both performed for the first time at the Aldeburgh Festival, and a set of *Five Flower Songs* for mixed chorus unaccompanied, dedicated to Leonard and Dorothy Elmhirst of Dartington Hall on the occasion of their 25th wedding anniversary (3 April, 1950). He also found time to prepare a new realization of *Dido and Aeneas* for the English Opera Group's Festival of Britain season.

About this period a number of ballets were adapted to existing scores of his. *Soirées Musicales* and *Matinées Musicales* (based on Rossini's music) were used once more for a ballet which was produced at the Théâtre de la Monnaie, Brussels, in 1948 under the title *Fantaisie Italienne*. The following June *The Young Person's Guide to the Orchestra* was used for a modern classical ballet *Oui ou Non?* (*Ballet de la Paix*) presented by the Association

74

des Amis de la Danse at their annual gala performance at the Théâtre National Populaire du Palais de Chaillot, Paris. On 26 April, 1949, the Ballets de Paris de Roland Petit gave the first performance at the Prince's Theatre, London, of *Le Rêve de Léonor*, a surrealist ballet with choreography by Frederick Ashton to the *Variations on a Theme of Frank Bridge* arranged for full orchestra by Arthur Oldham. The same *Variations* were used in Lew Christensen's *Jinx*, the revised version of which was included in the New York City Ballet's repertory for its spring 1950 season and brought to Great Britain when the company visited Covent Garden in July 1950. The same company performed *Les Illuminations* as a ballet at City Center, New York, on 2 March 1950, with choreography by Frederick Ashton and décor by Cecil Beaton; and this too was performed at Covent Garden in the summer of the same year. The *Sinfonietta* was used for a little drama in ballet form called *Die Versunkene Stadt* with choreography by Mara Jovanovits which was given at the Stadttheater, St Gallen, Switzerland, on 26 April, 1950.

The production of Purcell's *Dido and Aeneas* by the English Opera Group represented the fulfilment of a long cherished ambition. Britten was closely associated with Imogen Holst in this work. Together they collated and compared all the extant manuscript material and copied the whole of the transcript made by some copyist in the latter part of the eighteenth century and preserved in the Library of St Michael's College, Tenbury.* They then examined Nahum Tate's printed libretto and were particularly struck by the fact that no setting by Purcell exists for the witches' chorus and dance at the end of Act II. In a note dated 4 April, 1951, printed in the Lyric Theatre, Hammersmith, programme, Britten wrote: 'Anyone who has taken part in, or indeed heard a concert or stage performance, must have been struck by the very peculiar and most unsatisfactory end of this Act II as it stands; Aeneas sings his very beautiful recitative in A minor and disappears without any curtain music or chorus (which occurs in all the other acts). The drama cries out for some strong dramatic music, and the whole key scheme of the opera (very carefully adhered to in each of the other scenes) demands a return to the key of the beginning of the act or its relative major (i.e. D, or F major). What is more, the contemporary printed libretto (a copy of which is preserved in the library of the Royal College of Music) has perfectly clear indications for a scene with the Sorceress and her Enchantresses, consisting of six lines of verse, and a dance to end the act. It is my considered opinion that music was certainly composed to this scene and has been lost. It is quite possible that it will be found, but each year makes it less likely. It is to me of prime importance dramatically as well as musically to include this missing scene, and so I have supplied other music of Purcell's to fit the six lines of the

---

* See 'Dido and Aeneas', and article by George Malcolm in *Benjamin Britten: a commentary* (edited by Mitchell & Keller). Rockliff, 1952.

75

libretto, and a dance to end in the appropriate key . . . The realization of the figured bass for harpsichord is, of course, my own responsibility; in Purcell's time it was the custom for the keyboard player to work it out afresh at each performance. Therefore, no definitive version of this part is possible or desirable . . .'

This realization of *Dido and Aeneas* was given in a dual bill with Monteverdi's *Combattimento di Tancredi e Clorinda* at the Lyric Theatre, Hammersmith, on 1 May, 1951. Nancy Evans was Dido, and Joan Cross produced. Britten conducted from the harpsichord. After the Hammersmith season, it was played at the Aldeburgh, Holland, Cheltenham and Liverpool Festivals.

That summer he was created a freeman of his birthplace, Lowestoft.

At the end of September, while engaged on the final stages of the orchestral score of *Billy Budd*, Britten, together with Peter Pears and a group of friends, set out on a short pleasure cruise that took them direct from Aldeburgh, across the North Sea and up the Rhine as far as Bonn. Later that autumn, Peter Pears returned to Germany and sang the title part in *Oedipus Rex* in a performance for the Nordwestdeutscher Rundfunk, Cologne, that was conducted by Stravinsky, whose new opera, *The Rake's Progress*, had just been given in Venice.

Shortly after Britten's return, rehearsals started for *Billy Budd*. From discussions between the producer (Basil Coleman) and designer (John Piper), the general conclusion emerged that the sea and the ship would have to be suggested and not portrayed in realistic terms, and that the sets would have to be made 'so abstract that the absence or presence of particular details would not be noticed, so long as the shapes themselves were all intensely *ship-like*, and so long as the practical demands of the libretto and the score were all satisfied'.* This, on the whole, they succeeded in doing— particularly in the scene on the berth-deck with its skeleton of wooden ribs, low headroom, swung hammocks and illusion of claustrophobia.

The first performance of *Billy Budd* took place at the Royal Opera House, Covent Garden, on 1 December, 1951. Britten conducted; and the cast included Theodor Uppman in the title part, Peter Pears as Captain Vere, and Frederick Dalberg as Claggart.† Despite the possibility that the special nature of the subject and its all-male cast might restrict public interest, the first six performances were played to capacity houses. The opera was then taken on tour to Cardiff, Manchester, Glasgow and Birmingham; and when it returned to Covent Garden after Easter it continued to attract quite good attendances. In May the Covent Garden Opera Company gave two performances of it at the Théâtre des Champs-Elysées, Paris, during the

---

* 'Billy Budd on the Stage' by Basil Coleman and John Piper. *Tempo*, Autumn 1951.
† See Appendix B.

Peter Pears, Kathleen Ferrier and Britten, 1952

Festival of Twentieth-Century Art. The attitude of the Parisian audience appears to have been rather tepid.

This was not so with the first German production at Wiesbaden in March 1952. Despite severe cuts that amounted almost to mutilations, the opera was an unqualified success with the public. The Earl of Harewood, who attended the sixth performance, wrote that the German audience's enthusiasm was something he could not 'remember ever having seen exceeded elsewhere'.*

True to his usual custom, Britten celebrated the completion of his new opera by composing a smaller-scale work. The text of *Canticle II: Abraham and Isaac* is taken from the Chester Miracle Play and is set for alto, tenor and piano. The first performance was given by Kathleen Ferrier and Peter Pears at Nottingham (21 January, 1952) with the composer at the piano. The work is a dramatic scene, in which the tenor is cast as Abraham, the alto as Isaac; and the two voices combine, whether in two parts or in unison, to form the voice of Jehovah. Just as the second String Quartet is an instrumental sequel

* 'Foreign Diary' by the Earl of Harewood. *Opera*, May 1952.

77

to *Peter Grimes*, this is a parergon to *Billy Budd*. Abraham's dilemma over Isaac is similar to the problem that confronts Captain Vere and Billy Budd; but in the opera there is no *deus ex machina*.

Shortly after George VI's death in February 1952, Britten had the idea of writing an opera on the theme of Elizabeth I and Essex. This subject had made a strong impression on him when young; and when HM Queen Elizabeth II gave him permission to compose an opera on the occasion of her Coronation, he received just the stimulus he needed. In the event, *Gloriana* was produced at a gala performance at the Royal Opera House, Covent Garden, on 8 June, 1953, in the presence of the Queen and members of the Royal Family. It was a unique occasion. As Ralph Vaughan Williams pointed out in a letter to *The Times*,* this was 'the first time in history the Sovereign has commanded an opera by a composer from these islands for a great occasion'.

As his librettist, Britten chose William Plomer. Lytton Strachey's tragic history, *Elizabeth and Essex*, was the starting point of their collaboration; but, in the course of the opera's composition, both Plomer and Britten became 'less concerned than Strachey with the amatory motives of the two principal characters and more concerned with the Queen's pre-eminence as a Queen, a woman and a personality'.† The opera was planned so as to give considerable scope for pageantry—particularly the scenes of the Masque at Norwich and the dancing at the Palace of Whitehall. Basil Coleman was the producer and John Piper the designer; and between them they devised a magnificent and beautiful production, fully worthy of the great occasion for which the opera was written. John Pritchard conducted.‡ Joan Cross was cast as Queen Elizabeth and gave one of the finest performances of her career. No opera-lover who was present will ever forget the extraordinary effect of Elizabeth I on the stage directly addressing the glittering audience grouped round Elizabeth II in the Royal Opera House with the following words: 'I have ever used to set the last Judgment Day before mine eyes, and when I have to answer the highest Judge, I mean to plead that never thought was cherished in my heart that tended not to my people's good. I count it the glory of my crown that I have reigned with your love, and there is no jewel that I prefer before that jewel.'

But it must be admitted that the special audience at that gala was for the most part unmusical and disinterested, and the atmosphere in the Royal Opera House, when compared with that of a normal opera or ballet performance, distinctly frigid. This gave certain elements an excuse to denigrate the opera and its composer, and to express disapproval of the

---

* 18 June 1953.
† 'Notes on the Libretto of Gloriana' by William Plomer. *Tempo*, Summer, 1953.
‡ See Appendix B.

circumstances that had led to its commissioning. As Martin Cooper wrote in the *Spectator* (19 June), 'the work has been very generally over-blamed, with an almost sadistic relish or glee that has little to do with musical merit or demerit', and he noted the feelings of resentment that had arisen and the envy of 'special patronage and special conditions of work and performance not accorded to other performers'.* Despite its mixed reception at the première, subsequent performances of *Gloriana* at Covent Garden drew nearly as full houses as for *Billy Budd*. Its first performance outside London was given by the Covent Garden Company at Bulawayo, Southern Rhodesia (8 August 1953), at the Rhodes Centennial Exhibition; and early in 1954 it was toured to Cardiff, Manchester, and Birmingham.

The schedule for the composition of *Gloriana* had been a particularly onerous one. The idea of the opera had probably first been mooted during a skiing holiday with the Harewoods in Austria in March 1952. By the end of April the suggestion had received Royal approval; and the official announcement, naming the date of the gala performance of the Coronation opera as (probably) 8 June of the following year was made on 28 May. By July part of the first act existed in rough draft; and by the following February the work was sufficiently far advanced for a private run through to be held at the Royal Opera House. The actual composition had to be fitted in a very busy diary of festival appearances in the summer of 1952— Aldeburgh, followed by Copenhagen, Aix-en-Provence, Menton and Salzburg—and it is doubtful whether Britten would have managed to finish the score on time, had it not been for the devoted help of Imogen Holst. Subsequently she described this collaboration in some detail:† 'It was while he was still having discussions with William Plomer about the libretto that I first began working for him in Aldeburgh. My job was to copy out his pencil sketches of the music for each scene as soon as he had finished writing it, and to make a piano arrangement that the singers could use at rehearsals. I already knew that he was as practical a composer as my father had been, but even so it was astonishing to see how strictly he could keep to his time-schedule of work. He was able to say in the middle of October, when he was just beginning Act I, that he would have finished the second act before the end of January . . . When he began work on the full score of the opera, he wrote at such a tremendous speed that I thought I should never keep pace with him. He managed to get through at least twenty vast pages a day, and it seemed as if he never had to stop and think.'

In September 1953, Britten completed the composition of a set of eight lyrics and ballads by Thomas Hardy for high voice and piano. These were

---

* See also 'Let's Crab an Opera' by William Plomer. The *London Magazine*, October 1963.
† *Britten* by Imogen Holst. Faber, 1966.

dedicated to John and Myfanwy Piper, and *Winter Words*, as the song sequence was called, was performed for the first time at Harewood House on 8 October by Peter Pears with the composer at the piano, as part of the Leeds Triennial Festival programme.

While composing *Gloriana*, Britten had already been thinking about his next opera, which had been commissioned for the Venice Biennale in 1954; but in the autumn of 1953 he fell ill with acute bursitis in his right shoulder and had for the time being to cancel all conducting and recital engagements. He refused, however, to allow his illness to interfere with his composition schedule and started to write down the music for *The Turn of the Screw* with his left hand. The libretto had been prepared by Myfanwy Piper. Ultimately the score was finished in time; and the production of the opera by the English Opera Group at the Teatro la Fenice on 14 September 1954 with the composer conducting was a great triumph. Some of the critics confessed themselves puzzled by the subject matter of Henry James's story; but all praised the ingenuity of the score, the designs by John Piper and the production by Basil Coleman.* The work was first heard in London during a fortnight's season given by the English Opera Group at Sadler's Wells Theatre at the beginning of October.

That autumn he completed a third canticle. His choice was a poem by Edith Sitwell, *Still falls the rain*, which he set for tenor, horn and piano. It was dedicated to the memory of Noel Mewton-Wood, the brilliant Australian pianist, who had recently committed suicide; and the first performance was announced for a memorial concert at the Wigmore Hall on 4 December 1954. This particular concert had to be postponed till 28 January 1955, however, owing to the indisposition of Peter Pears.

During 1955 the English Opera Group gave further performances of *The Turn of the Screw* at Munich, Schwetzingen, and Florence (as part of the Maggio Musicale), at the Aldeburgh Festival in June, and at the Scala Theatre, London, in September. There was also an autumn tour and a Christmas Season of *Let's Make an Opera!*

Towards the end of the year Britten and Pears left England on an extended tour that took them via Yugoslavia and Turkey to the Far East, where they visited Bali, Japan and India before returning to England the following spring.

At this period Britten was much preoccupied with thoughts of writing a full-length ballet. He greatly admired the work of John Cranko, the young choreographer who was attached to the Sadler's Wells Theatre Ballet; and in fact Cranko had been chosen to direct a new production of *Peter Grimes* at Covent Garden in November 1953 with sets by Roger Ramsdell. A preliminary announcement was made in January 1954 that Britten would

* See Appendix B.

80

write a ballet for Cranko to choreograph for the Sadler's Wells Theatre Ballet during its 1954/5 season; but in the event the composition of the score was delayed for nearly two years.

Meanwhile, a new crop of ballets appeared, based on existing Britten scores. At the time of the Coronation of HM The Queen, a ballet called *Fanfare* was mounted by the New York City Ballet with choreography by Jerome Robbins to Britten's *Young Person's Guide to the Orchestra*. The same score was used for a ballet entitled *Variations on a Theme of Purcell* given by the Sadler's Wells Ballet at Covent Garden (6 January 1955) with choreography by Frederick Ashton and scenery and costumes by Peter Snow. (The programme carried the following quotation from Ovid: 'If you have a voice, sing: if soft arms, dance—and with whatever gifts you have for pleasing—please!') The *Variations on a Theme of Frank Bridge* were used for three different ballets. John Cranko created a ballet called *Variations on a Theme* for the Ballet Rambert, which was danced to a version specially scored for small orchestra by James Bernard (21 June 1954). In February 1955 Alan Carter mounted a new version called *Haus der Schatten* with the Bayerischer Staatsopernballett in Munich. And on 10 February 1957 the American Ballet Theatre produced *Winter's Eve* in New York with choreography by Kenneth MacMillan and settings by Nicholas Georgiadis. At the end of 1955, a new ballet by Zachary Solov called *Soirée* was presented at the Metropolitan Opera, New York, with décor and costumes by Cecil Beaton. The action represented a masked ball, and the music was chosen from Britten's arrangement of Rossini's music in *Soirées Musicales*. In the summer of 1957 the José Limon Ballet gave a season at Sadler's Wells Theatre, in the course of which they used Britten's String Quartets 1 and 2 as accompaniment for a ballet entitled *Ruins and Visions* after the collection of poems by Stephen Spender.

During 1954 and 1955 Britten gave much preliminary thought to the ballet score he intended to write. Cranko had already prepared a draft scenario for a three-act ballet and, knowing that what he wanted was 'a vehicle for creative choreography rather than "classical" pastiche', had decided that a 'mythological fairy-tale would supply the framework needed', his idea being 'to make a series of images from traditional fairy stories, linked by a thread of plot which was as important or unimportant as the audience chose to make it. These images would provide the various divertissements . . .'* Accordingly, he prepared a rough outline for *The Prince of the Pagodas*. When Britten formally agreed to write the music, Cranko's early draft scenario was reviewed, the composer making it clear that in his music he would introduce 'various themes on which he would make variations short enough to provide the episodic dances, but which

---

* 'Making a Ballet' by John Cranko. The *Observer*, 13 and 20 January 1957.

would give the work as a whole a sense of continuity.'* A revised script was prepared, on which the composer based his score; and the choreographer found the music so compelling that he had to revisualize the entire choreography.

Britten's interest in oriental music and his voyage to the Far East had yielded a timely musical harvest. Cranko agreed with the designer, John Piper, that some of the more exotic episodes in *The Prince of the Pagodas* would be related 'to the strange edifices of Steinberg and Paul Klee',† and the scene with the pagodas (in the second part of Act II) offered the composer a cue for introducing special colour elements into the orchestra — vibraphone, celesta, piano, xylophone, bells, tomtoms, gongs, all joining in outbursts of gay, festive sound, reminiscent of the Javanese gamelan.

The scenario by virtue of its deliberate construction out of stock characters and situations and various props supposed to be common to the conventions of fairy tales did not completely avoid the trap of pastiche. Nevertheless, the shape of the plot showed that Cranko had devised a suitable armature for a full-length ballet so that the right dances came at the right moments. For instance, the second act opens with a beautiful travelling sequence, where the heroine, Belle Rose, travels through three of the elements (air, water, fire) ultimately to track the Prince to earth in the Kingdom of the Pagodas; and in the third act after the transformation scene there is an elaborate *divertissement* consisting of eleven varied dance numbers containing some of the most original and inventive music in the score.

The Prince of the Pagodas contains about thirty-six different dances — some of them character dances, some of them *divertissements*. To differentiate these dances in the course of a full-length ballet score requires great ingenuity on the part of the composer. As Erwin Stein said,‡ 'It is not only a question of inventing many good and diverse tunes, but also of co-ordinating and balancing them. And as rhythm is the life blood of dancing, it is especially the rhythmic shapes of the tunes that must be well defined and diversified.' Here Britten was extraordinarily successful. The absence of words to set seemed to cause him no inhibitions; and the score has a vitality and a variety of melody, metre, texture and form that puts it in the same category as Prokofiev's full-length ballets *Cinderella* and *Romeo and Juliet*.

The first performance of *The Prince of the Pagodas* was given by the Sadler's Wells Ballet at the Royal Opera House, Covent Garden. At first, the advertised date was 19 September 1956; but as Britten found he was behind-hand with the instrumentation of the score, the première had to be postponed to 1 January 1957. The main roles of the Princess Belle Rose and

---

* Ibid.   † Ibid.
‡ From 'The Prince of the Pagodas: The Music' by Erwin Stein. *Tempo*, Winter 1956/7.

the Prince of the Pagodas were danced by Svetlana Beriosova and David Blair. The composer conducted; and at the end of the performance he was presented with a giant laurel wreath. The ballet was dedicated by composer and choreographer to Imogen Holst and Ninette de Valois.

The Prince of the Pagodas was performed at the Metropolitan Opera, New York, when the Royal Ballet visited America at the end of the summer of 1957; but shortly after that it was dropped from the company's repertory. It was mounted at the Scala, Milan, in the early summer of 1957.

A small group of occasional pieces preceded the composition of *The Prince of the Pagodas*. In the winter of 1955, Britten went skiing with friends at Zermatt, and one of them, Mary Potter, hurt her leg and was laid up. To divert her, he wrote an *Alpine Suite* for three recorders, which the two of them together with Peter Pears used to play in the evenings. To the same year belongs a *Scherzo* for recorder quartet (dedicated to the Aldeburgh Music Club). For the quincentenary of St Peter Mancroft, Norwich, 1955, he supplied a *Hymn to St Peter* for choir and organ; and for the centenary of St Michael's College, Tenbury, 1956, an *Antiphon*, also for choir and organ.

In April 1957, he was elected an honorary member of the American Academy of Arts and Letters and of the National Institute of Arts and Letters. In presenting him with the insignia, the American Ambassador in London read the following citation: 'Your compositions for voice, chamber groups and orchestra have been received with delight in many lands beside your own. Aware of the musical values and the history of your country, often collaborating with poets of distinction, as their contemporaries did with Shakespeare, Jonson, Milton, and Dryden, you have recaptured the great English tradition of word, song, and instrument. Your operas, *Peter Grimes*, *Albert Herring* and *The Rape of Lucretia* among others, are in the world repertory. They do honour to your country and to you.'

In August and September 1957 the English Opera Group visited Canada and played *The Turn of the Screw* at Stratford, Ontario. In October it gave performances of the same opera in Berlin as part of the Berlin Festival. A number of these performances were conducted by Britten himself.

On his return to Aldeburgh in October he decided to start work on a new children's opera. There had been several occasions in the past when he had thought of writing a successor to *The Little Sweep*. In the summer of 1951 he had plans to compose *The Tale of Mr Tod*, after the story by Beatrix Potter, for production by the English Opera Group in 1952; but copyright difficulties intervened, and the project never got off the ground. About 1954 he discussed with William Plomer an opera for children based on a space travel theme. Recalling this abortive venture years later he said:* 'I did in

* From an interview with Britten on his return from Russia in 1964 printed in *The Times*.

83

fact start on [an opera on astronauts], for children, about ten years ago, with William Plomer—who'd written a superb first scene; we occasionally look back at it and I may finish it sometime. It had a character with the magnificent name of Madge Plato.' This time, however, he chose a biblical theme—the episode of the Flood from the Chester Miracle Plays. *Noye's Fludde* was written at great speed. It was started at Crag House, Crabbe Street, on 22 October and finished a few weeks later at the Red House, Aldeburgh, to which he had just moved.

VIII
Aldeburgh: At the Red House

The move from Crag House to the Red House in November 1957 was brought about partly because the former house could be so closely overlooked by people strolling along Crag Path that Britten inevitably suffered from a lack of privacy. The occupier of the Red House, situated in a secluded part of Aldeburgh near the Golf Club, was the artist, Mary Potter; and she now moved into the house in Crabbe Street that had been vacated by Britten and Pears. The Red House, with its attractive garden, was larger and more convenient from the composer's point of view; and his life continued to follow a fairly regular routine of composition interspersed with performances as conductor and pianist, and festival direction.

Britten playing tennis at the Red House, Aldeburgh. He is partnered by Mary Potter the painter

Britten's studio at the Red House, Aldeburgh

The first composition to be completed at the Red House was *Noye's Fludde* (on 15 December 1957), and this was produced with great success as part of the 1958 Aldeburgh Festival programme at the church in the neighbouring village of Orford. The producer was Colin Graham, and the designer Ceri Richards.* This Festival also featured the first performance of *Songs from the Chinese*, a setting of Chinese poems translated by Arthur Waley for high voice and guitar, which had been completed before the move to the Red House the previous autumn. These were performed by Peter Pears and Julian Bream.

Although Britten had never been particularly keen on writing as such, feeling his true medium was music, nevertheless he agreed to collaborate with Imogen Holst on a book entitled *The Story of Music*, which was published by Rathbone Books, London, in 1958. This was a popular manual, lavishly illustrated and with collages specially designed by Ceri Richards, dealing first with the essential components of music—sound and rhythm, songs and singers, instruments and players—and then with drama

* See Appendix B.

The Library at the Red House

in music, styles in music, and western music as compared with eastern music. The final section touches on the work of the composer; and this leads to a reference to twentieth-century experiments in distortion, particularly the twelve-note or twelve-tone system which is basic for the construction of 'serial' music. Here the two authors write: 'It is impossible for anyone to say whether this is to be the recognized music of the second half of the twentieth century. Some musicians think it is. Some think that there are still unlimited possibilities in the seven-note scale and the chords that grow out of it. Some think that it does not matter what style a composer chooses to write in, as long as he has something definite to say and says it clearly.'

Other occasional articles by Britten published about the same time include a tribute to the virtuoso horn-player, Dennis Brain, killed in a car crash in 1957 (*Tempo*, Winter 1958), and an essay 'On Realising the Continuo in Purcell's Songs' which was published in *Henry Purcell (1659–1695)*, a symposium edited by Imogen Holst in honour of the tercentenary of Purcell's birth (Oxford University Press, 1959). The Purcell celebrations included a performance of his *Ode for St Cecilia* conducted by Britten at the Royal Festival Hall (10 June 1959), on which occasion Dame

87

Edith Sitwell read a new poem entitled *Praise We Great Men*, which she had written in honour of the occasion and dedicated to Britten.

The previous summer Britten had broadcast a programme, 'Personal Choice', on the BBC Home Service (16 July 1958), in which he selected a number of his favourite poems, which were then read over the air. He excluded poems which he had already set to music; and the following were his choice (in many cases prophetic of settings to come):

Crabbed Age and Youth	William Shakespeare
When most I wink	William Shakespeare
Vertue	George Herbert
The Chimney Sweeper	William Blake
'The World is too much with us'	William Wordsworth
'Ah! Sun-flower!'	William Blake
From *The Wanderings of Cain*	S. T. Coleridge
'Break, break, break'	Alfred Lord Tennyson
'If it's ever Spring again'	Thomas Hardy
Strange Meeting	Wilfred Owen
The Youth with Red-gold Hair	Edith Sitwell
'Lay Your Sleeping Head, My Love'	W. H. Auden

He also included *Hälfte des Lebens* by Hölderlin, together with Michael Hamburger's translation into English.

He forthwith proceeded to make a new short anthology of poems connected with night and sleep and dreams, with the intention of setting them on similar lines to his *Serenade*. For this purpose he chose a few lines from Shelley's *Prometheus Unbound*, Tennyson's *The Kraken*, Coleridge's *The Wanderings of Cain*, Middleton's *Blurt, Master Constable*, an extract from Wordsworth's *The Prelude*, *The Kind Ghosts* by Wilfred Owen, *Sleep and Poetry* by Keats and Shakespeare's 43rd Sonnet ('When most I wink'). He set these for tenor solo and string orchestra; but whereas in the case of the *Serenade* he had chosen a single *obbligato* instrument (the horn) for the whole work, in the *Nocturne* he picked seven. The opening poem (by Shelley) is set for strings alone. A bassoon *obbligato* is added to the strings for the Tennyson; harp for the Coleridge; horn for the Middleton; timpani for the Wordsworth; English horn for the Owen; flute and clarinet for the Keats; and all seven solo instruments join together with the string orchestra for the Shakespeare finale. Each *obbligato* instrument gives a different tone colour to the movement it appears in. Particularly impressive are the bassoon's submarine wallowings and burblings in *The Kraken*, and the menace of the timpani in the passage from *The Prelude*:

> *But that night*
> *When on my bed I lay, I was most mov'd*

And felt most deeply in what world I was . . .
I thought of those September Massacres,
Divided from me by a little month,
And felt, and touch'd them, a substantial dread . . .
And in such way I wrought upon myself
Until I seem'd to hear a voice that cried
To the whole City, 'Sleep no more!'

This *Nocturne* is in the sort of lyrical elegiac vein that is both characteristic of Britten and reminiscent of Mahler; and it seemed most appropriate that the work should be dedicated to Mahler's widow, Mrs Alma Mahler Werfel. The first performance was given on 16 October 1958 as part of the Leeds Centenary Festival by Peter Pears with the BBC Symphony Orchestra conducted by Rudolf Schwarz.

Another poem from the BBC 'Personal Choice' programme provided the cue for his next song cycle. It was Prince Ludwig of Hesse and the Rhine who brought Britten's attention to Hölderlin and his poetry, and he now chose *Hälfte des Lebens* and five other poems, to make *Six Hölderlin Fragments*, which he set for tenor and piano. These were first performed at Schloss Wolfsgarten, near Frankfurt-am-Main, on Prince Ludwig's fiftieth birthday (20 November 1958). The poems are short, and the songs represent Britten's style at its pithiest. Even though he is dealing with a foreign language, his inflexions are impeccable; and he also allows the words and their meaning to provide an occasional cue for musical transliteration, e.g. the drooping melodic tendrils in *Hälfte des Lebens* and the wonderful scalic ladders that slowly intersect in *Die Linien des Lebens*.

In 1959 Britten received a very attractive commission. It came from the University of Basle and was an invitation to write a cantata in honour of the 500th anniversay of the University's foundation which would be celebrated in 1960. He accepted and agreed to set a Latin text compiled from the Charter of the University and from older orations in praise of Basle.

After the 1959 Aldeburgh Festival, certain adaptation work was carried out to the Jubilee Hall. The stage and orchestral pit were enlarged, the dressing-room accommodation improved, the seating capacity increased to 316, and proper ventilation installed. In order to celebrate these improvements, Britten decided to compose a new opera for performance at the 1960 Festival. As usual, the production was to be entrusted to the English Opera Group; but in view of the increased facilities at the Jubilee Hall, he felt he could now write for a somewhat larger orchestra than the usual English Opera Group ensemble of about a dozen solo instruments. In 1960 he himself described the inception of the opera as follows:* 'Last August it was

* 'A New Britten Opera' by Benjamin Britten. The *Observer*, 5 June 1960.

decided that for this year's Aldeburgh Festival I should write a full-length opera for the opening of the reconstructed Jubilee Hall. As this was a comparatively sudden decision there was no time to get a libretto written, so we took one that was ready to hand . . . I have always loved *A Midsummer Night's Dream*. As I get older, I find that I increasingly prefer the work either of the very young or of the very old. I always feel *A Midsummer Night's Dream* to be by a very young man, whatever Shakespeare's actual age when he wrote it. Operatically, it is especially exciting because there are three quite separate groups—the Lovers, the Rustics, and the Fairies—which nevertheless interact . . . In writing opera I have always found it very dangerous to start writing the music until the words are more or less fixed. One talks to a possible librettist, and decides together the shape of the subject and its treatment. In my case, when I worked with E. M. Forster or William Plomer, for instance, we blocked the opera out in the way an artist might block out a picture. With *A Midsummer Night's Dream*, the first task was to get it into manageable shape, which basically entailed simplifying and cutting an extremely complex story . . . I do not feel in the least guilty at having cut the play in half. The original Shakespeare will survive . . . I actually started work on the opera in October, and finished it on, I think, Good Friday—seven months for everything, including the score. This is not up to the speed of Mozart or Verdi, but these days, when the line of musical language is broken, it is much rarer. It is the fastest of any big opera I have written, though I wrote *Let's Make an Opera!* in a fortnight.'

Although the work of composition proceeded speedily, a number of checks and difficulties were encountered—illness for one—but he could not afford to allow his regular working schedule to be affected. 'A lot of the third act', he admits,* 'was written when I was not at all well with flu. I find that one's inclination, whether one wants to work or not, does not in the least affect the quality of the work done.'

Anyway, the work was finished on time. John Cranko was appointed producer; John Piper designed the sets; and Carl Toms was responsible for the costumes.† The first performance at Aldeburgh (11 June 1960) was an unqualified success. Later that month the opera was given by the English Opera Group as part of the Holland Festival programme. When it arrived at Covent Garden the following winter (2 February 1961) a larger complement of strings was used in the orchestra. Georg Solti conducted; and the production was entrusted to Sir John Gielgud. Within a year of its first production, the opera had also been staged in Hamburg, Zurich, Berlin, Pforzheim, Milan, Vancouver, Göteborg, Edinburgh, Schwetzingen, and Tokyo. This made it clear that *A Midsummer Night's Dream* was likely to

---

* Ibid.   † See Appendix B.

prove the most popular in the international field of Britten's full-scale operas since *Peter Grimes*.

About this time a change occurred in the status of the English Opera Group. Originally it had been set up as a company limited by guarantee and had pursued an independent artistic policy, receiving an annual subsidy from the Arts Council of Great Britain. But now it was judged expedient for it to become closely associated with the Royal Opera House, Covent Garden, which henceforth undertook its direct management.

The first performance of *Cantata Academica* (*Carmen Basiliense*) took place in Basle on 1 July 1960 when it was performed by the Basle Chamber Choir and the Basle Chamber Orchestra under the direction of Paul Sacher. The work caught the imagination of the public with its infectious high spirits, good humour and academic ingenuities (e.g. Chorus *alla Rovescio*, *Tema Seriale con Fuga*, *etc.*).

During 1960 Britten completed a revision of *Billy Budd*, changing it from a four-act to a two-act opera and in the process shortening the end of the original Act 1. The new version was heard for the first time in a studio performance broadcast by the BBC (13 November 1960) and was staged at Covent Garden early in 1964.

In recent years it looked as if Britten had been composing more for the theatre and concert hall than for the church; but the time had now come to redress the balance. In 1959 he wrote the *Missa Brevis in D* as a parting present to George Malcolm, the retiring organist at Westminster Cathedral, who was well known as an excellent trainer of boys' voices and a doughty opponent of the artificial and unnatural sound known as 'Cathedral Tone'. He was a firm upholder of the belief that it was perfectly feasible 'to train a boy's voice—to refine it, and develop it, and turn it into a medium of real musical beauty—without destroying its natural timbre, and without removing from it the characteristics of the normal human Boy.'* In the *Missa Brevis*, the writing for boys' voices is particularly well fitted to the type of singing voice that Malcolm had trained so successfully in the Cathedral choir during his term as organist. The new work was first given during High Mass at Westminster Cathedral (22 July 1959).

A little later (1961) Britten wrote a *Jubilate Deo* for St George's Chapel, Windsor, at the request of HRH The Duke of Edinburgh, as a companion piece to the *Te Deum* of 1934.

But now the chance of a major church commission came his way, and he embraced it readily, though not without feeling slightly daunted at the magnitude of the task.

For some years the building of a new Cathedral in Coventry, designed by

---

* From 'Boys' Voices' by George Malcolm. *Tribute to Benjamin Britten on his Fiftieth Birthday*.

Sir Basil Spence and intended to complement the ruins of the bomb-damaged medieval Cathedral, had been under way; and many artists— including Graham Sutherland, Jacob Epstein, John Piper and John Hutton—had been engaged to beautify it. When the new Cathedral was seen to be approaching completion, it was agreed to celebrate its rededication with a special festival of the arts. A number of new works were required for the occasion, including a large-scale oratorio; and this commission was offered to Britten, who accepted it. He fully realized the importance of the occasion, for it would mark not only the phoenix-like resurgence of the new Cathedral at the side of the shattered shell of the old, but also the healing of many wounds. He wanted to make some public musical statement about the criminal futility of war, and this seemed a good opportunity to do so. Clearly a big work was called for—something for full orchestra and full chorus; something calculated for 'a big, reverberant acoustic'.* From the beginning the idea of the *Missa pro Defunctis* was in his mind—but this was to be a setting with a difference. When looking for something that would help him to contrast the spiritual pleadings and terrors and consolations of the *Missa pro Defunctis* with the realistic horror, suffering and desolation caused by twentieth-century war, he thought once again of his 1957 'Personal Choice' programme for the BBC and found what he required in Wilfred Owen's poem *Strange Meeting*. And in Owen's unpublished preface to his poems (*c.* 1918) lay the key to what was to be called *War Requiem*. '. . . I am not concerned with Poetry,' wrote Owen. '*My subject is War, and the pity of War. The Poetry is in the pity.* Yet these elegies are to this generation in no sense consolatory. They may be to the next. *All a poet can do to-day is to warn.* That is why the true Poets must be truthful.' (The italicized passages appear on the title page of the *War Requiem* as an epigraph.)

Britten now decided that on the big-scale level he would set part of the *Missa pro Defunctis* for chorus and full orchestra, while on a more intimate level (using soloists and chamber orchestra of twelve) he would set the text of nine of Owen's war poems, which would be inserted episodically into the six main movements of the Requiem. Two further musical decisions followed. The setting of the poem *Strange Meeting* seemed to postulate two male soloists, one symbolizing a British soldier (the poet) and the other a German soldier whom he had killed—here Britten chose a tenor and a baritone; and the setting of the rest of the song cycle was divided between them. This meant that, if only for purposes of contrast, it was preferable for the Requiem proper to have a female soloist (soprano); and a boys' choir was added in order to strengthen the high tessitura.

The Owen song cycle has the same sort of stature and importance as *Les*

* *On Receiving the First Aspen Award* by Benjamin Britten. Faber, 1964.

Illuminations or the *Nocturne*. Britten has set the following nine poems: 1. *Anthem for Doomed Youth*, 2. *Voices* (part only), 3. *The Next War*, 4. *Sonnet on Seeing a Piece of our Artillery Brought into Action* (six lines only), 5. *Futility*, 6. *The Parable of the Old Men and the Young*, 7. *The End*, 8. *At a Calvary near the Ancre*, and 9. *Strange Meeting* (thirty-three lines out of a total of forty-four). The songs are self-contained, with certain exceptions. *Futility* (which is treated as whispered recitative) is cross-cut with the *Lacrimosa dies illa* verse of the *Dies Irae* movement. A cue from the *Offertorium*—'*quam olim Abrahae promisisti, et semini ejus*'—leads into Owen's poem *The Parable of the Old Men and the Young* with its retelling of the Abraham and Isaac parable in twentieth-century terms:

> *When lo! an angel called him out of heaven,*
> *Saying, Lay not thy hand upon the lad,*
> *Neither do anything to him. Behold,*
> *A ram, caught in a thicket by its horns;*
> *Offer the Ram of Pride instead of him.*
> *But the old man would not so, but slew his son,—*
> *And half the seed of Europe, one by one.*

And here the music deliberately recalls the parallel passage in Britten's *Canticle II*. The penultimate song *At a Calvary near the Ancre* is cross-cut with the *Agnus Dei* movement, and at the end the tenor soloist is drawn out of the Owen poems with a final ascending scalic phrase '*Dona nobis pacem*'. The final song, *Strange Meeting*, is treated as a duologue in recitative for tenor and baritone; and their last line '*Let us sleep now*' becomes a coda which is absorbed into the final chorus '*In paradisum deducant te Angeli*'.

There are many striking features in the *War Requiem*;* and it undoubtedly succeeds in what it sets out to do, in moving between two planes—the setting of the *Missa pro Defunctis* in Latin for soprano solo with full chorus and full orchestra, and the setting of the Owen poems for the two male soloists with chamber orchestra. But at times one wonders whether it is really possible to achieve a conciliation between these elements. '*Requiem aeternam dona eis, Domine*' sings the chorus, leading directly to the tenor soloist's agonized question '*What passing-bells for these who die as cattle?*' and his answer '*Only the monstrous anger of the guns.*' In the *Sanctus* movement, the choral cries of *Hosanna in excelsis!* are followed by Owen's poem *The End*, where the baritone soloist asks:

> *Shall life renew these bodies? Of a truth*
> *All death will life annul, all tears assuage?*

---

* An excellent study of the score of *War Requiem* by Peter Evans will be found in *Tempo*, Spring/Summer 1962.

Britten receiving the Freedom of the Borough of Aldeburgh in the Moot
Hall, 22 October 1962

and Earth answers

My fiery heart shrinks, aching. It is death.
Mine ancient scars shall not be glorified,
Nor my titanic tears, the sea, be dried.

The first performance of *War Requiem* was given in St Michael's
Cathedral, Coventry, on 30 May 1962. The soloists were Heather Harper,
Peter Pears and Dietrich Fischer-Dieskau. The chorus and full orchestra
were conducted by Meredith Davies and the chamber orchestra by the
composer. The work made a profound impression both then and later when
it was heard in London at Westminster Abbey (6 December 1962). By the
end of 1963 it had been performed in more than a dozen cities abroad,
including Berlin, Boston, Paris, Perugia and Prague. The Decca gramophone recording sold over 200,000 sets in only five months.

In May 1962, just before the first performance of *War Requiem*, Britten
completed a setting of *Psalm 150* for two-part children's voices. This was
intended for the centenary celebrations of his preparatory school, Old
Buckenham Hall School (formerly South Lodge School) Lowestoft, and it
was performed there in July, the first public performance being given the
following year (1965) in the course of the Aldeburgh Festival.

To mark the fourteenth centenary of St Columba's missionary journey
from Ireland to Iona, a *Hymn of St Columba* was written in May 1963 and
given its first performance on 2 June 1963 at Gartan, County Donegal, the
Saint's birthplace.

94

Meanwhile, Britten had gradually formed a number of friendships with Russian musicians that were to have an important effect on his work. In September 1960, he met the great Russian cellist, Mstislav Rostropovich, in London, and it was agreed that he should compose a special Cello Sonata for him. This was dispatched to Russia in the winter of 1961, and rehearsed in London at the beginning of March when Rostropovich was passing through on his way to South America. The first public performance was given by Rostropovich and the composer at the Aldeburgh Festival on 7 July 1961. During this Festival visit, Rostropovich was joined by his wife Galina Vishnevskaya, who gave a vocal recital in honour of the Festival. Britten was so impressed by her lovely voice that he hoped she would be able to sing the solo soprano part at the first performance of his *War Requiem*; but unfortunately her engagements did not permit her to do so. Another Russian musician, who became a constant visitor to the Aldeburgh Festival, was the great pianist, Sviatoslav Richter, whom Rostropovich introduced to Britten in London in 1961. The following year Dmitri Shostakovich heard *The Turn of the Screw* at the Edinburgh Festival in company with Rostropovich, and the music of the opera made a tremendous impact on him. So it was not surprising when an invitation came for Britten and Pears to visit the USSR during a festival of British music in March 1963.

The programme of this festival in Moscow and Leningrad included orchestral performances of the *Sinfonia da Requiem*, the *Serenade*, and the Sea Interludes and Passacaglia from *Peter Grimes*, and chamber music recitals which included the Sonata in C for Cello and Piano, *Winter Words* and the *Six Hölderlin Fragments*. This was the moment when the Second Plenary Meeting of the Board of the USSR Union of Composers was sitting in Moscow, and special attention was being given to the work of the younger generation of composers. The current attitude to contemporary British music seemed propitious, especially when contrasted with the hostile atmosphere of the All-Union Congress of Soviet Composers only fifteen years previously. Britten and Pears received a very warm reception from their Russian audiences.

During this visit, Britten gave an interview to a *Pravda* correspondent in which he said: 'I must own that until my arrival in USSR I was assailed with doubts whether the Soviet audiences would understand and accept our musical art which had been developing along different national lines than the Russian. I am happy at having had my doubts dispelled at the very first concert. The Soviet public proved not only unusually musical— that I knew all along—but showed an enviable breadth of artistic perception. It is a wonderful public.'

This interview was widely quoted in the international press and, coming when it did, it may have played a not unhelpful part in the improvement in Anglo-Soviet cultural relations.

95

Britten in Moscow in 1963, with Peter Pears and Galina Vishnevskaya (*left*) and Mstislav Rostropovich and Marion, Countess of Harewood— now Mrs Jeremy Thorpe (*right*)

On his return to Aldeburgh, Britten completed two important works: the *Symphony in D for Cello and Orchestra*, which was dedicated to Rostropovich, and the *Cantata Misericordium* for tenor and baritone solos, small chorus and string orchestra, piano, harp and timpani, written to celebrate the centenary of the foundation of the Red Cross and dedicated to the Countess of Cranbrook, the Chairman of the Aldeburgh Festival Council. The first performance of the *Symphony for Cello* was scheduled for the 1963 Aldeburgh Festival, but had to be cancelled owing to Rostropovich's illness. The first performance of the *Cantata Misericordium* took place in Geneva on 1 September 1963 as planned, with Peter Pears and Dietrich Fischer-Dieskau as soloists, the Motet Genève and the Orchestre de la Suisse-Romande, conducted by Ernest Ansermet. The first performance in Great Britain occurred on 12 September at the Royal Albert Hall in the course of a Promenade Concert devoted entirely to Britten's music.

This special 'Prom' programme marked the opening of an autumn season of productions and performances intended to honour his fiftieth birthday on 22 November. Another new work appeared in September entitled *Night Piece* (Notturno) for piano solo written as a test piece for competitors in the first Leeds International Pianoforte Competition. *Peter Grimes* was revived in a new production by Basil Coleman at Sadler's Wells (16 October).

On the actual birthday, a special tribute from about three dozen of his friends—including writers, artists, musicians—was published by Faber and Faber. Anthony Gishford was the editor; and in his Foreword he explained that the 'invitation to those of Mr Britten's friends and collaborators who we thought would like to join in such a gift specified only that they should contribute something that they themselves liked and that they thought would give the recipient pleasure'. This loose formula covered such contributions as a chapter from an unfinished novel by E. M. Forster, a study of Edward Fitzgerald by William Plomer, substantial extracts from the diary written by Prince Ludwig of Hesse and the Rhine when he accompanied Britten and Pears on their journey to the Far East in 1956, and three works which had been specially dedicated to Britten—Edith Sitwell's poem 'Praise We Great Men' (1959), Hans Werner Henze's *Kammermusik 1958*, and Michael Tippett's *Concerto for Orchestra*, which bears the superscription 'To Benjamin Britten with affection and admiration in the year of his fiftieth birthday'. These last two tributes were represented by page facsimiles from the manuscript scores.

Many tributes appeared also in the national daily and weekly press, in periodicals, and on radio and television.

On the evening of Britten's birthday, a concert performance of *Gloriana* at the Royal Festival Hall was presented as a tribute to the composer by the Polyphonia Orchestra with Bryan Fairfax as conductor. Sylvia Fisher sang the part of Queen Elizabeth I for the first time, and Peter Pears was heard in his original part of the Earl of Essex. The revival of this opera was warmly welcomed by the composer and the audience, but the pleasure generated by the occasion was sadly affected as the news of the assassination of President Kennedy earlier that day in Dallas started to seep through the auditorium.

In some ways Britten's fiftieth birthday proved a climacteric. For nearly thirty years the firm of Boosey & Hawkes had been the exclusive publishers of his music. Now he decided to make a change. His contract was not renewed, and he threw in his fortunes with a new music publishing venture started by Faber and Faber, whose ambitious plan was 'to build up a catalogue of the highest quality, comprising both old and new music'. The first work of Britten's they published was *Nocturnal* (after John Dowland) for guitar, op. 70. The second was *Curlew River*, a new type of opera styled 'a parable for church performance'. Unlike most of Britten's works this was not written at home in Aldeburgh, but abroad in Venice, where he went for a change of air in February 1964. The way in which he and his librettist, William Plomer, succeeded in adapting stylistic methods characteristic of the Japanese Noh plays to a new kind of operatic convention for church performance marked a new departure in his operatic practice, not the least revolutionary feature being the abolition of the conductor.

After Venice, he paid a brief visit to the USSR in order to conduct the first performance of the *Symphony in D for Cello and Orchestra* in Moscow on 12 March 1964 with Rostropovich as soloist. He went over to Leningrad for a few days; and while he was there a group of music students performed part of the *War Requiem* in his honour. A little later the first performance of *Peter Grimes* in the USSR was given in concert form in the large hall of the Leningrad Conservatoire by opera soloists, the Leningrad Radio chorus, and the Leningrad Philharmonic Orchestra under Dzhemal Dalgat; and this led to two performances at the Kirov Theatre, Leningrad, and one in Moscow at the end of the season.

In May 1964 Britten was named the first winner of the Aspen Award in the humanities. This had been established by Mr Robert O. Anderson of Roswell, New Mexico, chairman of the Institute of Humanistic Studies at Aspen, Colorado, to honour 'the individual anywhere in the world judged to have made the greatest contribution to the advancement of the humanities'. The award (of $30,000) was handed over on 31 July at a ceremony held in the amphitheatre at Aspen and attended by about 1,500 guests. In his speech of thanks, Britten referred again to a subject that was always near his heart—that of 'occasional' music—and emphasized the importance of the 'occasion'. He reminded his audience that 'Bach wrote his *St Matthew Passion* for performance on one day of the year only—the day which in the Christian church was the culmination of the year, to which the year's worship was leading' and went on to say 'it is one of the unhappiest results of the march of science and commerce that this unique work, at the turn of a switch, is at the mercy of any loud roomful of cocktail drinkers—to be listened to or switched off at will, without ceremony or occasion'.* He expressed his dismay that musical performances could be made 'audible in any corner of the globe, at any moment of the day or night, through a loudspeaker, without question of suitability or comprehensibility. Anyone, anywhere, at any time, can listen to the B minor Mass upon one condition only—that they possess a machine. No qualification is required of any sort—faith, virtue, education, experience, age.' These were brave words, and possibly words that some members of his audience would not be prepared to accept; but anyone acquainted with the scientific advances during the last half-century that have made it possible to popularize music on such a universal scale must realize that this has not necessarily brought with it an automatic rise in the standards of listening and appreciation—in fact, everywhere the dangers of aural debasement are only too real.

Britten placed the money from the Aspen Award in a special Trust Fund which had been set up for the purpose of helping and encouraging young musicians.

During the summer of 1964 the English Opera Group toured the successful Aldeburgh Festival production of *Curlew River* (first performance, Orford Church, 13 June 1964) to the Holland Festival, and subsequently to the City of London and King's Lynn Festivals. In the autumn it embarked on a tour of the USSR (Riga, Leningrad, and Moscow) with a repertory consisting of *The Rape of Lucretia*, *Albert Herring* and *The Turn of the Screw*. The Russian audiences were most responsive; and special interest in the scale and style of the English Opera Group productions and their suitability for touring was expressed by Madame Furtseva, the Soviet Minister of Culture.

In 1965 Britten and Pears made two important trips abroad. First, there was a visit to India early in the year. Then in the summer they were invited to visit the USSR and to stay for a period in the Composers' Home for Creative Work in Dilidjan, Armenia. Two days after a performance of the *Symphony for Cello* by Rostropovich at the Royal Festival Hall (1 August), they flew with Rostropovich and his wife Galina Vishnevskaya to Yerevan via Moscow. From there they motored to Dilidjan where all four of them stayed for just over three weeks.† On 28 August a Britten Festival opened in Yerevan and the composer attended the various concerts, whose programmes included performances of the *Peter Grimes* Interludes, *The Young Person's Guide to the Orchestra*, the 1st String Quartet, three movements of the *Sonata for Cello and Piano*, the first *Suite* for unaccompanied cello (first

* *On Receiving the First Aspen Award* by Benjamin Britten. Faber, 1964.
† *Armenian Holiday*, an attractively informal account of this holiday in diary form by Peter Pears, was privately published in 1966.

performance in the USSR), and various songs, including two extracts from a new Pushkin song cycle.

Britten had bought a copy of the Penguin edition of Pushkin when he was passing through London Airport on his way to the USSR and fallen in love with some of the lyrics. He thought that setting some of these poems might help his Russian, so he got Rostropovich and Vishnevskaya to read aloud the poems he had chosen from the literal English translation in the Penguin, and they set about teaching him to pronounce them properly. He then worked out a transliteration of six of them, and began setting them to music. By the time the holiday party left Armenia the new cycle (to be called *The Poet's Echo*) had been sketched out in rough. A fascinating account is given in Peter Pears's *Armenian Holiday** of a visit they paid to Pushkin's birthplace in the course of a special journey between Novgorod and Leningrad by car. '. . . It was not until 8 p.m. that we reached our destination, Pushkin's home, just twenty-four hours later than we were expected. Our hosts had waited for us all night. Slava† had not thought to telephone; but we were welcomed and greeted as if they had been in no way inconvenienced or put out . . . Before we retired our host took a torch and showed us Pushkin's house and museum, and outside the front door was the clock tower and its cracked clock which was there in Pushkin's time and still struck its old hours . . . After a meal of soup and excellent cold leg of lamb, and a sort of barley with meat balls (so good), marvellous coffee and plum syrup, our host begged to hear the Pushkin songs. We moved into the lamp-lit sitting-room with an upright piano in the corner, and started on the songs after an introduction by Slava. Galya sang her two, and I hummed the others. The last song of the set is the marvellous poem of insomnia, the ticking clock, persistent night-noises and the poet's cry for a meaning in them. Ben has started this with repeated staccato notes high-low high-low on the piano. Hardly had the little old piano begun its dry tick-tock tick-tock, than clear and silvery outside the window, a yard from our heads, came ding, ding, ding, not loud but clear, Pushkin's clock joining in his song. It seemed to strike far more than midnight, to go on all through the song, and afterwards we sat spell-bound. It was the most natural thing to have happened, and yet unique, astonishing, wonderful . . .'

As a coda to this Russian visit, *A Midsummer Night's Dream* received its first Russian production at the Bolshoi Theatre, Moscow, on 28 October 1965 with Gennadi Rozhdestvensky as conductor, Boris Pokrovsky as producer, and Nicolai Benois as designer.

Meanwhile there had been a plentiful outpouring of other new works. At the 1965 Aldeburgh Festival, the first *Suite* for unaccompanied cello, the *Gemini Variations* for piano duet, flute and violin, and *The Songs and*

* Ibid.   † Rostropovich.

Proverbs of William Blake, an extended song-cycle for baritone and piano, had all received their first performances. Rostropovich played the *Suite*; the Hungarian twins, Gabriel and Zoltán Jeney, performed the *Variations*; and the composer accompanied Dietrich Fischer-Dieskau in the Blake cycle.

Somewhat earlier Britten had accepted a commission from the United Nations for an anthem to mark their twentieth anniversary. The closing section was a setting of Virgil's fourth Eclogue—the vision of the Golden Age to come. *Voices for Today* (as it was called) was given a simultaneous première in New York, Paris and London on 24 October 1965. In his address delivered in the General Assembly Hall, U Thant, the Secretary-General of the United Nations, said: 'To Benjamin Britten, the ideal of peace is a matter of personal and abiding concern. At the head of an earlier composition (also about war and peace), he once wrote this stark preamble: "All the artist can do . . . is warn." Today he speaks for all of us, with an eloquence we lack, in a medium of which he is a master.'

For the 1966 Aldeburgh Festival Britten planned to compose a second parable for church performance as a companion piece to *Curlew River*. Once again William Plomer was invited to write the libretto; and the subject chosen was Nebuchadnezzar and the burning fiery furnace. But Britten's composition schedule was upset in the winter of 1966 by illness; and he had to undergo an operation for diverticulitis. Nevertheless, *The Burning Fiery Furnace* was ready on time and, like *Curlew River*, received its first performance in Orford Church (9 June 1966).

Earlier in the year, the film of *Curlew River*, made at Louvain in 1965 by the Belgian Television, was awarded first prize in the Monte Carlo 'Unda' Festival; and *The Burning Fiery Furnace* was telerecorded at Louvain later in the summer. The English Opera Group tour of *The Burning Fiery Furnace* also covered Holland, and the City of London and King's Lynn Festivals.

In the autumn *Gloriana* was successfully revived at Sadler's Wells. For this production (by Colin Graham), the composer made a few rather minor revisions to the score.

Recitals in Moscow and Leningrad at the end of December 1966 gave Pears and Britten a chance to remain a few days longer in Moscow so that they could celebrate the Russian New Year with Rostropovich, Vishnevskaya, Shostakovich and others. (A sparkling account of this trip is given in the extracts from Peter Pears's diary printed in the 1967 *Aldeburgh Festival Programme Book*.) An interesting incident occurred after the recital at the Philharmonic Hall in Leningrad, when a group of young dancers mainly from the Kirov Ballet, formed and organized by the horn player Bouyanowski, gave Britten a private performance of a ballet that had recently been devised to his *Metamorphoses* for oboe solo.

In England the year 1967 was notable for the completion of a number of new concert halls. In London two halls were opened on the South Bank: the Queen Elizabeth Hall on 1 March, and the Purcell Room two days later. Britten conducted the greater part of the inaugural concert at the Queen Elizabeth Hall in the presence of HM Queen Elizabeth II; and the programme included the first performance of his *Hankin Booby*,* a folk dance for wind and drums, that had been commissioned by the Greater London Council.

On 2 June came the opening of the new Maltings Concert Hall, Snape, by HM The Queen, who was accompanied by HRH The Duke of Edinburgh. This was the first event of the 1967 Aldeburgh Festival. The inaugural concert programme also contained a new work by Britten, an Overture (with chorus) called *The Building of the House*. This had been 'inspired by the excitement of the planning and building—and the haste!' † and it proved to be one of his most successful pieces of 'occasional music'. The new concert hall was generally acclaimed as one of the finest in the country; and when a new production of *A Midsummer Night's Dream*, which the English Opera Group had presented in Paris a few weeks previously, was mounted there on 7 June, it was seen to be suited for opera too. The availability of the new hall made a great difference to the general Festival planning. In fact the sales of tickets for the 1967 Festival rose to a figure well over double that of 1966. In 1968 the hall won a Civic Trust award.

A vaudeville for boys and piano based on the old West Country ballad of *The Golden Vanity* was another new work that received its première at the 1967 Festival when it was performed by the Vienna Boys' Choir (3 June) for whom it had been specially written. The work was staged imaginatively by its librettist, Colin Graham—in costume, but without scenery—and the action was mimed, in a simple way, a few basic properties being provided. The result was like a miniature opera.

In September the English Opera Group played *Curlew River* and *The Burning Fiery Furnace* at Expo '67, Montreal; and afterwards Britten and Pears embarked on a tour that took them, first to New York, and then through various countries of Latin America including Mexico, Peru, Chile, Argentina, Uraguay and Brazil. Each of their two recital programmes contained works by Britten: in the first there was *The Poet's Echo* and some of Britten's folk song arrangements; in the second the *Seven Sonnets of Michelangelo*, *Winter Words*, and the *Six Hölderlin Fragments*. Their reception everywhere was most enthusiastic.

A few years previously, after seeing Rembrandt's great painting *The Return of the Prodigal* in the Hermitage Museum, Leningrad, Britten had decided that this was to be the theme of his next church parable. William

---

* *Hankin Booby* was later incorporated into the *Suite on English Folk Tunes: 'A time there was . . .'* (1974).

† Programme Note by B.B. in the *Aldeburgh Festival Programme Book*, 1967.

Britten (in the orchestra pit) rehearsing Mozart's *Idomeneo* with the
English Opera Group at Snape Maltings, June 1970

The Maltings Concert Hall, Snape

Plomer was once again chosen as librettist. A considerable part of the score was written during Britten's stay in Venice at the beginning of 1968. But on his return to Aldeburgh at the beginning of March he succumbed to an attack of fever which delayed the completion of the score for a few weeks. Fortunately he recovered in time for the new work (*The Prodigal Son*) to be rehearsed and produced at the 1968 Aldeburgh Festival, where it repeated the success of the two earlier church parables.

Early in January 1969 he completed *Children's Crusade*, a setting of Brecht's poem *Kinderkreuzzug* for boys' choir and orchestra; and on 19 May it was performed by the boys of Wandsworth School in the impressive setting of St Paul's Cathedral, London, to commemorate the fiftieth anniversary of the Save the Children Fund. But the events of the year were overshadowed by the disastrous fire that destroyed the Maltings Concert Hall, Snape, after the opening performance of the Aldeburgh Festival on 7 June.

Since its opening the hall had acquired an international reputation. In addition to Festival performances, it had been used for the following purposes: a Bach week-end, an Antique-Dealers' Fair, summer orchestral and band concerts, and a series of 'Jazz at the Maltings' for BBC Television. Recordings had been made there with Rostropovich, Vishnevskaya, Britten himself, Peter Pears, Philip Ledger, the English Chamber Orchestra, the Ambrosian Singers, and the Katchen-Suk-Starker Trio. The most ambitious operation, however, had been the television film of *Peter Grimes* made by the BBC during February 1969. And plans were under discussion for extending its use in other ways.

The fire put a temporary stop to all that. But it was clear that the same spirit of indefatigable determination that produced a revised programme for the 1969 Aldeburgh Festival in less than twenty-four hours and saw to it that the performances originally scheduled for the Maltings (including the stage production of Mozart's *Idomeneo*) were transferred elsewhere would be applied also to the rebuilding plans. In the event, after a year of intensive rebuilding activity the hall rose again like a phoenix from its ashes; and the fact that the opening concert on 5 June 1970 (entitled 'Music for a Royal Occasion') was attended by both HM The Queen and HRH The Duke of Edinburgh showed the continuity of royal interest. And later that summer, Britten and Pears were invited to give a recital at Sandringham House in honour of the seventieth birthday of HM Queen Elizabeth The Queen Mother.

Peter Pears, Britten and HM Queen Elizabeth, The Queen Mother (Patron of the Aldeburgh Festival) at the Red House, Aldeburgh, 13 June 1975

Britten and Peter Pears at Snape, 1974

About this time Britten began to show an increasing interest in television. Early in 1969 he agreed to conduct the BBC 2 colour production of *Peter Grimes*; and subsequently he accepted a commission to compose a new opera for television. As subject he chose *Owen Wingrave*, a short story by Henry James, which had much impressed him when he first read it twenty years or more ago, and as in the case of James's *The Turn of the Screw* he asked Myfanwy Piper to provide the libretto. Part of the music was written in Venice and Hessen during the winter of 1969/70; and the opera was 'shot' at the Maltings, Snape, in November 1970. The world première broadcast took place on 16 May 1971; and the first 'live' performance was given at Covent Garden (10 May 1973).

During his late fifties, he continued to tour extensively. He visited Australia in March 1970 with the English Opera Group, which presented all three of his church parables at the Adelaide Festival. In April the following year he returned to Moscow and Leningrad, this time with the London Symphony Orchestra, which he conducted in a programme that included his Piano Concerto and Symphony for Cello and Orchestra with his Russian friends, Richter and Rostropovich respectively, as soloists. The distinguished audience at the Moscow concert included Madame Ekaterina

106

Furtseva, Dmitri Shostakovich, and Madame Prokofiev. In later years, Sir Duncan Wilson, who acted as Britten's host in Moscow on this occasion, recalled how 'perhaps the most moving experience of all for an Englishman was to attend next day in his company part of a performance of *A Midsummer Night's Dream* at the Bolshoi Theatre, and to see, as he walked through the corridors at the interval, how the mainly young audience recognized and with obvious spontaneity applauded him'.*

By 1971 an important new opera was under way. Wishing to write something that would provide a specially varied and challenging role for Peter Pears, he chose Thomas Mann's story *Death in Venice* as his subject matter; and once again the work of adaptation was entrusted to Myfanwy Piper. By the beginning of 1973 the score was nearing completion, but Britten was beginning to feel seriously ill. His doctors examined him, and their diagnosis showed that one of the valves of his heart was defective and would have to be replaced. An operation took place in May; but unfortunately the replaced valve was not wholly successful, and the patient's condition was complicated by a slight stroke that occurred during the actual operation and paralysed his right side.

The last three and a half years of his life were those of an invalid, who was easily exhausted by the slightest exertion, and could summon up enough strength to devote only a very limited amount of his time to composition. Nevertheless, during this period he managed against all odds to complete over half a dozen important new works, including *The Death of St Narcissus*, a canticle for tenor and harp, *Phaedra*, a dramatic cantata for mezzo-soprano and small orchestra, and his Third String Quartet. He was engaged on a large-scale cantata, a setting of *Praise We Great Men*, a poem that had been written for him by Edith Sitwell some years previously, when death intervened.

He died at the Red House, Aldeburgh, on 4 December 1976 and was buried in the churchyard, within sight and sound of the sea he loved.

* *The Times*, 9 December 1976.

IX
Personal Postscript

Britten was supremely a professional musician. Composer, pianist, viola-player, conductor, research-scholar and musical editor—in the course of his career he was engaged in multifarious activities connected with music, and if he carried them out successfully and well, it was because he took the trouble to acquire the necessary skills. He always believed in the importance of technique. In his broadcast talk *The Composer and the Listener* (1946) he said: 'Obviously it is no use having a technique unless you have the ideas to use this technique; but there is, unfortunately, a tendency in many quarters today to believe that brilliance of technique is a danger rather than a help. This is sheer nonsense. There has never been a composer worth his salt who has not had supreme technique. I'll go further than that and say that in the work of your supreme artist you can't separate inspiration from technique. I'd like anyone to tell me where Mozart's inspiration ends and technique begins.'

For technical reasons, among others, he was always prepared to work to order. He did not believe in allowing his talents to rust. As an artist he wanted to serve the community and showed himself ready to accept commissions of every kind. He found virtue in serving all sorts of different persons and believed that even 'hackwork will not hurt an artist's integrity provided he does his best with every commission'.*

To have as many ideas as he had and to work as hard as he did argued not only extraordinary fertility and fluency but also great sensitivity. If an artist lacks feeling, he loses much of the impetus towards expression. Britten showed himself sensitive in many ways—particularly to cruelty and suffering. Many of his operas contain (or imply) scenes of almost sadistic cruelty, but they inevitably lead to episodes of warm compassion and pity. This intense sympathy with the victims of oppression lay at the heart of his pacifism. Hans Keller, in an interesting essay,† went so far as to suggest that

* Quoted in 'Benjamin Britten: Another Purcell' by Phoebe Douglas. *Town and Country*, December 1947.
† 'The Musical Character' by Hans Keller. *Benjamin Britten: a commentary*. Rockliff, 1952.

108

'what distinguishes Britten's musical personality is the violent repressive counterforce against his sadism; by dint of character, musical history and environment, he has become a musical *pacifist* too'.

He was also sensitive to critical misunderstanding, or lack of understanding, especially where it appeared to be the result of wilfulness or stupidity.* In his own case, it was not as if his music were particularly obscure or revolutionary. His style was eclectic; his idiom modal; his musical metrics often echoed the more or less familiar structure of English poetical metrics. The surface value of his music was quite easy to understand; but an appreciation merely of its superficial qualities would not reach the heart of the matter.

As an occasional composer he had a flair for the various elements that made an occasion unique, and his works were often supremely effective in the setting and circumstances for which they were designed. It was this feeling for what was likely 'to come off' in performance that stood him in such good stead in his music for the opera-house, theatre, cinema, radio and television.

Occasionally the effect on listeners was so strong that normal critical standards seemed to be swept away. E. M. Forster's comment on the first performance of *Saint Nicolas* at Aldeburgh has already been quoted—'It was one of those triumphs outside the rules of art which only the great artist can achieve.' The same work had a very similar effect on another critic. Donald Mitchell wrote:† 'I was so confused by its progressively overwhelming impact that all I could find to say was: "This is too beautiful".' Lord Clark, who spent the first fifteen years of his life on the other side of the river Alde from Aldeburgh, was similarly moved by *Noye's Fludde*.‡ 'To sit in Orford Church, where I had spent so many hours of my childhood dutifully awaiting some spark of divine fire, and then to receive it at last in the performance of *Noye's Fludde*, was an overwhelming experience.' Perhaps Auden in his poem 'The Composer'** found the best way of putting into words the inexplicable thrill that floods the mind and senses at such a moment:

Pour out your presence, O delight, cascading
The falls of the knee and the weirs of the spine,
Our climate of silence and doubt invading:
You alone, alone, O imaginary song,

---

* See 'Variations on a Critical Theme' by Benajmin Britten. *Opera*, March 1952.
† 'A Note on *Saint Nicolas*: Some points on Britten's Style' by Donald Mitchell. *Music Survey*, Spring 1950.
‡ 'The Other Side of the Alde' by Kenneth Clark from *Tribute to Benjamin Britten*, Faber 1963.
** *Another Time*, XXII.

Are unable to say an existence is wrong,
And pour out your forgiveness like a wine.

Britten always responded deeply to words and loved setting them. When he was a small boy of just ten years old, he chose the best of his juvenile songs and wrote them out neatly in a special manuscript book. Amongst his juvenilia there are 'settings of anonymous poems, poems by Tennyson, Longfellow, Shelley, Shakespeare, Kipling, a rather obscure poet writing under the pseudonym of "Chanticleer", pieces from the Bible, Thomas Hood, bits of plays, and poems in French too.'* Subsequently he turned to poems by Auden, Donne, Hardy, Blake for some of his song cycles. Although he had no great knowledge of any foreign language, he enjoyed reading foreign poems, sometimes with the help of cribs, and at various times set texts in French, German, Italian, Russian, and Latin.

He was sensitive to the relationship of words and music. He was not inhibited about words like some composers, but was capable of assessing the different values of the syllable, the word, and the idea behind the word, and knew how to give them a musical gravity of their own. Sometimes one had the feeling that his instrumental music aspired to the condition of vocal or choral music; and sometimes in the vocal and choral music one had the conviction that the word had been made music and the music had taken on a new dimension. If critics object that the issue is being confused by the presence of an extra-musical element, the answer must be that precisely this combination of disparate elements lies at the heart of the problem of opera, and Britten approached excitingly near to one of the possible solutions.

There were many aspects of his character that could be pursued if one felt inclined. His sense of humour (or perhaps one should say his sense of proportion); his brisk fancy and ambivalent imagination; his fondness for children; his deep religious conviction.

As a man he recalled with pleasure his youth in East Anglia—fêtes and obstacle races, bicycle rides, tennis tournaments, bathing parties, making friends and making music—and projected himself without difficulty into the minds and hearts of young people of a later age. That was why he wrote such good music about children and for them to listen to and play. He was always interested in their problems and prepared to go out of his way to give them advice and show sympathy. At the same time he did not lower his sights, but expected work of the highest quality from them.

His religious beliefs were central to his life and his work. As a devout and practising Christian, he was keen, wherever possible, to work within the framework of the Church of England, and many of his compositions were planned accordingly.

* From *Personal Choice*, a broadcast talk by Benjamin Britten, BBC Home Service, 16 July 1958.

His talent for music manifested itself at an astonishingly early age, and his precocity as a composer startled audiences when he was in his twenties. He reached complete musical maturity in his early thirties; and by his thirty-fifth year (1948) the full extent of his remarkable gifts had been revealed— the fluency, the protean variety, the feeling for effect, the love of setting words to music, and the deceptive simplicity of the melodic and harmonic means employed.

Although his success occasionally excited opposition and jealousy, his career did not go unhonoured.

The first university to honour him was Queen's University, Belfast (1951). The initiative shown by Northern Ireland was soon followed up by England, and honorary degrees (all D.Mus.) were conferred on him by the Universities of Cambridge, Nottingham, Hull, Oxford, Manchester, London, Leicester, East Anglia, Wales and Warwick. In 1957 he was made an honorary member of the American Academy and the National Institute of Arts and Letters; and in 1961 he was awarded the Hanseatic Goethe Prize. The following year he was created Commander of the Royal Order of the Pole Star (Sweden); the Aspen Award followed in 1964, and the Wihuri-Sibelius Prize the following year. On 25 November 1964 he was presented with the Gold Medal of the Royal Philharmonic Society. He was made an Honorary Fellow of Magdalene College, Cambridge and an Honorary Member of Worcester College, Oxford, and he received the Freedom of the Worshipful Company of Musicians in 1966. The Mahler Medal (awarded by the Bruckner and Mahler Society) and the Leone Sonning Prize (Denmark) followed in 1967 and 1968 respectively. In 1973 he became the first recipient of the Ernst Siemens–Musikpreis; and the following year he received the Ravel Foundation Prize, which was awarded to composers who revealed in their work 'qualities recalling the scrupulous attention to detail, the search for beauty of expression, sonority and invention which characterize the music of Ravel'. The very last honour he received was the Mozart Medal for 1976, awarded by the Mozartgemeinde, Vienna.

Suffolk was his home county. He was born there, and lived there for the greater part of his life. Some of his operas were set there—*Peter Grimes*, *Albert Herring*, *The Little Sweep* and *Curlew River*. Of all the numerous honours he received, he probably valued most highly the compliments paid him when he received the freedom, first in 1951 of the Borough of Lowestoft, the town where he was born, and then in 1962 of the Borough of Aldeburgh, the town where he chose to reside. In his speech of thanks on the former occasion, he took the opportunity of confirming his allegiance to that part of England. He said: 'Suffolk, the birthplace and inspiration of Constable and Gainsborough, the loveliest of English painters; the home of Crabbe, the most English of poets; Suffolk with its rolling, intimate countryside; its heavenly Gothic churches, big and small; its marshes, with those wild seabirds; its grand ports and its little fishing villages. I am firmly rooted in this glorious county. And I proved this to myself when I once tried to live somewhere else.'

His services to English music were outstanding; and it was largely due to him that today it is better known and stands higher in the esteem of countries abroad than was ever the case before. The tributes were well merited, therefore, when he was made a Companion of Honour in the Coronation Honours of 1952, awarded the Order of Merit in 1965, and created Baron Britten of Aldeburgh in the Birthday Honours of 1976.

Part Two
THE OPERAS

The opening chorus of the Prologue to *Paul Bunyan* in Britten's original composition sketch

I
Paul Bunyan

At the time of its first performance *Paul Bunyan* was described by its authors as a choral operetta with many small parts rather than a few star roles.

For his subject Auden took the legend of Paul Bunyan. This giant lumberman, who was reputed to stand forty-two axe handles high and to sport a twist of chewing tobacco between his horns, was among the pioneers working in the American wilderness who helped prepare the way for the advance of civilization westwards. Auden considered America to be unique in being the only country to create myths after the industrial revolution, and this particular legend to be not only American but universal in its implications. He looked on Paul Bunyan as 'a projection of the collective state of mind of a people whose tasks were primarily the physical mastery of nature'* and intended that the operetta should present 'in a compressed fairy-story form the development of the continent from a virgin forest before the birth of Paul Bunyan to settlement and cultivation when Paul Bunyan says goodbye because he is no longer needed, i.e. the human task is now a different one, of how to live well in a country that the pioneers have made it possible to live in.'†

In a newspaper article,‡ he set out his attitude to Bunyan and his friends in some detail: 'Appearing so late in history, Paul Bunyan has no magical powers; what he does is what any man could do if he were as big and as inventive; in fact, what Bunyan accomplishes as an individual is precisely what the lumbermen managed to accomplish as a team with the help of machinery. Moreover, he is like them as a character; his dreams have all the native swaggering optimism of the nineteenth century. . . .' Babe the Blue Ox, who gave Bunyan advice, Auden found something of a puzzle—'I conceive of her quite arbitrarily as a symbol of his anima'—and he omitted her as a character. He went on to say: 'Associated with Bunyan are a number of satellite human figures, of which the most interesting are Hel Helson, his

---

* Quotation from the *Paul Bunyan* programme.   † Ibid.
‡ From 'Opera on an American Legend' by W. H. Auden. *New York Times*, 4 May 1941.

Swedish foreman, and Johnny Inkslinger, his book-keeper. These eternal human types; Helson, the man of brawn but no brains, invaluable as long as he has somebody to give him orders whom he trusts, but dangerous when his consciousness of lacking intelligence turns into suspicion and hatred of those who possess it; and Inkslinger, the man of speculative and critical intelligence, whose temptation is to despise those who do the manual work that makes the thought possible. Both of them learn a lesson in their relations with Paul Bunyan; Helson through a physical fight in which he is the loser, Inkslinger through his stomach.'

In writing an operetta about Bunyan, Auden found three main difficulties confronted him. In view of his previous record as poet and playwright, he was hardly likely to approach his subject from a literal or realistic angle; and it is not surprising that he tried to surmount these difficulties in ways consistent with the didactic style of epic drama as advocated by Brecht. 'In the first place [Bunyan's] size and general mythical characteristics prevent his physical appearance on the stage—he is presented as a voice and, in order to differentiate him from the human characters, as a speaking voice. In consequence some one else had to be found to play the chief dramatic role and Inkslinger seemed the most suitable, as satisfying Henry James's plea for a fine lucid intelligence as a compositional centre. Inkslinger, in fact, is the only person capable of understanding who Paul Bunyan really is, and, in a sense, the operetta is an account of his process of discovery. In the second place, the theatrical presentation of the majority of Bunyan's exploits would require the resources of Bayreuth, but not to refer to them at all would leave his character all too vaguely in the air. To get round this difficulty, the librettists interposed simple narrative ballads between the scenes, as it were, as solo Greek chorus. Lastly, an opera with no female voices would be hard to produce and harder to listen to, yet in its earlier stages at least the conversion of forests into lumber is an exclusively male occupation. Accordingly the collaborators introduced *a camp dog* and *two camp cats* sung by a coloratura soprano and two mezzo-sopranos respectively.'*

In his article from which the above quotations are taken, Auden went on to say: 'The principal interest of the Bunyan legend today is as a reflection of the cultural problems that occur during the first stage of every civilization, the stage of colonization of the land and the conquest of nature. The operetta, therefore, begins with a prologue in which America is still a virgin forest and Paul Bunyan has not yet been born, and ends with a Christmas party at which he bids farewell to his men because now he is no longer needed. External physical nature has been mastered, and for this very reason can no longer dictate to man what they should do. Now their task is one of their human relations with each other and, for this, a collective mythical

* Ibid.

Paul Bunyan: The lumberjacks, Moppet and Poppet (the camp cats) and Fido (the camp dog) in Act I from the English Music Theatre revival in 1976

figure is no use, because the requirements of each relation are unique. Faith is essentially invisible.'

Finally, he pointed out that the implications of the Bunyan legend were not only American, but also universal.

Paul Bunyan was presented on 5 May 1941,* for a week's run by the Columbia Theater Associates of Columbia University, with the co-operation of the Columbia University Department of Music and a chorus from the New York Schola Cantorum. It was financed by a grant from the Alice M. Ditson Fund. The producer was Milton Smith, and the conductor Hugh Ross.

CHARACTERS
In the Prologue

Old Trees	Young Trees	Three Wild Geese

* There had been a preview the previous evening for members of the League of Composers.

THE OPERAS

In the Interludes   Narrator

In the Play   The Voice of Paul Bunyan
Cross Crosshaulson   John Shears   Sam Sharkey
Ben Benny   Jen Jenson   Pete Peterson
　　Andy Anderson   Other Lumberjacks
　　Western Union Boy   Hel Helson   Johnny Inkslinger
Fido*   Moppet*   Poppet*
          The Defeated
Slim   Tiny
          The Film Stars and Models
          Frontier Women

Scene: A Grove in a Western Forest
　　　　Prologue　　—　　Night
　Act I　　Scene i　　—　　A Spring Morning
　　　　　Scene ii　　—　　Summer
　Act II　　Scene i　　—　　Autumn
　　　　　Scene ii　　—　　Christmas

On the whole the work was received by the New York critics with dismay. *Time*, suspicious of this 'anemic operetta put up by two British expatriates', complained that it was 'as bewildering and irritating a treatment of the outsize lumberman as any two Englishmen could have devised'. *The New Yorker* said that though on paper or in conference there may have been certain items that 'looked like the makings of something pretty exciting . . . in the theatre *Paul Bunyan* didn't jell'. A more revealing description of the music came from Robert Bagar in *World Telegraph*. 'Mr. Britten, who is an up and coming composer, has written some worth-while tunes in this score. It ranges, in passing, from part-writing to single jingle. Its rhythms are often interesting and the harmonies fit rather well. There are arias, recitatives, small ensembles and big choral sequences. Most of the last named are good. The music makes occasional reference to *Cavalleria Rusticana* and one item, a stuttering bit, goes back to *The Bartered Bride*.'

Under the heading 'Musico-Theatrical Flop', Virgil Thomson, writing in the *New York Herald Tribune*, attacked the form of the work, which he considered fell into the category of 'the Auden semi-poetic play', going on to assert that on the stage the Auden style had 'always been a flop. It is flaccid and spineless and without energy.' Turning to Britten, he showed similar lack of enthusiasm over the music: 'Benjamin Britten's music, here as elsewhere, has considerable animation. His style is eclectic though not without savour. Its particular blend of melodic "appeal" with irresponsible

* These are the camp dog and two camp cats mentioned above.

118

counterpoint and semi-acidulous instrumentation is easily recognisable as that considered by the BBC to be at once modernistic and safe. Its real model is, I think, the music of Shostakovich, also eclectic, but higher in physical energy content than that of Mr. Britten. Mr. Britten's work in *Paul Bunyan* is sort of witty at its best. Otherwise it is undistinguished. . . .'

But the most perceptive review came from Olin Downes writing in the *New York Times*. Like Virgil Thomson, he found Britten's style 'eclectic' and thought his sources ranged widely 'from Prokofiev to Mascagni, from Rimsky-Korsakov to Gilbert and Sullivan'. Although he had a few reservations to make, he admitted that Britten 'knows how to set a text, how to orchestrate in an economical and telling fashion, how to underscore dialogue with orchestral commentary'. He added: 'What is done by Mr. Britten shows more clearly than ever that opera written for a small stage, with relatively modest forces for the presentation, in the English language, and in ways pleasantly free from the stiff tradition of either grand or light opera of the past, is not only a possibility but a development nearly upon us.'

At first neither score nor libretto was published. In fact, it was hinted that during *Paul Bunyan*'s brief run in New York, the work had been subjected to so many changes, cuts, and revisions that no definitive version could be said to exist. But after Auden's death in 1973 this attitude began to change. A number of people expressed interest in the work and asked for the material to be made accessible; a recording of one of its original performances was deposited in the Brander Matthews Dramatic Museum at Columbia University; and the value of this operetta could be assessed more justly in the light of the subsequent achievements of both composer and librettist in the operatic field. In 1975 Britten agreed to release it for performance. The score was edited and revised for stage performance and was produced by the English Music Theatre Company on 4 June at the 1976 Aldeburgh Festival. It was immediately recognized as an enjoyable, carefree entertainment that was fun to listen to and fun to perform.

Whether there could have been a future to this particular collaboration if the two partners had decided to pursue it is difficult to say. Although Auden appeared to be successful with some of his subsequent librettos, such as *The Rake's Progress* for Stravinsky and *The Bassarids* for Henze, one has the feeling that in the long run Britten and Auden would not have worked happily together. Auden, the elder character, was fairly dogmatic in his attitude to opera and usually took the view that he knew most of the answers to most of the questions. It is unlikely that a person of Britten's shy and sensitive nature could have flourished in this rather overbearing atmosphere.* Anyway, surmise is useless, because after *Paul Bunyan* the partnership was not resumed.

* For a fascinating insight into the problems in this relationship see *Britten and Auden in the Thirties* by Donald Mitchell. Faber, 1981.

II
Peter Grimes

I

When Britten approached Montagu Slater in 1942 and asked him to write a libretto for the opera that had been commissioned by the Koussevitzky Music Foundation, the theme was already fixed—Aldeburgh was to be the scene and the subject Peter Grimes, the story of whose life is told by Crabbe in *The Borough*.

Aldeburgh was Crabbe's birthplace. He was born there on 1 January 1755. His son described it as a poor and wretched place lying between 'a low hill or cliff, on which only the old church and a few better houses were then situated, and the beach of the German Ocean. It consisted of two parallel and unpaved streets, running between mean and scrambling houses, the abodes of seafaring men, pilots and fishers. The range of houses nearest to the sea had suffered so much from repeated invasions of the waves, that only a few scattered tenements appeared erect among the desolation.' As for the beach, then as now it consisted of 'large rolled stones, then loose shingle, and, at the fall of the tide, a stripe of fine hard sand. Vessels of all sorts, from the large heavy troll-boat to the yawl and prame, drawn up along the shore—fishermen preparing their tackle, or sorting their spoil—and, nearer the gloomy old town-hall (the only indication of municipal dignity) a few groups of mariners, chiefly pilots, taking their quick, short walk backwards and forwards, every eye watchful of a signal from the offing—such was the squalid scene that first opened on the author of *The Village*.'* And such was the place and community that in 1810 Crabbe described so vividly in his poem *The Borough* by means of a series of twenty-four letters written in heroic couplets.

In the first of these letters he gives a general description of the Borough. He mentions the River Alde, which (as his son explains) 'approaches the sea close to Aldeburgh, within a few hundred yards, and then turning abruptly continues to run about ten miles parallel to the beach, until it at length finds

* The Life of George Crabbe, by his Son, 1834.

120

its embouchure at Orford'; the craft on the river—'hoys, pinks and sloops; brigs, brigantines and snows'—and also the quayside with its clamour of sailors and carters and lumber of 'package and parcel, hogshead, chest, and case.' After night-fall some of the inhabitants of the Borough pass their times at parties, whist-drives, concerts, plays or taverns, while—

> *Others advent'rous walk abroad and meet*
> *Returning Parties pacing through the Street . . .*
> *When Tavern-Lights flit on from Room to Room,*
> *And guide the tippling Sailor staggering home:*
> *There as we pass the jingling Bells betray,*
> *How Business rises with the closing Day:*
> *Now walking silent, by the River's side,*
> *The Ear perceives the rimpling of the Tide;*
> *Or measur'd cadence of the Lads who tow*
> *Some enter'd Hoy, to fix her in her row;*
> *Or hollow sound, which from the Parish-Bell,*
> *To some departed Spirit bids farewell!*

Crabbe then proceeds to deal with the various professions and trades in the Borough. Letter XI enumerates the inns—particularly *The Boar*.

> *There dwells a kind old Aunt, and there you see*
> *Some kind young Nieces in her company;*
> *Poor village Nieces, whom the tender Dame*
> *Invites to Town, and gives their Beauty fame.*

No fewer than ten of the later letters are devoted to the inhabitants of the almshouse and to the poor; and among the latter figure Abel Keene, a clerk in office (Letter XXI), Ellen Orford, the widowed school mistress (Letter XX), and Peter Grimes, a fisherman (Letter XXII). From the first, Slater borrowed no more than his surname, which he attached to Ned, the quack (who is described in Letter VII); the second he elevated to the principal female part; and the third became the protagonist of the opera.

Edward Fitzgerald, who was a friend of Crabbe's son, has left it on record that the Peter Grimes of the poem was based on an actual fisherman named Tom Brown, who lived in Aldeburgh in the middle of the eighteenth century. According to Crabbe, there were few redeeming features about Peter Grimes. As soon as he was out of his teens, he became impatient of parental control and started to knock his father about. Then he went to live on his own and 'fished by water and filched by land'; but he was dissatisfied so long as there was no unfortunate victim living with him on whom he could wreak his strength at any hour of the day or night. Presently he heard of workhouse-clearing men in London who were prepared to bind orphan parish-boys to needy tradesmen. He obtained such an apprentice for himself and was at last able to give full rein to his sadistic instincts and his lust for power. The first apprentice, Sam, lived for three years and then was found lifeless in his bed. The second fell one night from the main-mast of the fishing boat and was killed. The third died in the course of a stormy voyage from Aldeburgh to London. After his death, the conscience of the Borough was thoroughly roused; and the Mayor himself forbade Grimes to take any more apprentices to work for him. Thenceforward he was ostracized. Gradually his mind began to fail and, as he sailed up and down the river, he was haunted by the spirits of his father and two of the dead boys. In raving delirium shortly before his death he described one such scene:

> *In one fierce Summer-day, when my poor Brain*
> *Was burning-hot and cruel was my Pain,*
> *Then came this Father-foe, and there he stood*
> *With his two Boys again upon the Flood;*
> *There was more Mischief in their Eyes, more Glee*
> *In their pale Faces when they glar'd at me:*
> *Still did they force me on the Oar to rest,*
> *And when they saw me fainting and opprest,*
> *He, with his Hand, the old Man, scoop'd the Flood,*
> *And there came Flame about him mix'd with Blood;*
> *He bade me stoop and look upon the place,*
> *Then flung the hot-red Liquor in my Face;*
> *Burning it blaz'd, and then I roar'd for Pain,*
> *I thought the Daemons would have turn'd my Brain.*

In the Preface to *The Borough*, Crabbe embarks on a brief analysis of the character of Peter Grimes. 'The mind here exhibited', he says, 'is one untouched by pity, unstung by remorse, and uncorrected by shame; yet is this hardihood of temper and spirit broken by want, disease, solitude and disappointment; and he becomes the victim of a distempered and horror-stricken fancy. . . . The corrosion of hopeless want, the wasting of unabating disease, and the gloom of unvaried solitude, will have their effect on every nature; and the harder that nature is, and the longer time required to work upon it, so much the more strong and indelible is the impression.'

If Peter Grimes was to become the hero of a twentieth-century opera and win the sympathy of a modern audience, some of these eighteenth-century values would have to be altered and adjusted. Slater accordingly embarked on a reinterpretation of the character, as a result of which Crabbe's grim fisherman became something of a Borough Byron, too proud and self-willed to come to terms with society, and yet sufficiently imaginative to be fully conscious of his loss. A clue to this new reading is perhaps to be found in an episode of Grimes's childhood, which becomes even more poignant when it is remembered that Crabbe too as a boy had been bitterly hostile to his own father. Grimes recalled

> *How, when the Father in his Bible read,*
> *He in contempt and anger left the Shed:*
> *'It is the Word of Life', the Parent cried;*
> *—'This is the Life itself', the Boy replied.*

To fit his more modern interpretation, Slater decided to post-date the action of the drama from the latter part of the eighteenth-century, when the stories related in *The Borough* actually took place, to 1830 when the tide of Byronism was in full flood. In view of the usual time-lag between metropolitan and provincial fashions, little or no injury was thereby done to the accuracy of the general description of the Borough and its inhabitants as based on Crabbe; but the new date accentuated the rift between Grimes and the rest of the community—between, on the one hand, the comparatively modern type of the psychopathic introvert, divided against himself and against the world, and, on the other, reactionary extrovert society.

For the purpose of his plot, Slater omitted Peter's father and reduced the number of his apprentices from three to two, the first of whom has just died at sea when the opera opens. Ellen Orford, the widowed school-mistress, is promoted to the position of Peter's friend and confidante—in fact, there is a moment when Peter deludes himself into thinking his problems would be solved if he could marry her. At the end of the inquest into the apprentice's death, Ellen asks Peter to come away with her; but he feels he cannot accept until he has rehabilitated himself in the eyes of Borough—and to him rehabilitation means money, wealth. He explains this to Captain Balstrode, a retired sea-captain, during the storm in Act I:

> *These Borough gossips*
> *Listen to money,*
> *Only to money.*
> *I'll fish the sea dry,*
> *Sell the good catches.*
> *That wealthy merchant*
> *Grimes will set up*
> *Household and shop.*
> *You will all see it!*
> *I'll marry Ellen!*

Balstrode replies:

> *Man—go and ask her,*
> *Without your booty,*
> *She'll have you now.*

But when Peter demurs at the idea of being accepted out of pity, Balstrode realizes that it is too late to remedy the defects in his character and that sooner or later the fatal pattern of the former tragedy is bound to be repeated. And so it turns out. The new apprentice arrives; but although he is ill-treated by Peter, it is accident rather than deliberate cruelty that ultimately brings about his death. By then, however, the Borough conscience has been thoroughly aroused—the man-hunt is up—and Balstrode realizes that the best thing for Peter will be to disappear. But how? At an earlier point in the action, Peter, asked why he didn't leave the Borough to 'try the wider sea with merchantman or privateer', replied:

I am native, rooted here . . .
By familiar fields,
Marsh and sand,
Ordinary streets,
Prevailing wind.

As exile is out of the question, the only alternative appears to be suicide; and on Balstrode's advice, he sails his fishing boat out to sea and scuttles it.

Through his imagination Peter is aware of wider universal issues at stake at the same time as he wrestles with the immediate problems caused by the flaws in his nature. This is made clear from his soliloquy in the crowded pub during the storm in Act I:

Now the Great Bear and Pleiades
where earth moves
Are drawing up the clouds
of human grief
Breathing solemnity in the deep night.
But if the horoscope's
bewildering . . .
Who can turn the skies back and begin again?

And this intensity of vision helps to raise to the tragic plane what might otherwise have been merely a sordid drama of realism.

Here is a synopsis of the libretto:

Prologue. The interior of the Moot Hall. At the end of the inquest into the death of Peter Grimes's apprentice, Mr Swallow, the coroner, brings in a verdict of death in accidental circumstances; but Peter complains that this verdict does not really clear him of the charge, for the case will still go on in people's minds. Act I, scene i. A street by the sea a few days later, showing the exterior of the Moot Hall and *The Boar*. Peter is already experiencing difficulty in working his fishing boat single-handed; but Ned Keene, the apothecary, tells him he has found another apprentice boy, whom Ellen Orford, despite the general disapproval of the Borough, agrees to fetch by

124

Peter Grimes: Peter Pears as Grimes and Joan Cross as Ellen Orford in their duet from the Prologue

the carrier's cart. Shortly after her departure, a storm breaks, which is all the more to be feared because it comes with a spring tide. The boats are made fast, the nets brought in and the windows of the houses shuttered. After a dialogue between Peter and Captain Balstrode, the scene changes to *The Boar* (Act I, scene ii) on the evening of the same day. Although it is past closing time, the pub is full, and people are still coming in out of the storm for shelter and refreshment. News is brought that the coast road has been flooded and a landslide has swept away part of the cliff up by Peter Grimes's hut. A quarrel or two break out among the topers; a round is sung; and when at last Ellen Orford arrives back with the boy, Peter—to everyone's consternation—insists on taking him away at once to his desolate hut through the storm.

Act II, scene i. The scene is the same as in Act I, scene i; the time a Sunday

125

morning a few weeks later. Ellen and Peter's new apprentice sit in the sun on the beach, while morning service goes on in the Parish Church. By chance she discovers the boy's clothes are torn and his body bruised; and when Peter, who has just caught sight of a shoal, arrives to take him out fishing, her reproaches lead to an open quarrel between the two, which is overseen and overheard by some of the neighbours. By the time the church service is over, the news has spread round the Borough that '*Grimes is at his exercise!*' and a party of men sets out to investigate. Meanwhile, Peter and the apprentice have reached Peter's hut, which is made out of an old upturned boat (Act II, scene ii). Here he gathers together his fishing gear; but the boy's blubbering delays him and when, after a clumsy attempt to soothe the lad, he hears the sound of the neighbours coming up the hill, he suddenly decides to make a quick get-away. He flings his nets and tackle out of the cliff-side door; but the boy, as he starts to climb down the cliff, slips and is dashed to death. Peter scrambles down after him. On arrival, the search party—to its surprise—finds the hut empty, neatly kept and reasonably clean, but there is no sign of its recent occupants.

Act III, scene i. The scene is the same as in the first scenes of Acts I and II; the time, two or three nights later. A subscription dance is taking place in the Moot Hall, and there is considerable traffic between the Hall and *The Boar*. Though neither Peter nor his apprentice has been seen during the last few days, it is assumed that both are away fishing, until Mrs Sedley, one of the leading gossips in the Borough with a keen nose for scenting out crime as well as scandal, overhears Ellen telling Balstrode that the jersey she embroidered for the boy some time ago has been found washed up on the beach. Seeing that Peter's boat is now back, Mrs Sedley imparts her suspicions to Swallow, who in his capacity as Mayor summons the constable of the Borough and bids him take a posse of men to apprehend Grimes. A few hours later (Act III, scene ii) when the dance in the Moot Hall is over, a fog has crept up from the sea, and only the occasional cries of the man-hunt and the moan of a fog-horn break the stillness of the night as Peter creeps back to his hut. There Ellen and Balstrode find him, hungry, wet, exhausted, almost insane. It is Balstrode who proposes the way out—that he take his boat out to sea, scuttle it and sink with it—and this Peter does, as dawn breaks. Gradually the Borough reawakes to life. Lights appear at windows. Shutters are drawn back. The coastguard station reports a boat sinking far out at sea, but the news is dismissed as an idle rumour; and as the light of the morning waxes, the people of the Borough start to go about their daily tasks. It is the beginning of another day.

This outline is sufficient to show how far Slater's libretto is removed from Crabbe and *The Borough*, and how fundamentally different a character Slater's Grimes is from Crabbe's. Slater himself explained that 'the story as worked out in the opera uses Crabbe's poem only as a starting-point. Crabbe

produced character sketches of some of the main persons of the drama. I
have taken these character sketches as clues and woven them into a story
against the background of the Borough: but it is my story and the
composer's (the idea was originally not mine but Britten's), and I have to
take the responsibility for its shape as well as its words.'*

In writing his libretto, Slater avoided the heroic couplet as used by
Crabbe and blank verse, because he felt that the five-stress line was 'out of
key with contemporary modes of thought and speech'.† Instead, he adopted
a 'four-stress line with rough rhymes for the body of the drama', while the
Prologue was written in prose and various metres used for the set numbers.

There are at least five different published versions of the text. The first is
in Montagu Slater's *Peter Grimes and other poems* published by The Bodley
Head in 1946.‡ This may be called the literary text—as Slater explains in his
Preface, it is 'to all intents and purposes the one to which the music was
composed', but omits 'some of the repetitions and inversions required by
the music'. The second is a text that has not been published in full, but
extensive passages are quoted from it in an essay of Slater's included in the
Sadler's Wells Opera Book devoted to *Peter Grimes* (The Bodley Head,
1945). The third is the libretto of the opera published by Boosey & Hawkes
in 1945 (and reissued with minor revisions in 1961); but this does not
contain all the amendments and corrections that appeared in the fourth and
near-final text as printed in the vocal score published by Boosey & Hawkes
also in 1945. The final version of the libretto *as performed* was not published
by Boosey & Hawkes until 1979.

The main divergencies between the literary text and the others are to be
found in Peter's monologues in the second scenes of Acts II and III; and
these changes were clearly made to meet the composer's musical exigencies.
It may be of interest, however, to take a different passage and to compare the
various versions of Ellen's monologue at the beginning of Act II, scene i as
they appear in the four different texts:

Text I   *The sun in*
             His own morning
             And upward climb
             Makes the world warm.
             Night rolled
             Away with cold.
             The summer morning
             Is for growing.

* *Peter Grimes* (*Sadler's Wells Opera Books No.* 3 edited by Eric Crozier). The Bodley
Head, 1945.
† *Peter Grimes and other poems* by Montagu Slater. The Bodley Head, 1946.
‡ Ibid.

Text 2   *The sun in*
          His fair morning
          And upward climb
          Makes the world warm . . .

                                        [Cetera desunt]

Text 3   *Glitter of sun*
          On curling billows,
          The earth is warm
          Old ocean gently flows.
          Man alone
          Has a debt to pay
          In this tranquility
          *Mindful of yesterday.**

Text 4   *Glitter of waves*
          and glitter of sunlight
          Bid us rejoice
          And lift our hearts on high.
          Man alone
          has a soul to save,
          And goes to church
          to worship on a Sunday.

Whatever the poetic merits of these different versions, there is no doubt that the final one was the best suited to its context. It clinches the impression made by the orchestral interlude at the beginning of the act and puts the forthcoming church service into perspective.

II

When Britten started to set this libretto, he was confronted by various problems.

In the first place, the division of each of the three acts of the opera into two scenes, the action of which was continuous or nearly continuous or partly overlapping, made it possible for him to decide to compose each act as an unbroken piece of music; but as there were scene changes between scenes i and ii of Acts I and II, interludes would be needed there, and a further interlude of some sort was indicated between scenes i and ii of Act III to mark the passing of time. There was also the formal problem of the Prologue to consider. Prosaic though it might be, it gave such back history as was needed, provided an exposition of the theme and introduced the main

---

* But in the 1961 revision of this libretto, text 4 has been substituted for text 3.

128

Peter Grimes: Peter Pears as Grimes in the pub scene (Act I, scene ii) from the BBC television production recorded at Snape Maltings in 1969

characters of the opera by name—all this so expeditiously and succinctly that it could hardly be expected to stand alone. Clearly, it ought to be joined to Act I; and this would entail another interlude to cover the necessary scene change. To complete the scheme, he added introductions to Acts II and III, making a total of six orchestral 'interludes' in all.

And then he had also to take into account the fact that each act of Slater's libretto contained cues for actual sound or song effects as opposed to the music to be composed in accordance with operatic convention. For instance, the scene in *The Boar* works up to a moment when a song is suggested and someone spontaneously starts up a round, '*Old Joe has gone fishing*', in which the rest of the company joins. The following Act opens with Ellen talking to the boy apprentice on the beach, while from the neighbouring church are overheard strains of the Sunday morning service. Later that morning, some of the men of the Borough form a procession and, led by Hobson the carter playing a tenor drum, go off to Peter's hut, chanting a sinister marching song. In the last Act, the dance band at the Moot Hall, consisting of fiddle, double-bass, two clarinets and percussion, is heard playing fragments of a barn dance, waltz *alla Ländler*, hornpipe and galop. And later that night, when the man-hunt is up, the search for Grimes is punctuated by the slow booming of a fog-horn.

In the scene in *The Boar*, not only was it important for the round to stand

129

out properly in its context; but there was the added complication of storm without and warmth and drink-happy company within. Here he profited by his experience in writing for radio drama. The technique of the mixing panel had shown him how varied were the possibilities of using music at different levels—background, foreground or intermediate—and how with two or more distinct streams of sound, one could be brought up into the foreground while the other was faded out, or (if necessary) the two streams could be mixed together. He accordingly decided to depict the storm in its full fury in Interlude ii and shut it out as soon as the curtain went up. Most of the scene in *The Boar* is accordingly set to an animated form of free recitative, punctuated by brief fragments of the storm that burst through the doorway as various characters enter from outside; and this provides an excellent setting for the round. Thanks to the cross-fading device, the music of the storm, having been heard in full in Interlude ii, continues by implication unbrokenly throughout this scene.

Cross-fading is also used for the church service, the song chanted by the procession that visits Peter's hut, and the Moot Hall dance.

As for the fog-horn, Britten realized that here was an unique opportunity for dramatic effect. As the first half of the second scene of Act III is virtually a soliloquy by Grimes, he let the orchestra be silent after the *fortissimo* shouts of the chorus at the end of the previous scene and let Grimes's monologue be accompanied only by the fog-horn and occasional cries from the distant man-hunt. Then when he has sailed out to sea to drown himself and life returns to the Borough with the dawn of another day, the repetition of the orchestral music from the opening of Act I is particularly impressive.

The first Interlude, joining the Prologue to Act I, is based on three motifs:

Ex. 1

(*a*) The high unison strings that faintly outline the key of A minor cling hard and long to each holding note, and the tension is emphasized by the grace notes; (*b*) in the middle register, *arpeggii* of diatonic thirds from the harp reinforced by clarinets and violas, describe fragmentary arcs of sound; and this musical superstructure is underpinned by (*c*) a sequence of slowly shifting bass chords from the brass—A major against the shrill A minor of the upper strings. It is not over-fanciful to find these three motifs evocative of (*a*) the wind that 'is holding back the tide', as it blows through the rigging of the boats on the beach and over the chimney-pots of the Borough, (*b*) the lapping of the water, and (*c*) the scrunch of the shingle beneath the tide. The clash between major and minor gives an extraordinarily salty tang to the scene.

Interlude ii is a storm of almost symphonic stature, which might take the following lines of Crabbe as motto:

> *But nearer Land you may the Billows trace,*
> *As if contending in their watery chace . . .*
> *Curl'd as they come, they strike with furious force,*
> *And then re-flowing, take their grating course,*
> *Raking the rounded Flints, which ages past*
> *Roll'd by their rage, and shall to ages last.*

It follows directly on the unresolved cadence of Peter's monologue at the end of Act I, scene i, and its four main episodes are: (*a*) a theme (*presto con fuoco*), which is treated fugally at its first appearance, and whose periodic recurrence in different forms gives this Interlude something of the character of a rondo; (*b*) an altered form of the brass groundswell theme from the first Interlude, with a particularly grinding passage of close imitation at the interval of a minor ninth; (*c*) a grotesque bitonal passage in triplets (*molto animato*), where the gale indulges in particularly malicious pranks in the keys of D natural and E flat simultaneously; and (*d*) a reprise and development of the music of Peter's unresolved monologue from the end of the previous scene.

This Interlude is broken off short by the rise of the curtain on the scene in the interior of *The Boar*; but, as explained above, its continuance is implicit in the fragmentary bits of storm that burst into the pub each time someone opens the door. These make it clear that the episodes of the storm Interlude are following each other in the same order as above; but when a strangely altered phrase from (*d*) ushers in Peter's arrival, it serves as direct introduction to his soliloquy, '*Now the Great Bear and Pleiades*', which is thereby given a wider significance.

The Third Interlude is an impressionist description of the sea on a warm Sunday morning—the greatest possible contrast to the storm of the previous Act (A major after E flat minor). An *ostinato* by the horns playing contiguous but overlapping thirds gives a kind of blurred background to a merry toccata-like theme for woodwind. This with the syncopated

Ex. 2

reiteration of its notes recalls the animated glitter of sunlight on water. After a brief episode consisting of a sustained tune from cellos and violas rising and then falling back through an octave, the toccata material is repeated and leads directly to the rise of the curtain and Ellen's opening *arioso*, '*Glitter of waves and glitter of sunlight*', which is set to the tune from the preceding episode, but now appearing a fourth higher and in the key of D.

As in Shostakovich's *Lady Macbeth of Mtsensk*, the central Interlude is a passacaglia. The turning-point of the opera has been reached in the middle of Act II, scene i, when, after striking Ellen at the end of their quarrel, Peter is overcome by the full realization of their failure and cries out '*God have mercy upon me!*' The musical phrase to which these words are sung becomes

Ex. 3

a key motif for the rest of the opera. After providing the main theme of the following chorus, '*Grimes is at his exercise!*' it serves (in augmented form) as

Ex. 4

the ground-bass of the passacaglia and later will be found inverted.

This large-scale passacaglia Interlude, in which it may be claimed that Peter's apprentice who has been mute throughout the opera at last becomes musically articulate, consists of a poignant air for solo viola:

Ex. 5

followed by nine variations, developed freely over this unchanging ground-bass (key of F). The last variation, a fugal *stretto*, leads directly into the next

132

scene, which opens with a series of disjointed ejaculations from Peter, punctuated by scrappy orchestral references to eight of the foregoing variations. In this way, Britten makes it clear that just as Interlude ii depicted the fury of the storm as it impinged on the senses, so Interlude iv reflects the agony that is undermining Peter's mind. The true close of the passacaglia is deferred until the end of Act II, just after the apprentice's death. Then, through a whispering *bisbigliando* figure for celesta, the solo

Ex. 6

viola repeats the original air inverted, at the end of which the ground-bass returns for a single final statement (key of C).

The two remaining Interludes are not so fully developed from the musical point of view. Like those introducing Acts I and II, Interlude v is a descriptive piece. By skilful placing, a sequence of almost static, carefully punctuated swell chords (mainly in their first and second inversions), strung together on a thread of quietly moving inner parts, is made to suggest the tranquil beauty of the sea and Borough under the moon. A secondary theme played by flutes and harp, which gives the impression of an occasional glint of reflected moonlight from wave or slaty roof or weather-vane, should be contrasted with the gayer and more glittering daylight toccata theme of Interlude iii.

The last Interlude is a cadenza freely improvised by various instruments over, under and through a ghostlike chord of the dominant. This chord arises like a faint overtone between the mighty shouts of '*Peter Grimes!*' at the end of the previous scene, is sustained by the muted horns *pianissimo* throughout the movement, and at the beginning of the last scene melts into the distant voices (off) still shouting '*Grimes!*' The effect of this *ostinato* is to emphasize the all-pervasive featureless fog, while the free improvisation of the orchestra, based on snatches of many previous themes, shows something of the raging turbulence and agony of Peter's mind.

The main purpose of these Interludes was to serve as impressionist and expressionist introductions to the realistic scenes of the opera, in much the same way as Virginia Woolf used the device of prose poems about the sea to introduce each different section and period of her novel *The Waves*, and also to secure continuity within the acts.

As for vocal presentation, Britten decided (in his own words) to embrace 'the classical practice of separate numbers that crystallize and hold the emotion of a dramatic situation at chosen moments'.* He did not, however, make each number complete and self-contained, but by following a method of construction similar to that used (for instance) by Verdi in *Falstaff* succeeded in reconciling the classical practice of separate numbers with an uninterrupted musical action. *Peter Grimes* shows with what remarkable skill he managed the transitions between the various degrees of intensity needed for recitatives, airs, *ariosi* and concerted numbers, and how he usually allowed the emotion engendered by each number to lead on to its sequel before the musical construction could reach a full close. The music flows accordingly without check or hiatus from the beginning to the end of each act, and this continuity of development is achieved at the expense of any sense of interim relaxation.

The nearest approach to the fully developed air is to be found in Ellen's solos in Act I and III; '*Let her among you without fault cast the first stone*' and '*Embroidery in childhood was a luxury of idleness*'. Her song in Act II, '*We planned that their lives should have a new start*', would fall into this category too, were it not for the fact that it excites the comments and interruptions of so many bystanders that after a few bars it becomes an eleven-part concerted ensemble and then the full chorus joins in.

As for Peter, his solos are in the nature of monologues or soliloquies; and their construction is looser and more rhapsodic accordingly. It has already been shown how his solo at the end of Act I, scene i is at first interrupted and then resumed by the storm Interlude, and how it is (as it were) completed by his soliloquy, '*Now the Great Bear and Pleiades*', in the middle of the subsequent scene. It may also be argued that his air in the hut, '*And she will soon forget her schoolhouse ways*', is a further instalment of the same large utterance, for it shows its strong family likeness to the earlier passages by its use of diatonic idiom and its consistent loyalty to the tonality of sharp keys (e.g. A and E).

There is one other fully developed musical number that deserves special mention, and that is the impassive quartet of women's voices that occurs at the end of Act II, scene i and forms such an excellent contrast to the all-male vocal writing of the following scene. When the procession of men has marched off to Peter's hut, Ellen, Auntie (the proprietress of *The Boar*) and her two nieces remain behind and dejectedly reflect on women's lot. A *ritornello* of bedraggled, trailing diatonic seconds played by the flutes separates each phrase of this trio, and the vocal parts, which are generally a major second lower than the notes of the accompaniment, betray a sluttish weariness—which conforms well with the character of Auntie and her nieces.

Most of the minor characters have an opportunity of singing memorable

* *Peter Grimes* (*Sadler's Wells Opera Books No.* 3).

snatches of song, generally in stanza form and sometimes with a refrain, and these can perhaps best be described as half-numbers. Such are Hobson's song (Act I, scene i), '*I have to go from pub to pub*', in which the second verse is sung by Ellen; Auntie's song (I, ii) with its refrain '*A joke's a joke and fun is fun*'; Balstrode's song in the same scene, '*Pub conversation should depend*' with its refrain '*We live and let live and look we keep our hands to ourselves!*'; and in Act III, scene i, Swallow's tipsy '*Assign your prettiness to me*' and the Rector's goodnight '*I'll water my roses*' with its male-voice sextet accompaniment. In addition, there is the capstan shanty in Act I, scene i, '*I'll give a hand*', which is started up by Balstrode, who is subsequently joined by Keene as helper and Auntie and Bob Boles, the Methodist, as lookers-on.

Stimulated but also kept on tenterhooks by these half-measures, the listener longs for the satisfaction of a fully completed musical number and in this state of tension welcomes any chance of relief however slight. Such a moment comes during Ellen's quarrel with Peter in Act II, scene i. The music leading up to this has been superimposed on the church service (off) with its in-going voluntary, morning hymn, responses, Gloria and Benedicite. As the Credo is reached, Peter and Ellen start to quarrel; and at this point the rift between the two musical streams widens, the Credo being intoned by the Rector and congregation to an F held by the organ, while Ellen cross-examines Peter (in the key of D flat) about the boy's bruises:

Ex. 7

After a particularly obstinate clash has developed between the subdominant of Ellen's key and the organ's F, her two-bar *dolce* phrase, '*Were we*

Ex. 8

135

mistaken?' in which her G flat appears to have modulated to F, has the surprise and relief of a final reconciliation. This relief is only momentary, however, for the passage leads directly to Peter's cry of despair, '*God have mercy upon me!*' (Ex. 3) and all the tumult released by that pregnant musical phrase.

As for the recitative in the opera, Britten's purpose can best be expressed in his own words:* 'Good recitative should transform the natural intonations and rhythms of everyday speech into memorable musical phrases (as with Purcell), but in more stylized music the composer should not deliberately avoid unnatural stresses if the prosody of the poem and the emotional situation demand them, nor be afraid of a high-handed treatment of words, which may need prolongation far beyond their common speech length, or a speed of delivery that would be impossible in conversation.' In *Peter Grimes*, there are numerous examples of both natural and unnatural intonation and rhythm. For instance, as the orchestra is playing the final bars of Ellen's air '*Let her among you without fault*', Ellen drops her voice to a *parlando* level and, turning to the carter, says: '*Mister Hobson, where's your cart? I'm ready.*' The unforced naturalness of this passage should be

Ex. 9

contrasted with another piece of recitative. At the beginning of Act II, after her short *arioso*, '*Glitter of waves*', she also drops her voice to a *parlando* level and asks Peter's apprentice '*Shall we not go to church this Sunday?*' Her words are set to the glittering toccata-like theme just heard in Interlude iii (Ex. 2) whose wide intervals and syncopated measure are utterly at variance with natural conversational idiom.

Ex. 10

The chorus plays an important part in the opera. The opening chorus of Act I, scene i, sung partly in unison and partly in parallel thirds, is sufficiently stolid with its diatonic hymnlike tune to bring out the drab as well as the picturesque aspect of life in a little fishing port; and its recapitulation at the end of Act III, where the whole musical structure is cut through in cross-section at its climax so that the opera ends with the same

* *Peter Grimes* (*Sadler's Wells Opera Books No.* 3).

abruptness as the East Anglian coast with its eroded cliffs facing the sea, constitutes an essential element of construction. Later in the opening scene of Act I, at the approach of the storm, the chorus is given a simple but moving appeal, '*O tide that waits for no man, spare our coasts!*' But its most impressive moment—and in some ways the climax of the whole opera—comes at the end of the first scene of Act III with its unaccompanied fortissimo shouts of '*Peter Grimes! Peter Grimes! Grimes!*' If this passage is to obtain its full effect, the chorus must be sufficiently strong for its cries to resound through the theatre and 'lift the roof'. Otherwise, the device of complete musical silence broken only by Balstrode's spoken words at the climax of the following scene (Grimes's suicide) loses some of its power by contrast.

Some critics, after comparing *Peter Grimes* with *Boris Godunov*, have suggested that *Grimes* is an opera whose protagonist is the chorus. But the analogy is misleading. The statement may well be true of Mussorgsky's opera, for the Russians tend to exalt the collective or communal ideal at the expense of the individual, but not of Britten's. Although the majority of the inhabitants of the Borough are prejudiced bigots, they nevertheless remain closely defined individuals who are absorbed into the general community only when their finer feelings are submerged by the herd instinct—as on the occasion of the man-hunt. The changes of focus whereby Balstrode, Boles, the Rector, Swallow, Keene, Hobson, Mrs Sedley, Auntie and her two nieces appear sometimes as individuals (with short solos to sing), sometimes as neighbours (with parts in an ensemble), and sometimes as members of the general chorus, are deliberately designed by Slater and Britten as a means of obtaining a degree of characterization in depth.

Whereas in his musical dramas Wagner was thinking first and foremost of his orchestral texture and his peculiar form of symphonic development led to an apparently unbroken flow of melody into which the vocal parts fitted like additional instruments, Britten's prime concern in *Peter Grimes* is to display the voices of protagonists, minor characters and chorus to the best possible advantage. This means, not that the vocal writing is necessarily easy and uncomplicated, but that, with the exception of the six Interludes, the orchestra is definitely used in a subordinate position as a means of accompaniment. How important this is can be seen from the scene in *The Boar* where, thanks to the preceding Interlude, the orchestra is able to allude to the furious storm outside without drowning the singers or, indeed, making it necessary for them (for the most part) to lift their voices above the level of normal recitative. Despite the use of a full symphony orchestra and of certain symphonic effects in the Interludes, *Peter Grimes* owes its characteristic movement and idiom to Britten's imaginative treatment of the voices.

Peter Grimes himself as portrayed in Slater's libretto is what might be called a maladjusted aggressive psychopath. There is a chasm, which he fails to bridge, between himself and the external world; and Britten has shown much ingenuity in finding appropriate devices to express this maladjustment in musical terms. Peter's disturbed state of mind leads, not only to the fragmentary style of utterance on which his monologue in Act I scene ii and his soliloquy in Act II scene ii are built up, but also to a disjunct motion in his vocal line and a tendency to use intervals wider than the octave. The minor ninth seems to be particularly symptomatic of his difficulty in adjusting himself to the outside world, and an upward leap of this interval occurs several times in his narration of the events that led up to the death of his first apprentice. But when he sees in Ellen a possible solution

Ex. 11

of his troubles, this interval does not resolve on the octave, but widens to the *major* ninth:

Ex. 12

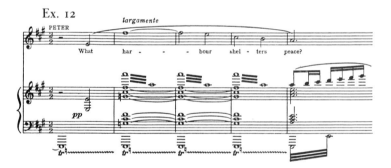

In so far as Peter is different from the rest of the Borough, augmentation and inversion are particularly associated with his music—augmentation when during the inquest in the Prologue he takes the oath and, later (I, ii), when he joins the round in *The Boar* (Ex. 13); inversion in the hut scene (II, ii) when to an inverted version of Ex. 3 he turns on his apprentice and accuses him of being the cause of all his troubles (Ex. 14). Another example of inversion deserves special mention. In the Prologue, the chorus is accompanied by the woodwind with a simple staccato chattering figure

(Ex. 15 (*a*)), which clearly becomes associated in Peter's mind with the persecution of Borough gossip. It reappears, inverted, in his soliloquy at the end of Act I, scene i (Ex. 15 (*b*)).

Occasionally Britten feels justified in using bitonality to emphasize Peter's maladjustment, and then he sometimes tries to reconcile the simultaneous use of hostile keys by enharmonic means. An excellent example occurs at the end of the Prologue, when Peter and Ellen, left alone in the Moot Hall after the inquest, sing an unaccompanied duet. At first, Peter's key is F minor and Ellen's E major; but as the voices intertwine, Peter, thanks to the enharmonic mediation of A flat and G sharp, is won over

Ex. 16

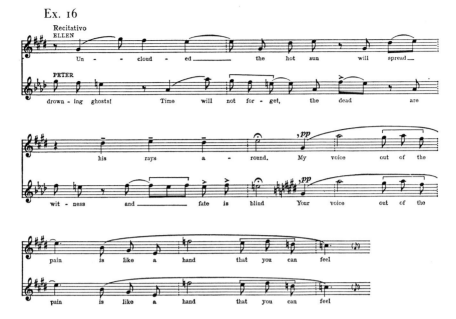

to Ellen's key, and the duet ends with both singing in unison. Exactly the same bitonal clash (F minor and E major) is to be found in the violent tremolo passage for full orchestra that punctuates the chorus's shouts at the end of the first scene of Act III.

Ex. 17

A similar example of bitonality reconciled by enharmony occurs in the first scene of Act II. There a simple sequence of notes—E flat, F natural, E flat, D natural—is harmonized by organ chords in the key of C minor and

Ex. 18

becomes the sung Gloria in the church service (off). The enharmonic equivalent of this theme—D sharp, E sharp, D sharp, C double sharp—then appears in the theatre orchestra, harmonized in the key of B major, and becomes the accompaniment to Ellen's *arioso*, '*Child, you're not too young to know where roots of sorrow are*':

Ex. 19

These enharmonic and bitonal devices which are used to express Grimes's maladjustment, together with the various passages where by cross-fading two musical streams impinge implicitly, if not explicitly, upon the ear, at first cause a kind of auditory dichotomy on the part of the listener. But as this new idiom became more familiar and acceptable, it was realized that in *Peter Grimes* Britten had certainly widened the boundaries of opera by introducing new and stimulating ideas culled from cinema and radio technique and had shown himself a remarkably subtle delineator in musical terms of complex psychological states of mind.

The originality of the work remains unimpaired, even after one has made allowances for such unconscious or subconscious echoes as the resemblance between the A Lydian of Peter's '*And she will soon forget her schoolhouse ways*' and the D Lydian of Ping's '*Ho una casa nell' Homan*' in *Turandot*, and the unresolved cadence at the end of Act I scene i repeated faster and faster until it almost becomes a trill just like the curtain to the first act of *Wozzeck*. Britten himself considered it the most realistic of all his operas.* This East Anglian story comes from an East Anglian heart, and in its operatic form it is never likely to lose the exciting force of its original impact.

* See 'Producing the Operas' by Basil Coleman. The *London Magazine*, October 1963.

III
The Rape of Lucretia

I

When *The Rape of Lucretia* was first performed at Glyndebourne, the programme note explained that Ronald Duncan's libretto had been written 'after the play *Le Viol de Lucrèce* by André Obey and based on the works of Livy, Shakespeare, Nathaniel Lee, Thomas Heywood and F. Ponsard'.

The main Latin sources for this story of Roman virtue outraged by Etruscan lust and treachery are Livy and Ovid; but at the end of the Renaissance, Shakespeare made this story so much his own, partly by direct narration in his early poem *The Rape of Lucrece* and partly by references in some of his later plays (notably *Macbeth* and *Cymbeline*), that echoes of his voice are likely to be heard in any subsequent attempt to dramatize it. This is particularly true of *Le Viol de Lucrèce*, which André Obey wrote in 1931 for Jacques Copeau's Compagnie des Quinze. In fact, the play quotes from Shakespeare's poem such passages as the description of Lucrèce asleep in her bed, her arraignment of Opportunity after the rape, and the invocation of Philomel ('*Poor bird . . .*') as she contemplates suicide. To comment on the action, Obey decided to have a Chorus of both sexes, but reduced it from plural to singular numbers, the Male and Female Chorus being endowed with special insight into the characters of Tarquin and Lucrèce respectively. In this way, he was able to expand the Shakespearean device of the soliloquy and throw a revelatory beam on the subconscious workings of the two protagonists' minds. At the same time, both Choruses were free to comment on the action; and this they did from a more or less contemporary angle, occasionally quoting fragments of Shakespeare or Livy when it suited their purpose.

Obey arranged his tragedy in four acts, according to the following scheme. Act I, scene i. During the siege of Ardea, two sentinels overhear the Etruscan and Roman generals carousing in a tent and describing how, in the course of a surprise visit to Rome the previous night to make trial of their wives, only the virtue and chastity of Collatinus' wife Lucrèce had been triumphantly vindicated. Inflamed by this account, Tarquin steals from the

142

The Rape of Lucretia: Kathleen Ferrier as Lucretia in the original 1946 production by the English Opera Group

tent and sets off for Rome on his steed. Scene ii. The same evening, Lucrèce is discovered, spinning, with her maids at home. Tarquin arrives unexpectedly and is offered hospitality for the night. Act ii. Lucrèce's bedchamber: the same night. Tarquin enters her room, wakes and ravishes her. Act iii. The same scene: the following morning. Lucrèce awakens and sends for Collatinus. Act iv, scene i. A commentary, mainly by the Male and Female Chorus, on the revolutionary state of feeling in Rome. Scene ii. The same scene as Act i, scene ii. After telling the story of her rape, Lucrèce stabs herself. Brutus, on behalf of the other Romans there present, swears revenge on the Etruscans.

The presence of the Male and Female Chorus on the stage and their commentary on the action made it necessary for the actors to develop a style that could merge almost imperceptibly from acting into mime as occasion demanded and produced a kind of extra dimension that seemed to transcend the normal limitations of realistic stagecraft. As produced by Michel St Denis for Copeau's company in the early years of the 1930s, *Le Viol de Lucrèce* lives in the memories of those who saw it as one of the masterpieces of twentieth-century theatre.

For his libretto, Duncan kept fairly closely to Obey's play. He slightly reduced the number of *dramatis personae* by doing away with a few of the servants and a pair of sentinels, but adopted without change the device of the Male and Female Chorus. He compressed the action into two acts, instead of four. Act I: *Prologue*. Male and Female Chorus: general exposition. Scene i. The generals' tent in the camp outside Rome. (This loosely follows Obey's Act I, scene i.) *Interlude*. Male Chorus: description of Tarquinius' ride to Rome. Scene ii. A room in Lucretia's house in Rome the same evening. (This closely follows Obey's Act I, scene ii.) Act II: *Prologue*. Male and Female Chorus: further exposition. Scene i. Lucretia's bedroom. (This follows Obey's Act II.) *Interlude*. Male and Female Chorus: chorale. Scene ii. A room in Lucretia's house the next morning. (This partly amalgamates Obey's Act III and Act IV, scene ii.) *Epilogue*. Male and Female Chorus: final commentary.

It will be seen that, according to Duncan's scheme, the rape comes three-quarters instead of half-way through the action, with the result that, whereas in *Le Viol* there was a long gradual diminution of tension during the last half of the play—especially in Act III where Lucrèce soliloquizes alone for a considerable part of the scene—in *The Rape* dramatic interest is built up comparatively slowly during the first act and the action gathers impetus during the second act.

In an essay describing his method of writing the libretto,* Duncan maintains that 'the legend of Lucretia has much in common with Etruscan mythology'. He adds: 'Just as fertility or life is devoured by death, so is spirit defiled by fate. Lucretia is, to my mind, the symbol of the former, Tarquinius the embodiment of the latter.' In a further attempt to magnify the symbolic significance of this simple tale and to universalize its values, he has placed the Male and Female Chorus outside the temporal framework of the action and allowed them to submit it to Christian interpretation.

> *Whilst we as two observers stand between*
> *This present audience and that scene;*
> *We'll view these human passions and these years*
> *Through eyes which once have wept with Christ's own tears.*

And from this lofty religious viewpoint, they offer (in the Epilogue) the consolation of general absolution.

FEMALE CHORUS: *Is this all? Is this it all?*
MALE CHORUS: *It is not all.*
> *. . . yet now*
> *He bears our sin and does not fall*
> *And He, carrying all,*
> *turns round*

---

* *The Rape of Lucretia* (commemorative volume), ed. Eric Crozier.

144

Stoned with our doubt and then forgives
us all.

This is a gloss that is always likely to provoke critical comment. The severely dogmatic tone of Duncan's Chorus is far removed from the pagan spirit of Shakespeare and Obey; but in the Epilogue it certainly provides the composer with a cue for a musical coda of great solemnity.

Fortified by his experience in writing *Peter Grimes*, Britten laid great store by the close collaboration of poet and composer and in his preface to the published libretto maintained that this seemed to be 'one of the secrets of writing a good opera. In the general discussion on the shape of the work — the plot, the division into recitatives, arias, ensembles and so on — the musician will have many ideas that may stimulate and influence the poet. Similarly when the libretto is written and the composer is working on the music, possible alterations may be suggested by the flow of the music, and the libretto altered accordingly. In rehearsals, as the work becomes realized aurally and visually, other changes are often seen to be necessary.' Duncan's libretto reveals nearly as many variations in its various texts as Slater's libretto for *Peter Grimes*. The initial text is to be found in the libretto published by Boosey & Hawkes in 1946, a corrected version of which was printed the same year in the vocal score. The following year, a revised edition of both score and text was issued by Boosey & Hawkes, together with a German translation by Elizabeth Mayer. This is substantially the same text as was reprinted in the commemorative volume of 1948* and the library edition of 1953.†

The effect of some of these changes has been to make the information imparted by the Male and Female Chorus rather less didactic in tone. For instance, in the first version of the Prologue to Act II, the Female Chorus, reading from a book, explained that:

The prosperity of the Etruscans was due
To the richness of their native soil,
the virility of their men,
and the fertility of their women—
See Virgil, Book Eleven, verse five three three,
'Sic fortis Etruria crevit' etcetera.
All authorities agree
that the Etruscan conquest of Rome
dates from six hundred B.C.—
that is, approximately.

In the revised edition, the last six lines are omitted entirely. Other changes are probably due to the need for simplification. Duncan learned that in his

---

* Published by The Bodley Head. † Published by Faber and Faber.

145

role as librettist he must avoid writing complicated sentences. As he says: 'The poet must drive his metaphor to the point of clarity and contain in one image the condensation of a mood. He must never forget that the audience is listening to both words and music, and that their concentration, thus divided, cannot be imposed on.' An example of such simplification is to be found at the beginning of Act I, scene i, where the Male Chorus's original comment on the generals' drinking bout:

> *The grape's as wanton as the golden boy*
> *Whom the Naiads drew to the whispers of the well*
> *But these generals drink to drown his wanton echo.*

was later changed to

> *The night is weeping with its tears of stars*
> *But these men laugh—for what is sad is folly.*
> *And so they drink to drown their melancholy.*

Here, a rather intricate, if not confused, simile has given way to a clearer and simpler contrast between the natural sorrow of the unclouded night and the forced laughter of the generals flushed with wine.

But the main alterations affect two of the characters: Junius and Lucia. In the revised version, Junius' devouring jealousy of Collatinus is established more firmly in Act I, scene i, and this helps to render more credible his sudden irruption with Collatinus just before Lucretia's suicide in the last scene. Meanwhile, Lucia who appeared in the earlier version as a young, frivolous and slightly over-sexed maid-servant, regains a measure of her not yet lost innocence and is given a charming *arioso*, '*I often wonder whether Lucretia's love is the flower of her beauty*'—her only one in the opera—which helps to establish her character musically.

II

As has already been shown, both natural predilection and economic necessity influenced Britten in deciding to choose an instrumental contingent of chamber orchestra dimensions for the *The Rape of Lucretia*. A quarter of a century previously, when Stravinsky, partly for reasons of wartime economy, had felt a similar urge in planning *The Soldier's Tale*, he had chosen representative outer-range instruments from each of the main orchestral groups and had found that seven was his irreducible minimum. *The Soldier's Tale* was accordingly written for violin and double-bass, clarinet and bassoon, cornet and trombone, and percussion. Britten's solution was different: a string quintet, a woodwind quartet (with the flute doubling piccolo and bass-flute, the oboe doubling cor anglais, and the clarinet doubling bass-clarinet) and a miscellaneous trio consisting of horn,

146

harp and percussion—a dozen players in all—with the *recitativo secco* accompanied by the conductor on a piano. (A similar ensemble was chosen by Menotti about the same time for his operas, *The Telephone* and *The Medium*.)

Although this meant that *The Rape* would be a chamber opera in the sense that every executant (vocalist or instrumentalist) would be a solo performer, it did not necessarily presuppose intimate chamber conditions of performance. The doublings in the instrumentation of works for symphony orchestra increase the volume of sound to only a limited extent—as Stravinsky says in the sixth lecture in his *Poetics of Music*, they thicken without strengthening, and above a certain point the impression of intensity is diminished rather than increased and sensation is blunted. Their main purpose is to blend and bind instrumental tone colour; and this may be compared with the custom of covering oil paintings with a heavy layer of varnish. Just as modern taste is in favour of removing these varnishes to reveal the original colours of a painting in all their unsubdued vigour, so one would like to think that modern audiences, suffering a revulsion from the inflated performances of super-orchestras, are prepared to demand a scaling-down of forces and to accept the convention whereby instrumental tone colours are presented directly in their primary state without any attempt to blend them before they reach the ear.

In the light of the experience gained in touring *The Rape* in England and abroad, it appeared (according to Eric Crozier) that 'the quality and vitality of the voices with instruments were much better in large theatres than in small. There were no complaints of thinness or raggedness in texture.'* On the other hand, there were moments when the voices, instead of being sustained by the blended accompaniment that comes from a full body of strings, had to fight for audibility on equal terms with each solo instrument and were in danger of being drowned or rendered insignificant by the unsubdued tone colour of a handful of instruments. As Edward Sackville-West wrote in the Preamble to *The Rescue*, 'it is, paradoxically, impossible to produce an overall orchestral pianissimo without using a considerable body of instruments, whereas a double *forte* requires only the minimum'.

In short, chamber opera demands the highest virtuoso standards from each individual executant—the slightest lapse is liable to prejudice the total effect. It makes similar demands of its audience too. The equipoise between voices and instruments is too precarious for the listener ever to be lulled into a sense of complete relaxation and security; and many listeners— particularly those reared on a rich diet of lush orchestral fare—are apt to resent the necessity of making this special effort. This is one of the problems that Britten and his collaborators had to face when producing *The Rape of*

* In *The Rape of Lucretia* (commemorative volume), ed. Eric Crozier.

THE OPERAS

Lucretia, Albert Herring, The Beggar's Opera, The Turn of the Screw, Curlew River and the other church parable operas, and one that they surmounted with extraordinary success.

III

Whether the conflict in *The Rape of Lucretia* is spirit defiled by fate or, more prosaically, Lucretia ravished by Tarquinius, Britten needed two simply contrasted musical ideas to express it. He found them by taking the melodic contents of the interval of the diminished fourth and arranging the notes in two different ways. The first (*a*), a descending scale-passage, is identified with Tarquinius; the second (*b*), a sequence of two thirds in contrary motion, is the kernel of the Lucretia motif. By extension, the idea of a scale-

Ex. 20a   Ex. 20b

passage (whether descending or ascending) becomes identified with the male element. This is aptly summarized in the Prologue, where the Male and Female Chorus sing a solemn hymn, which combines both elements.

Ex. 21

With this key, it is possible to unlock many of the secrets of the score.

Lucretia's motif, moving both forwards and backwards, serves as the generals' flamboyant toast in Act I, scene i. In diminution, it is frequently

Ex. 22

148

used as an accompanying figure—for example, in Junius' outburst *'Lucretia! I'm sick of that name!'*, also in the following comment by the Male Chorus,

> *Oh, it is plain*
> *That nothing pleases*
> *Your friends so much*
> *As your dishonour*

and in the Male Chorus's subsequent apostrophe, when he perceives what is going on in the privacy of Junius' heart and how like an empty vessel it is suddenly flooded with jealousy. Here in the swirling accompaniment, the

Ex. 23

minor third interval becomes widened progressively to major third, fourth and even fifth. During the same scene, Tarquinius' motif is not much in evidence, though it is clearly referred to in the Male Chorus's opening air which depicts the sultry atmosphere of the evening outside Rome, and three notes of it, repeated by the harp in diminution, become identified with the noise of the crickets (Ex. 24). In the Interlude, however, the male element dominates the furious ride of Tarquinius and his steed (Ex. 25). It is only when horse and rider, after being momentarily checked in their course by the sudden obstacle of the Tiber, plunge into the river and swim across that

Ex. 24

Ex. 25

the Lucretia motif returns, the music of Ex. 23, now accompanied by the cool metallic hiss of a cymbal tremolo, being repeated to the following words with their prophetic symbolism:

> *Now stallion and rider*
> *Wake the sleep of water*
> *Disturbing its cool dream*
> *With hot flank and shoulder.*

In the second scene, with Lucretia sewing, Bianca and Lucia spinning, and the Chorus interpolating her commentary, interlinked feminine thirds run riot in the harp accompaniment to the vocal quartet. The rising

Ex. 26

intonation of Lucretia's apostrophe to her absent husband may be thought of as an altered version of the Tarquinius motif, which naturally plays an important part a little later when Tarquinius arrives unexpectedly. A particularly subtle touch is to be found in the passage where this motif, preluded by its inversion, is accompanied by a figure made up of both the Tarquinius and Lucretia motifs in diminution. The final series of

Ex. 27

goodnights, when Lucretia leads Tarquinius to his bedchamber, contains the Tarquinius motif augmented and harmonized with thirds over a ground-bass of compound thirds rising by thirds:

Ex. 28

At the beginning of the second Act, the interval of the third—and also its inversion, the sixth—are predominant in Tarquinius' air, '*Within this frail crucible of light*', sung while Lucretia is still asleep; but when she awakens and in her agitation asks what he wants from her, the music gives an explicit answer, the brassed horn frenziedly attempting to fill the open interval of the

Ex. 29

cor anglais's minor third. The new figure formed thereby (*c*) becomes closely identified with Tarquinius' lust during the remainder of the scene, until near its climax Lucretia interrupts him with a broad and dignified rebuke based on the original Tarquinius motif in augmentation (Ex. 30). In the unaccompanied quartet with which the scene ends ('*See how the rampant*

151

Ex. 30

centaur mounts the sky') will be found a recapitulation of the jealousy theme (Ex. 23) sung by Lucretia, together with fragments of the Lucretia, Tarquinius and lust motifs.

For the opening of the last scene, where Bianca and Lucia greet the sunny morning and then start to arrange the flowers the gardener has brought,

both accompaniment and voices describe festoons of rising and falling thirds. When Lucretia enters and is told by Bianca that they have left the orchids for her to arrange, she bursts out hysterically, and the music at once refers to the lust theme. Just before her suicide there are two passages where use of the Lucretia motif deserves special notice. The first comes when she asks Bianca if she remembers teaching her as a child to weave garlands of wild flowers. *'Do you remember!'* In a brief aria, Bianca replies:

Ex. 33

and nothing could sound gayer than this carefree scherzo, with its sparkling accompaniment based on the Lucretia motif in diminution and in a major mode. Later when Collatinus arrives and Lucretia dressed in mourning makes her confession, the Lucretia motif returns to the minor mode, but is tenderly harmonized with a major chord. Meanwhile, her silent entry has

Ex. 34

been made all the more pitiful by the way in which both the regularly moving ground-bass and the hesitant sobbing phrases of the cor anglais are

Ex. 35

formed out of scale passages. As she stabs herself, her last words fall in thirds through the compass of two octaves:

Now I'll be forever chaste
With only death to ravish me.
See, how my wanton blood
Washes my shame away!

A funeral march in passacaglia form follows, with a ground-bass (Ex. 36) that brings to mind the Chorus's solemn hymn (Ex. 21) that has been heard in both Prologues. Over this ostinato, a magnificent sextet is built up, in

153

Ex. 36

which each vocal part is developed with full regard for the characters of the various persons concerned. At the end, special prominence is given to the repeated demi-semiquaver figure; and this figure persists throughout the Epilogue so that when the opera ends with a recapitulation of Ex. 21, the accompaniment is powdered with major and minor thirds—the minor predominating—spread over the entire orchestral compass, like stars that come out in the firmament after sunset.

Although these musical germs or cells have been spoken of as motifs, and their permutations and transformations traced in some detail, it must not be thought that in *The Rape* Britten has adopted a comprehensive *leitmotif* system of construction. Many beauties lie outside the passages quoted above—to take a single example, the marvellous lullaby at the beginning of the second Act, one of his most original and memorable passages. But just as in the *Symphony of Psalms*, Stravinsky used the device of two interlinked ascending thirds (rather similar to the Lucretia motif) to unify the thematic material in each of its three movements, so Britten has used his motifs to emphasize the persistence of the fundamental conflict in the action and to achieve a remarkably homogeneous texture in his score.

IV
Albert Herring

Albert Herring was intended to be both a companion piece and a contrast to *The Rape of Lucretia*: a companion piece since it is written for the same vocal and instrumental contingent; a contrast in the sense that its subject is comic as opposed to the tragedy of *The Rape*.

Eric Crozier adapted Guy de Maupassant's short story, *Le Rosier de Madame Husson*, with considerable skill. He transferred the action from Normandy to East Anglia; and although in the process something of the malicious sparkle of the Gallic original may have been lost, what remains is in most essentials faithful to the letter, if not the spirit, of the original.

The main outline of the plot is soon told. Lady Billows, virtuous herself and the self-appointed guardian of virtue in others, is anxious to select a May Queen in Loxford; but, in default of suitable female candidates, she decides on a May King, and her choice falls on young Albert Herring, who works in a greengrocer's shop and has a reputation for unassailable innocence and chastity. During the May Day celebrations, he is fêted and plied with lemonade that has been surreptitiously laced with rum. So fortified, he breaks out and escapes that evening from the stifling atmosphere of his home. When his absence is discovered the following morning, search parties are sent out. At first, it is feared he may have been killed; but just as his death is being lamented, he arrives back, dirty, dishevelled and defiant after a bibulous night out.

This is the point where opera and short story begin to diverge. According to Crozier, Albert Herring's night out consisted only of a pub-crawl, in the course of which he was thrown out of *The Dog and Duck* and *The Horse and Groom*. Although Herring himself refers to it as 'a night that was a nightmare example of drunkenness, and dirt, and worse', it seems mildly innocuous when contrasted with the virtuous Isidore's escapade in *Le Rosier de Madame Husson*. 'Isidore was drunk, dead drunk, besotted after a week of dissipation, and not merely drunk, but so filthy that a dustman would have refused to touch him. . . . He smelt of the sewer and the gutter and every haunt of vice.' This would have set too serious a note for the simple honest fun of Crozier's libretto. Nevertheless, the character of Herring remains slightly embarrassing; and when the curtain falls, one does not feel convinced that the experience gained in the course of this or any other drinking bout would have been sufficient to free him from the shackles of his painful inhibitions. Isidore eventually died of *delirium tremens*; Albert Herring, one suspects, lived down the momentary scandal of his May Day intoxication and became a respected citizen of Loxford.

Crozier aimed at clarity and simplicity in his libretto and wrote 'to be sung, not to be read from a printed page'. In his series of lectures on *The Poetic Image*, C. Day Lewis, speaking from the poet's point of view, explained the distinction as follows: 'The writing of words for music demands an entirely different technique from the writing of lyric poetry as we now understand the term. Words for music are like water-weed: they live only in the streams and eddies of melody. When we take them out of their element, they lose their colour, their grace, their vital fluency: on paper they look delicate perhaps, but flat and unenterprising.' It is to Crozier's credit that he understood this, and consequently his libretto is most successful when it is most self-effacing.

Writing in 1938 of Britten's early works, Henry Boys gave it as his opinion that if Britten chose, 'he could undoubtedly become the most original and probably the most successful maker of light music in England since Sullivan'.* In *Albert Herring* this prophecy is fulfilled. The music is light and loose, mercurial and full of fun. It recaptures something of the boyish high spirits of his earlier works. '*Avanti!*' cries the composer to his woodwind in the Interlude between the two scenes of Act I, just as eleven years before he had written the same direction in the first movement ('*Rats away!*') of *Our Hunting Fathers*. Sid, the butcher's assistant, laces Albert Herring's glass of lemonade with rum; and a sinister coil of chromatics arises from the orchestra to remind one of the love potion motif in *Tristan und Isolde*. Police Superintendent Budd complains in the last act, when the hunt for Albert Herring is on:

> *Give me a robbery with force*
> *Or a criminal case of rape,*
> *But God preserve me from these disappearing cases!*

And as he mentions rape, the orchestra refers slyly (but *fortissimo*) to the Lucretia motif (cf. Ex. 20b).

One of the score's most distinctive characteristics is to be seen in the handling of the recitative. Not only are there examples of every shade of *recitativo secco* and *recitativo stromentato*, but many of these show an extraordinary freedom in rhythm, the voices (*ad libitum*) often going a

---

* 'The Younger English Composers: V. Benjamin Britten' by Henry Boys. *Monthly Musical Record*, October 1938.

Albert Herring: Lady Billows (Joan Cross), The Vicar (William Parsons), and Miss Wordsworth (Margaret Ritchie) in the committee meeting from Act I in the original 1947 English Opera Group production

completely different way from the accompaniment—as in the *recitativo quasi ballata* of Act I, scene i—while others combine in contrapuntal patterns to form recitative ensembles of great complexity. There are three such ensembles in Act II, scene i, where each character is directed to sing his or her line at the natural speed of the diction without paying any regard to the other voices or the accompaniment; and the last one, which occurs as the coronation feast begins and just before the curtain falls, contains six independent recitative solos, two recitative duets (between Nancy and Sid, and the Mayor and the Police Superintendent), and a canon for the three village children.

In a three-act comic opera like *Albert Herring*, where the music is continuous, the composer has an enormous canvas to fill. As Erwin Stein says, in comic opera 'there are fewer opportunities for slow movements and lyrical expansion than in musical drama'.* Although by the use of recitative the composer is able to get over much of the ground at a spanking pace, it becomes all the more necessary for him to strengthen the tension and

---

* 'Form in Opera: *Albert Herring* Examined' by Erwin Stein. *Tempo*, Autumn 1947.

specific gravity of his music at the nodal points. This is done mainly by increasing the contrapuntal interest of the musical texture. For instance, there are fugal choruses, '*We've made our own investigations*' and '*May King! May King!*' in Act I, scene i; in the following scene, the 'pleasures of love' duet, where Nancy and Sid singing in unison are accompanied by a perky woodwind canon at the octave; the fugal Interlude between scenes i and ii of Act II; and, in particular, the magnificent nine-part threnody in the last Act. This is built up on a ground chorus in the minor mode. Above this

Ex. 37

the individual verses of lament are freely developed. The Vicar's contribution is marked *espressivo*, Nancy's *piangendo*, the Mayor's *marcato ed eroico*, Lady Billows's *brillante*, her housekeeper's *con forza*, the Superintendent's *pesante*, the schoolmistress's *lamentoso*, Sid's *con gravità* and Mrs Herring's *appassionato*. Towards the end, all these nine characteristic laments, after alteration to fit the major mode, are repeated together over a roll from the timpani and fused into a wonderfully intricate knot of polyphony.

It is the opera's special glory that frequently a character or episode is treated with a mixture of satire and sentiment that produces as unforgettable a vignette as a drawing by Rowlandson or a poem by Betjeman. There is the bland and hesitant Vicar trying to reassure Lady Billows on the subject of virtue:

Ex. 38

the twittering Miss Wordsworth nervously rehearsing the school children in the festive song:

Ex. 39

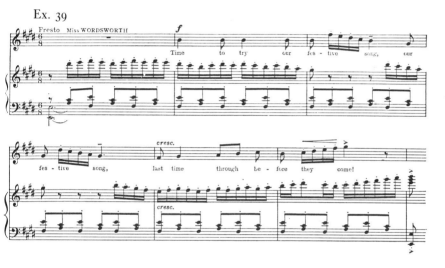

Mrs Herring clutching the faded framed photo of her son as a little boy:

Ex. 40

and Albert when, returning in the evening from the feast, he enters the
greengrocer's shop in the dark and looks round for matches to light the gas
to the accompaniment of the exquisite nocturne for bass flute and bass
clarinet first heard in the previous interlude:

Ex. 41

It might be thought that in foreign countries *Albert Herring* would be
handicapped by the slang element in its libretto; but this does not seem to be
the case. In translation the opera has proved a great success in Germany and
other countries; and in the original nowhere has it been more enthusiasti-
cally acclaimed than in Aldeburgh, which is only a few miles as the crow flies
from Loxford itself.

V

The Beggar's Opera

Planned by John Gay as a Newgate comedy, *The Beggar's Opera* was an immediate hit when produced at the Theatre in Lincoln's Inn Fields on 29 January 1728. Its record run of sixty-two performances, of which the first thirty-two were consecutive, remained unbroken until the production of *The Duenna* at Covent Garden in 1775. It made the fortunes of a number of persons connected with it. Lavinia Fenton, the original Polly, became the toast of the town and ultimately married the Duke of Bolton. It is calculated that from his four benefit nights (third, sixth, ninth and fifteenth performances), John Gay, the author, received between £700 and £800. Details of the emoluments of Dr Pepusch, who adapted the music, are not known; but Rich, the manager, seems to have netted a profit of over £6,000 on the first season, and £1,000 or more the following (1728–9) season.* This must have contributed substantially to the capital needed a few years later when he promoted the building of the new Covent Garden Theatre.

The fame of *The Beggar's Opera* spread quickly through the provinces and there were performances in Dublin, Glasgow and Haddington in 1728. It was the second piece to be produced at Covent Garden (16 December 1732). The following year a company took it to the West Indies, where Polly had already gone in Gay's published but for many years unacted sequel, and subsequently it was played in North America. It set a fashion for ballad opera which swept London and persisted for about ten years—at least 120 ballad operas were produced during the period 1728–38—and later the type developed into *pasticcio* opera which flourished during the latter part of the eighteenth century. In both ballad and *pasticcio* operas the predominating role was the playwright's, while the composer had the function of a musical director, who arranged and scored the music selected. But whereas the majority of the tunes used in the ballad operas were popular and anonymous—for instance, fifty-one out of the sixty-nine in *The Beggar's Opera*—the numbers in the *pasticcio* operas had nearly all been composed by

---

* These figures are based on the calculations in Sir St Vincent Troubridge's article 'Making Gay Rich, and Rich Gay'. *Theatre Notebook*, October 1951.

contemporary musicians, many of them probably by the arranger himself.

It is not known for certain whether Dr Pepusch was solely responsible for the music in *The Beggar's Opera* or whether Gay himself played some part in the choice of tunes. The music as published with the libretto consisted of Dr Pepusch's Overture printed in four-part score and the tunes of the sixty-nine airs given without any indication of harmony. When the songs were separately engraved as Dr Pepusch's composition, an unfigured bass was added. Other musical editors soon started to make additions and alterations to Dr Pepusch's score. Dr T. A. Arne was one of the first; and his dance of prisoners in chains (Act III, scene xii), for which Dr Pepusch had specified no music, was later incorporated into the Frederic Austin version. Many others followed.

In default of an authoritative version of the original score, each revival has had to solve afresh the problem of musical presentation. After the first world war, when Nigel Playfair produced *The Beggar's Opera* at the Lyric Theatre, Hammersmith, the text was carefully revised by Arnold Bennett and all offensive matter removed. Frederic Austin, basing his score fairly closely on a previous version by J. L. Hatton, set about two-thirds of the original tunes in a style that was an elegant *pastiche* of the eighteenth century, scoring the numbers for string quintet, flute, oboe and harpsichord with occasional use of the viola d'amore and viola da gamba. He inserted a number of dances and instrumental interludes, mainly of his own composition; and Lovat Fraser designed a gay colourful setting. The production was pretty and had panache. It proved to be exactly suited to the taste of the post-war public and ran for 1,463 performances, being frequently revived later in the 1920s and during the early 1930s.

By 1940, when Glyndebourne produced *The Beggar's Opera* on tour and later brought it to the Haymarket Theatre, London, the Austin version was becoming outmoded. John Gielgud, who on this occasion was the producer, moved away from the emasculated prettiness of the Playfair-Fraser entertainment towards a more realistic treatment of the miseries of London; but he chose the early Victorian period depicted in the style of Cruikshank, and in this inappropriate setting the Austin score was a complete misfit. About the same time Edward J. Dent made a new version of the score at the request of Sadler's Wells; and this was based as far as possible on the original edition. In the event, the Dent version was performed for the first time, not by the Sadler's Wells Opera Company, but by the Clarion Singers, Birmingham, and there have been several subsequent amateur productions.

When Britten decided to make a new adaptation of *The Beggar's Opera* for the English Opera Group in 1948, he turned to Tyrone Guthrie as his collaborator. Guthrie wished to restore the opera's original pungency. He saw it literally as a *beggar's* if not a *beggars'* opera—in his own words, 'the expression of people made reckless, even desperate by poverty, but in whose

162

The Beggar's Opera: Polly Peachum (Nancy Evans), Macheath (Peter
Pears) and Lucy Lockit (Rose Hill) in the original 1948 production by the
English Opera Group

despair there is none the less a vitality and gaiety that the art of elegant and
fashionable people often misses'—and he edited the text and planned his
production accordingly. The same intention had underlain *Die Dreigroschenoper* which Brecht wrote in 1928 for production at the Theater am
Schiffbauerdamm, Berlin: but whereas Brecht retained the essential
framework of the characters and the plot (although he transferred the period
from the early eighteenth century to the last years of Queen Victoria's reign
during the Boer War), Kurt Weill removed all the original airs from his
score, except the *Morgenchoral des Peachum*, which was based on Peachum's
opening song *'Through all the Employments of Life'*, and even this was
dropped after the dress rehearsal.

As far as Britten was concerned, there was no temptation to cut the
original airs. Quite the contrary. He himself said: 'These tunes to which
John Gay wrote his apt and witty lyrics are among our finest national
songs. These seventeenth- and eighteenth-century airs, known usually as
"traditional tunes", seem to me to be the most characteristically *English* of
any of our folk-songs. They are often strangely like Handel and Purcell:
may, perhaps, have influenced them, or have been influenced by them. They
have strong, leaping intervals, sometimes in peculiar modes, and are often

163

strange and severe in mood. While recognizing that the definitive arrangement of them can never be achieved, since each generation sees them from a different aspect, I feel that most previous arrangements have avoided their toughness and strangeness, and have concentrated only on their lyrical prettiness. For my arrangements of the tunes I have gone to a contemporary edition of the original arrangements by Dr Pepusch. Apart from one or two extensions and repetitions, I have left the tunes exactly as they stood (except for one or two spots where the original seemed confused and inaccurate).'

In actual fact he used sixty-six out of the sixty-nine airs of the original 1728 version, as against the forty-five set by Austin. Twice he combined two of these airs: the first time to create a duet between Lucy (Air XXXI*—'*Is then his Fate decreed, Sir?*') and Lockit (Air XXXII—'*You'll think e'er many Days ensue*'); the second to create a trio between Macheath (Air XXXV— '*How happy could I be with either*') and Polly and Lucy (Air XXXVI—'*I'm bubbled—I'm bubbled*'). Britten's solution of the scene in the condemned hold where Macheath tosses off glasses of wine and bumpers of brandy in the hope of working up sufficient courage to face being hanged upon Tyburn tree is a *tour de force*. The original 1728 version specified a sequence of ten different airs, several of them fragmentary. Austin quite frankly shirked the problem by omitting six of the ten airs. Dent tackled it by giving the original tunes to the orchestra complete as far as possible and allowing Macheath to come in with his fragmentary ejaculations as and when they occurred. Britten's solution was to set the final air '*Green Sleeves*' in such a way that its bass could be used as a ground linking the previous airs. Seven times this ground bass is broken at different points to allow the interpolation of fragmentary airs; twice the fragments fit it exactly; and the tenth time it sinks into its proper place as the bass to the concluding '*Green Sleeves*'.

Although, as the Beggar boasts in the Prologue, there is nothing in the opera so unnatural as recitative, occasional passages of melodrama occur where music underlines spoken dialogue and eases some of the transitions from song to speech and speech to song. There is a particularly ambitious passage of this kind in the form of a series of nine cadenzas introducing eight ladies of the town and a harper in the scene of the tavern near Newgate.

A somewhat similar musical introduction of characters was carried out in Britten's Overture, which replaced Dr Pepusch's. Guthrie's intention was that after the Beggar had spoken his introduction the curtain would rise and the orchestra would play the Overture while the actors were seen getting ready for the performance and setting the scene of Peachum's lock. Britten accordingly composed an episodic overture introducing tunes connected with the main characters, starting with Lucy Lockit (fugue) and ending with Mr Peachum's air '*Through all the employments of life*', which is also the

* The numbering of the airs is that of the original 1728 edition.

opening number of Act I. The effect is attractive so long as the producer can ensure that the end of the Overture and beginning of the Opera are clearly differentiated so that the actors can get the ictus they need for the start of the action proper.

Another change made by Guthrie was the lengthening of Act I to include the first scene (A Tavern near Newgate) of Gay's Act II. The effect of this was to bring the curtain down on Macheath's arrest and to confine the next act to the Newgate scenes covering Macheath's imprisonment and escape.

Careful attention was paid by Britten to the keys of the different numbers. Thirty-six airs were presented in their original keys. The transpositions of the remaining thirty were carried out mainly with an eye to emphasizing two important tendencies—of numbers concerned with Macheath and Polly, and their love for each other, to gravitate round B flat major and its related keys, and of numbers concerned with Newgate to gravitate round E minor and A major and their related keys. The opera's central tonality is F. The Overture opens in F major; Act I starts with Peachum's first air in F minor; Macheath's scene in the condemned hold is in F minor; and the following trio '*Would I might be hanged*' is in F major. This marks the end of the opera proper, the dance in G minor which comes after the reprise being of the nature of an afterthought, a kind of jig to wind up the proceedings.

Throughout the opera Britten takes special pains to set off the modal characteristics of the airs and not to force them into the straight-jacket of academic major/minor harmonization. An interesting example of his treatment is to be found in the duet between Polly and Lucy '*A curse attends a women's love*'. The original tune was '*O Bessy Bell*' and the first half of it is given in Gay's original version of *The Beggar's Opera* as follows:

Ex. 42

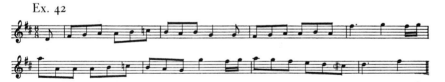

Austin, being apparently nervous of the 'strong, leaping intervals' in the vocal part, altered the tune and set it as follows:

Ex. 43

Britten's version is not only faithful to the original, but captures the real non-modulatory spirit of the flattened seventh and sets it in proper relief by the oboe phrase with its false relations:

Ex. 44

Quotations illustrating the numerous ingenuities of Britten's settings of these tunes could easily be multiplied—the augmentations and diminutions, the metric subtleties, the daring harmonic progressions that occasionally seem to modulate against the vocal line, the startling stab of bitonality in the chorus '*Let us take the road*', the use of canon and imitation, the free development of counter themes, the cunning instrumentation, and so on. But even more important than any of these is the fact that Britten still knows when mere cleverness is out-of-place and when it behoves him to be absolutely unaffected and simple. For instance, his setting of the duet between Mrs Peachum and Polly '*O Polly, you might have toy'd and kist*' is most lovely and haunting. It will be found that here too Britten's vocal line differs from the Austin version, the reason being that Britten has literally followed the original tune '*O Jenny, O Jenny, where hast thou been*' as given in the original editions.

166

The orchestra for this version of *The Beggar's Opera* consists of flute (doubling piccolo), oboe (doubling cor anglais), clarinet, bassoon, horn, percussion, harp, string quartet and double-bass. From it Britten obtains a maximum variety of instrumental effects. Examples that stick in the mind are the tremolo flute motif in Polly's air '*The Turtle thus with plaintive crying*'; the lace-like edging provided by the harp to the sturdy cotillon '*Youth's the season made for joys*'; the sensitive flutterings of the flute in Polly's air '*Thus when the swallow seeking prey*'; the savage unison strings with their dotted quaver motif in Lockit's air '*Thus Gamesters united in friendship are found*'; the rumbling of the low timpani bass accompanying the air sung by that deep drinker Mrs Trapes '*In the days of my youth*'; the sinister liquid gurgling of the clarinet *arpeggii* as Lucy urges Polly to '*take a chirping glass*'; and the vibration of the tremolo bell chords, the tolling of which punctuates the trio '*Would I might be hanged!*'

Britten's version of *The Beggar's Opera* is not the definitive version—no version will ever be that*—but it is to date not only the most brilliant musical version, but also the one in which the operatic nature of the work has been successfully emphasized. For the first time in its long history, *The Beggar's Opera* needs singers rather than actors to interpret it; and the result is that, whereas none of the earlier versions made much headway abroad, except in the United States and British Commonwealth, *The Beggar's Opera* in Britten's version has been played in many of the opera houses in Europe, and the work is at long last becoming universally recognized as one of the masterpieces of comic opera.

* In 1953 another version of the score was made by Sir Arthur Bliss for the film directed by Peter Brook with Sir Laurence Olivier as Captain Macheath.

VI
The Little Sweep

The Little Sweep is the opera that forms part of the entertainment for young people called *Let's Make an Opera!* Whereas in *Peter Grimes* the apprentice boy was a mute character, here Sam, the sweep-master's eight-year-old apprentice, becomes the hero of the opera, and a fully vocal hero too.

The story of how a number of children living at Iken Hall, Suffolk, in 1810 meet Sam, a new sweep-boy, while he is sweeping one of the chimneys, and succeed in rescuing him from his bullying master, Black Bob, was based by Eric Crozier, not on Lotte Reiniger's silhouette film *The Little Chimney Sweep* (1935), nor Charles Kingsley's *The Water Babies* (though Britten himself was brought up on Kingsley's book as a boy), nor Elia's mannered essay, *The Praise of Chimney-Sweepers* (1822), but on William Blake's Songs of Innocence, *The Chimney Sweeper*:

> *When my mother died I was very young,*
> *And my father sold me while yet my tongue*
> *Could scarcely cry ''weep! 'weep! 'weep!'*
> *so your chimneys I sweep, and in soot I sleep.*

Despite its surface gaiety, the atmosphere of this little opera is surcharged with intense pity and indignation for the plight of the hapless sweep-boy sold into servitude because of his parents' poverty. If to readers of Lamb and Kingsley the note of cruelty implicit in this tale seems rather exaggerated, they will find more than sufficient justification for Crozier's attitude if they consult *A People's Conscience** where the case of the climbing boys is presented on the basis of evidence contained in the reports of various Select Committees of the House of Commons. No wonder that as Blake wandered through the London streets he heard

> *How the chimney-sweeper's cry*
> *Every black'ning church appals.*

The practice of using young boys to sweep the more difficult and dangerous

---

* By Strathearn Gordon and T. G. B. Cocks. Constable, 1952.

168

flues was not really abolished until the latter part of the nineteenth century; and as late as June 1949, the month and year in which *Let's Make an Opera!* was first performed, the last of the chimney boys—Mr Joseph Lawrence of Windlesham, Surrey—celebrated his one hundred and fourth birthday.

It is to the credit of Britten and Crozier that, while not shirking the fundamental problem of this opera, they presented it in such a cheerful and sympathetic guise as completely to beguile the audience; and this audience, it must not be forgotten, included children and young people as well as adults.

Sam, the little sweep, is eight years old. A party of half a dozen children, three of them living at Iken Hall and three of them on a visit there, rescue him from the flue where he has stuck and eventually deliver him from his master. As against these seven child actors—all the boys with unbroken voices—there are six adult parts, two of which can be doubled. The cast of eleven is accordingly a mixture of amateurs and professionals; and Britten and Crozier had the audacious but brilliant idea to increase the amateur element by implicating the whole of the audience in the performance. This they did by giving the audience four songs to sing in the course of the opera: *The Sweep's Song* by way of overture, *Sammy's Bath* and *The Night Song* as interludes between scenes i and ii, and scenes ii and iii, and the *Coaching Song* that forms the finale.

Such an unorthodox plan could be carried out successfully only if the audience had an opportunity to rehearse their songs; and this gave Britten and Crozier the cue they needed for the earlier part of the entertainment. Clearly the opera must be shown in rehearsal; and this in fact was how *Let's Make an Opera!* was first presented at Aldeburgh. Later, however, Crozier revised this first part, expanding it into two acts so as to show quite clearly the conception, writing and composition of the opera as well as its stage production.* Although these two acts are in the form of a play, they contain a certain amount of incidental music. For instance, the first scene ends with a musical toast in the form of a round on the words '*Let's Make an Opera!*' and later on the local builder who is roped in to play the roles of the Sweepmaster's assistant and the gardener takes part in an audition at which he sings '*Early One Morning*'. Earlier, there is a glimpse of the composer at work on the scene where Sam, just after his rescue from the chimney, begs the children '*Please don't send me up again!*' During the second act, the

---

* Subsequently (1965) he revised it again, condensing the two acts into one. But in the most recent edition of the published libretto for *The Little Sweep* Eric Crozier writes in the Foreword: 'This preliminary play served its purpose at the time, but it does not wear so well as the opera. Ideally, it should be rewritten to suit the local circumstances and characters of any group performing *The Little Sweep*. We have therefore decided not to reprint it any longer, but copies will continue to be available from the Hire Library at Boosey & Hawkes Ltd. for you to adopt or adapt at your pleasure. But if the opera is to be performed, without the play, it should still be preceded by a short rehearsal of the audience songs.'

The Little Sweep: Black Bob (Norman Lumsden), Clem (Max Worthley) and Sam the sweep boy (John Moules) from the original 1949 production by the English Opera Group

audience is given a chance not only to learn its four songs, but also to see three of the children's ensembles being rehearsed: so by the time the opera proper is reached, the work of preparation has been so thorough that much of the music is greeted with the thrill of recognition.

The orchestra is smaller even than that used for *The Rape of Lucretia*, *Albert Herring* and *The Beggar's Opera*, and consists of a string quartet, piano (four hands) and percussion (single player).

The music is not continuous. There is spoken dialogue between the numbers, and practically no recitative.

The score is distinguished by the straightforward appeal of the vocal line; but the setting and presentation of the melodies often show a considerable degree of sophistication. The basic tunes of all four audience songs are so simple that some of them (particularly the Night Song) bid fair to attain an independent popularity; but in their setting they are distinguished by metrical and harmonic touches of considerable subtlety. For instance, each verse of the first audience song is prefaced by the cry 'Sweep!' and the opening phrase of the catchy tune is:

Ex. 45

But Britten sets it in 5/4 time, which gives a fascinating contrast between the groups of nine hurrying triplet quavers and the more measured tread of the crotchets, and he makes ingenious use of the fact that a large concourse of people of all ages singing in unison often produces a slightly indeterminate pitch:

Ex. 46

The cry 'Sweep!' is harmonized differently on every appearance, with the result that although the tune never varies, the song in the course of its five verses modulates from D minor through B flat major, G minor, E flat major to D minor (ending on a chord of the major). Ex. 47 shows the transition between verses three and four:

171

Ex. 47

Another example of the special effects to be obtained from shifting harmonies while the tune remains constant is to be found in the Ensemble *'O why do you weep through the working day?'* At the end of each verse a three-part phrase sung by the children leads to Sam's repeated, unforgettably poignant cry *'How shall I laugh and play?'* The first and last verses are given in Ex. 48; and it will be seen how natural and moving is the clash between the F natural and the F sharp in the final cadence:

Ex. 48

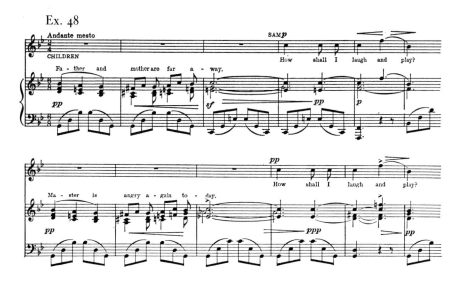

Although Britten takes considerable pains to protect the children's voices from anything too complicated in the way of melody, he does not hesitate to allow them to take part in a contrapuntal movement like the finale to scene ii, which is really a passacaglia on a ground bass formed by the rising scale of D major:

Ex. 49

Since Purcell produced *Dido and Aeneas* for Josias Priest's girls' school at Chelsea, no more beautiful opera for child performers has been written. Its immediate popularity as an entertainment, not only in this country but all over the world, has been phenomenal; but what is more difficult as yet to assess is the extent of its influence in introducing children to opera. In some countries a new generation is growing up that, thanks partly to its example, is prepared to accept opera and its conventions as a natural and familiar art form.

The Little Sweep is an opera of innocence—of innocence betrayed and rescued—and it is fully worthy of the singer of innocence who inspired it.

VII
Billy Budd

I

At the first Aldeburgh Festival in 1948, E. M. Forster gave a lecture on George Crabbe and Peter Grimes. The crowded audience, which included Benjamin Britten, sat on the hard benches of the Baptist Chapel that warm summer afternoon and heard the lecturer talk about the far-reaching alterations Montagu Slater had made when constructing his opera libretto on the basis of Crabbe's poem. Having enumerated some of these changes, he went on to say: 'It amuses me to think what an opera on Peter Grimes would have been like if I had written it. I should certainly have starred the murdered apprentices. I should have introduced their ghosts in the last scene, rising out of the estuary, on either side of the vengeful greybeard, blood and fire would have been thrown in the tenor's face, hell would have opened, and on a mixture of *Don Juan* and the *Freischütz* I should have lowered my final curtain.'*

Britten didn't forget these words, and shortly afterwards when he contemplated writing another full-scale opera and found his imagination irresistibly kindled by Herman Melville's posthumous novel, *Billy Budd, Foretopman*, it was to Forster as collaborator that he instinctively turned. With characteristic courage, this septuagenarian novelist and essayist agreed to set out on a new and hazardous career as opera librettist. Hitherto, his experience of dramatic writing had been confined to a couple of open-air pageants; but how deep was his love of music could be seen from the part it had played in his novels and essays. (Who, having read *Where Angels Fear to Tread*, could ever forget his account of the performance of *Lucia di Lammermoor* at Monteriano?) The libretto of *Billy Budd* was the joint creation of himself and Eric Crozier; and, as might be expected after his remarks on *Peter Grimes*, it was conscientiously faithful to the spirit and even the letter of the original.

Melville was an old man of sixty-nine when after twenty years' silence as a

---

* From 'George Crabbe and Peter Grimes' in *Two Cheers for Democracy*. Arnold, 1951.

174

novelist he began to write his last story. The first draft of the various Billy Budd manuscripts was dated 16 November 1888; the main revision begun on 2 March 1889; and the story finished on 19 April 1891, a few months before Melville's death. F. Barron Freeman has shown* how the initial short story of sixteen sections, *Baby Budd, Sailor*, was expanded into the novel of thirty-one sections generally known as *Billy Budd, Foretopman.*

'An Inside Narrative' Melville called it, going out of his way to suggest that this was the reason for some of its crudities—'The symmetry of form attainable in pure fiction cannot so readily be achieved in a narration essentially having less to do with fable than with fact.' Indeed, it appears almost as if he were bent on a work of rehabilitation, for he quotes a completely distorted account of the events that led to the execution of Budd, purporting to come from a contemporary naval weekly chronicle at the end of the eighteenth century, and explains that so far this had been the only record of this strange affair. But it would appear that the actual events on which *Billy Budd, Foretopman* is based took place, not in the British Navy at the time of the mutinies of Spithead and the Nore which is the period Melville has chosen for the setting of his story, but in the American Navy nearly half a century later; and the intensity of his interest in what came to be known as the Mackenzie Case can be appreciated from the fact that his first cousin, Guert Gansevoort, was one of the lieutenants on board the brig-of-war *Somers* in 1842 at the time of the so-called mutiny.

The Mackenzie Case is specifically referred to in *Billy Budd, Foretopman.* In explaining the difficulty the drumhead court experienced in trying Billy Budd, Melville wrote: 'Not unlikely they were brought to something more or less akin to that harassed frame of mind which in the year 1842 actuated the commander of the U.S. brig-of-war *Somers* to resolve, under the so-called Articles of War, Articles modelled upon the English Mutiny Act, to resolve upon the execution at sea of a midshipman and two petty-officers as mutineers designing the seizure of the brig. Which resolution was carried out though in a time of peace and within not many days sail of home. An act vindicated by a naval court of inquiry subsequently convened ashore. History, and here cited without comment.' Melville's care in explaining the reasons that actuated the captain of the *Indomitable* in giving his evidence and the drumhead court in reaching their verdict seems indirectly to be aimed at vindicating Gansevoort who as the chief aide of Captain Mackenzie against the 'mutineers' on the *Somers* came in for a good measure of subsequent criticism. It is perhaps worth noting that in June 1888, just five months before Melville started to write *Baby Budd, Sailor*, the case was reopened by the publication of an article by Lieutenant H. D. Smith in the American Magazine entitled *The Mutiny on the Somers*. As by then

* *Melville's Billy Budd*, edited by F. Barron Freeman. Harvard University Press, 1948.

Billy Budd: Billy Budd (Theodor Uppman) and Captain Vere (Peter Pears) in the original Covent Garden production of 1951

Gansevoort had been dead for twenty years, Melville must have felt that no time was to be lost if he wished to present the events in their proper perspective.

But this material could be used only after it had been digested and assimilated to its fictional purpose. As Freeman says, 'the "inner" drama of Billy Budd uses the Mackenzie Case in the same way that it uses *The Naval History of Great Britain* by William James and Melville's own *White Jacket*: as a factual source on which to build an "interior" drama of the forces of fate'.*

Here, as in *Peter Grimes*, the grit of actuality—actual persons and actual incidents—was apparently the prime cause of the long process of artistic secretion that built up, first, an independent literary work and, subsequently, an opera; and it would seem that, consciously or subconsciously as far as Britten was concerned, the actual element behind the fictional treatment still retained a powerful germinating force.

The action in *Billy Budd* takes place on the seventy-four HMS *Indomitable* during the summer of 1797. This was the moment when the mutinies of Spithead and the Nore had just occurred. Conditions in the English Navy were particularly brutalizing, cruel and horrible then, and

* F. Barron Freeman, op. cit.

there is little doubt that the men's grievances were justified; but the authorities were naturally appalled by the hint of revolution at a time when every ounce of energy had to be put into the life-and-death struggle with France.

The *Indomitable* is at sea alone, on her way to join the Mediterranean fleet. Like so many of the ships in the Navy at that time, she is short of her full complement of men: so when a passing merchantman named *The Rights of Man* is sighted, a boarding party is sent off, and among others Billy Budd is impressed from her crew. Billy is a handsome, pleasant fellow, sound in heart and limb, whose only physical defect is an occasional stammer. He gets on well with everyone on board ship, and everyone likes him—except the master-at-arms, John Claggart. From the outset Claggart pursues him with hidden but implacable malevolence. Natural depravity is bent on corrupting natural innocence and encompassing its downfall. But in plotting Billy's destruction, he reckons without the ship's captain, Edward Fairfax Vere, a bachelor of about forty, known throughout the Navy as 'Starry Vere'. A cultured, aristocratic man, popular with both officers and men, and a shrewd judge of character, Vere immediately sees through Claggart's vamped up charges against Billy. He confronts the sailor with the master-at-arms in his cabin; but the horror of Claggart's false accusations so staggers Billy that it brings out his lurking defect and his power to answer is momentarily taken from him by his stammer. He stands there, tongue-tied, until a kindly gesture from the Captain releases something inside him, and he strikes Claggart a fierce blow with his naked fist. Claggart falls—dead. Captain Vere immediately summons a drumhead court. Billy is tried under the articles of war for striking and killing his superior in rank and is condemned to be hanged from the yardarm, a sentence that is carried out at dawn the following day.

The most important respect in which Forster and Crozier have departed from Melville's story is by letting Vere live to a ripe old age—according to Melville, he was mortally wounded in a sea-fight a few years later, just before Trafalgar—and introducing him in a Prologue and Epilogue where he appears as an old man ruminating on the past. At first he finds it difficult to recall the details of those early days; but what remains indelibly impressed on his mind is that in the faraway summer of 1797 someone had blessed him, someone had saved him. Who was it? As recollection comes crowding back, he remembers his old ship the *Indomitable* and the strange, sad story of Billy Budd.

This device of a Prologue and Epilogue is important for a number of reasons. In the first place, the sea in *Billy Budd* is an isolating medium, the element on which the all-male microcosm, the crew of the *Indomitable*, depends for its temporary suspension. To see the ship from without, alone in the centre of its circumscribing horizon, it is necessary to provide the

Billy Budd: The full muster on the main deck and the quarter deck before Billy Budd is hanged at the yardarm, in the original Covent Garden production of 1951

audience with a view-point outside the field of immediate action. Secondly, the final scene (the hanging) is so tense and its impact so overwhelming that time is needed for the audience to recover. Thirdly, the fact that the action is seen through the eyes of an old man who calls it to mind after the lapse of many years parallels the case of Melville himself composing the story at the age of seventy a short time before his death. Finally, it places the Captain at the centre of the action, showing him caught on the horns of a cruel dilemma. As sole witness of Claggart's death, he finds fate has thrust on him the power of life or death over a human being of whose essential innocence he is fully convinced. The iron discipline of duty prevails over the dictates of his heart—what Melville calls 'the feminine in men'—and he allows Billy to die. The fact that Billy understands and forgives him may be the means of his ultimate salvation, but cannot completely reassure him, for (as he cries out in the Epilogue) he could have saved him, and Billy knew it.

This focusing on Vere puts him into what Henry James called 'the compositional centre'* of the drama; and the heroic aspect of the character is accentuated by the vocal casting. Vere is a tenor; Claggart, as befits the villain of the piece, a bass; Billy, a baritone. (In this connection it is interesting to note that in his one-act opera, *Billy Budd* (1949), Ghedini cast Vere as bass, Claggart as tenor and Billy as baritone.)

Some of the overtones in Melville's descriptions of the characters have been dropped. For instance, he more than once implies that the relationship

* Cf. p. 116.

178

of Vere to Budd is like that of father to son—this is particularly stressed when Vere communicates the finding of the court to the condemned sailor, and Melville suggests that at that moment Vere may have 'caught Billy to his heart even as Abraham may have caught young Isaac on the brink of resolutely offering him up in obedience to the exacting behest'*—but Forster and Crozier omit this, and in any case such an implied relationship would be contradicted by the vocal casting.

There certainly appear to be sufficient indications of unconscious or latent homosexuality in Melville's description of Claggart to justify Freeman's statement that 'the psychological key to Claggart's antipathy is not conscious suppression but unconscious repression of a perverted desire for the boy whose downfall he plotted';† but most of these indications are missing from the opera libretto, though Forster and Crozier plumb the depths of Claggart's character in the introspective monologue they give him in Act I, scene iii, '*O beauty, O handsomeness, goodness!*' No such soliloquy occurs in Melville; but there is excellent operatic precedent for it in Iago's Credo in Verdi's *Otello*. Claggart sings, '*If love still lives and grows strong where I cannot enter, what hope is there in my own dark world for me?*' So beauty, handsomeness, goodness, as personified by the Handsome Sailor, must be destroyed. In the libretto the clash between the two men is shown to be one facet of the eternal struggle between the powers of good and evil. As W. H. Auden wrote in his poem *Herman Melville*:‡

> *Evil is unspectacular and always human,*
> *And shares our bed and eats at our own table,*
> *And we are introduced to Goodness every day,*
> *Even in drawing-rooms among a crowd of faults;*
> *He has a name like Billy and is almost perfect*
> *But wears a stammer like a decoration:*
> *And every time they meet the same thing has to happen;*
> *It is the Evil that is helpless like a lover*
> *And has to pick a quarrel and succeeds,*
> *And both are openly destroyed before our eyes.*

One further change made by Forster and Crozier should here be noted. It was necessary for them to include in their libretto both the novice who receives a flogging for dereliction of duty (a passage in *Billy Budd, Foretopman* that is much more reserved and moderate in tone than the comparable passage in *White Jacket* describing the scourging of young Peter of the mizzentop) and the afterguardsman who comes to Billy at night to

---

* It is worth remembering that the first work Britten composed after *Billy Budd* was his *Canticle II: Abraham and Isaac*.
† F. Barron Freeman, op. cit. ‡ From *Another Time*.

bribe him to join the mutineers. But whereas in Melville these two characters, though rather shadowy, are quite distinct, in the libretto they are fused together. Humiliated and cowed by his experience, the novice is driven by the fear of further punishment to agree to become Claggart's cat's paw and by tempting Billy to compromise and betray him. A minor character admittedly, but one that Forster and Crozier portray 'in the round'.

The *Billy Budd* libretto is written almost entirely in prose. The only exceptions are the words of some of the interpolated shanties, a few free verse passages, Claggart's blank verse denunciation of Billy to the Captain, and the Ballad of Billy in the Darbies, twenty lines chosen from the thirty-two-line poem with which Melville's narrative ends. The realistic treatment of a subject at one or two removes from actual fact seems to have imposed a predominantly prosaic idiom on the librettists. Some of Melville's dialogue has been preserved intact: e.g. phrases like *Handsomely done, my lad. And handsome is as handsome did it, too!—Jemmy Legs is down on you!—A man-trap may be under his ruddy-tipped daisies.—Fated boy, what have you done!—Struck by an angel of God! Yet the angel must hang!* And there are even one or two passages where Forster and Crozier have based their text on some of Melville's discarded drafts. For instance, in the gun-deck scene, Billy tells his friend Dansker of the Chaplain's visit. '*Chaplain's been here before you—kind—and good his story, of the good boy hung and gone to glory, hung for the likes of me.*' This is a direct citation from one of the discarded fragments of the Ballad of Billy in the Darbies:

> *Kind of him ay, to enter Lone Bay*
> *And down on his marrow-bones here and pray*
> *For the like of me. And good his story—*
> *Of the good boy hung and gone to glory,*
> *Hung for the likes of me.*

It might have been thought that the predominance of prose would have tended to make the opera ponderous and heavy; but this is not so. Those prose sections that are dealt with on the near-declamation level proceed quickly and lightly: those that are set as passages of *arioso* or as ensembles take on a lyrical quality. Persons who have not seen the text printed as prose in the libretto would find it difficult to believe that such numbers as the post-flogging trio for the Novice, his friend and a small chorus of sailors with its refrain '*Lost for ever on the endless sea*' (I, i), or the duet between the Sailing Master and the First Lieutenant '*Don't like the French*' (I, ii), or the '*We've no choice*' trio between the First and Second Lieutenants and the Sailing Master at the end of the court-martial (II, ii), or Billy's farewell to life (II, iii) are not written to a definite metrical scheme embellished with rhyme or assonance. The prose is so supple and lissom in texture that it seems to take on some of the attributes of verse; and, in fact, as pointed out above, verse fragments are occasionally to be found embedded in it.

The original version of 1951 was written in four acts, each act being a complete and continuous movement, whether or not it was subdivided into scenes; and Britten was careful to bind Acts I and II closely together, and also Acts III and IV, by arranging almost literal musical joins between the end of Act I and the beginning of Act II, and the end of Act III and beginning of Act IV. But the action falls naturally into two parts—exposition and catastrophe—rather than four; and the impact of the work in its original form was considerably weakened by the necessity to break the action at no fewer than three different points. In 1960 he decided to make a revision (first heard in a BBC broadcast on 13 November 1960) in which the four acts were reduced to two. The main change came at the end of the original Act I, where in the 1951 version there had been a Captain's muster on deck, at the climax of which 'Starry Vere' appeared and harangued his men. In the 1960 version, this is substantially altered. Instead of a muster, the whistles sound for a changing of the watch; and instead of being roused by Vere's rhetorical 'Death or Victory' speech, the men discuss their captain in his absence, their enthusiasm proving so infectious that Billy catches fire and apostrophizes him as 'Star of the morning'—

Starry, I'll follow you . . .
Follow thro' darkness, never you fear . . .
I'd die to save you, ask for to die . . .
I'll follow you all I can, follow you for ever.

This means that Vere does not make his first appearance until the following scene where he is discovered reading Plutarch in his cabin. The other changes in the 1960 version are minimal, consisting mainly of a few cuts in the Vere/Claggart dialogue when the Master-at-arms denounces Billy as a mutineer. The music in each of the two revised acts is continuous; and this is an undoubted improvement. The following table makes these changes clear:

1951 *Version*		1960 *Version*		
		Prologue		
		Prologue		
Act I		Act I	Scene i	The Main-deck and Quarter-deck
Act II	Scene i		Scene ii	The Captain's Cabin
	Scene ii		Scene iii	The Berth-deck
Act III	Scene i	Act II	Scene i	The Main-deck and Quarter-deck
	Scene ii		Scene ii	The Captain's Cabin
Act IV	Scene i		Scene iii	A bay of the upper gun-deck
	Scene ii		Scene iv	The Main-deck and Quarter-deck
	Epilogue		Epilogue	

181

II

There is an analogy, though it should not be pushed too far, between the form of this opera and that of a symphony. The first scene of Act I is expository, most of the characters being introduced singly or in groups. The two following scenes are the equivalent of a slow movement—a reflective serenade below deck in two scenes, the first showing the Captain and some of his officers enjoying an after-dinner conversation, during which they hear distant singing from the Berth-deck, and the second showing the sailors singing shanties before slinging hammocks and turning in for the night. The first scene of Act II serves as a scherzo, namely the whole of the chase after the French frigate and the call to action stations. The remainder of the act provides the climax and *dénouement*.

The opposition between what Melville (in a discarded draft) called 'innocence and infamy, spiritual depravity and fair repute' lies at the heart of Britten's score. The opening bar of the Prologue with its quiet rippling movement contains a clash between B flat and B natural which typifies this struggle:

Ex. 50

As the two melodies unfold, they seem to be establishing a bitonality of E flat major and C major; but in reality the clash is much sharper and more powerful—B flat major against B minor—and this is confirmed by another passage in the Prologue where a bitonal B flat major/B minor chord thrice repeated encloses a musical phrase later to be associated with Claggart, the first part of which is related to the first of the two keys and the second part (the descending fourths) to the second.

Ex. 51

At the beginning of Act I when the lights go up on the *Indomitable* with the men holy-stoning her maindeck, the same notes are there; but now the B flat

182

has suffered an enharmonic change to A sharp and the chord has lost its
bitonal implications.

Ex. 52

The bitonal clash between two keys with their roots a semitone apart is
closely associated with Claggart's hatred for Budd. Before Claggart's first
entry (I, i) the orchestra is playing a theme in G major connected with the
boarding party; but as soon as he is addressed by the First Lieutenant, the
music clouds over and the key changes to G sharp minor. The same key
relationship governs the initial airs of the two characters: Billy's exultant
'*Billy Budd, king of the birds*' opens in E major; Claggart's '*Was I born
yesterday?*' in F minor. Similarly, the A major fight between Billy and
Squeak in the third scene of Act I is interrupted by Claggart's B flat arrival;
and a few bars later as the men sling their hammocks in the bays, a sailor (off)
sings a shanty '*Over the water, over the ocean*' in E major, while the orchestra
is still meditating in F minor on the way Claggart has just lashed out at one
of the boys with his rattan. In the following act the temperature of the music
drops each time Claggart approaches the Captain in order to denounce
Budd, and the key drops too—this time from G minor to F sharp minor, not
major.

The enharmonic change from B flat to A sharp that was noted at the
beginning of Act I (see Ex. 52 above) occurs in reverse at the end of Act II
scene iv just after Billy has been hanged at the yardarm. This is one of the
most astonishing and frightening moments in the opera. Melville describes
how at that moment an inarticulate noise like 'the freshet-wave of a torrent
suddenly swelled by pouring showers in the tropical mountains' burst forth
from the men massed on the ship's open deck. In a wordless fugal stretto
sung so quickly that they sound like a pack of wild beasts, the sailors seem for
the moment to be on the verge of mutiny. (This passage is in E major.) But
the officers order all hands to pipe down, and the command is given on B flat.

Ex. 53

For some bars an attempt is made to assimilate this note enharmonically as an A sharp into the Lydian mode of E: but ultimately force of habit is too much for the men, they obey the command and disperse, and the music resignedly modulates to the key of B flat major. At this point the music pivots on A sharp/B flat as if on the hinge of fate.

The final resolution of this bitonal conflict comes at the end of the Epilogue. Once again the phrase formerly associated with Claggart is enclosed by the bitonal chord (as in Ex. 51); but this time Claggart's descending fourths have disappeared, and in their place there is a fanfare—fanfares in this score are usually associated with Billy Budd—spread over the common chord of B major. At the same time Vere's voice is accompanied by a figure earlier associated with Billy's song of farewell (ii, iii), with the result that the final cadence contains a double clash—B major against B flat major, and B flat major against A major—and the ultimate resolution on B flat major comes with all the force and finality of a double resolution:

Ex. 54

Just as there is an occasional enharmonic ambivalence about the music, so there is sometimes an ambivalence between the music and the action, and the music and the words. A problem posited by the action may receive its solution in purely musical terms; and the listener must be prepared to recognize and accept this.

For instance, it is rewarding to examine the musical and psychological metamorphoses of a simple musical phrase of a fifth and a semitone. In the

184

Prologue, Vere as an old man is bewildered and troubled as he tries to remember what he did in the far-off case of Billy Budd, and why he did it.

Ex. 55

Right at the beginning of Act I the phrase appears in the shanty sung by the sailors holy-stoning the deck:

Ex. 56

When Billy Budd is impressed, he shouts a farewell to his old comrades on *The Rights of Man* and his words fall naturally into the simple tune of the shanty:

Ex. 57

Some of the officers on the *Indomitable*, however, misunderstand this spontaneous greeting of Billy's and take it as an unbecoming reference to the liberal and suspect views expressed by Thomas Paine in his book *The Rights of Man*; and as the chorus echoes Billy's phrase, they order the decks to be cleared. Henceforward this innocuous phrase is associated with the idea of mutiny. The officers, when discussing a glass of wine with the Captain after dinner, identify the phrase with Spithead and the Nore (I, ii):

Ex. 58

The oboes bring it back in sinister guise when the Novice tempts Billy to join the gang which he says is plotting mutiny on the ship (Ex. 50a); the muted trumpets transmute it into glittering guineas to bribe him (Ex. 59b):

Ex. 59a

Ex. 59b

When Claggart confronts Billy before Captain Vere and accuses him of mutiny, the motif has become the skeleton of the phrase:

Ex. 60

Its final appearance, apart from a repetition of Ex. 55 in the Epilogue, is in the scene already mentioned that takes place immediately after the hanging; and there it forms the subject of the wordless fugal stretto:

Ex. 61

It has already been suggested that fanfares and sennets are closely associated with Billy Budd himself. The air is filled with these calls when the boarding party returns from *The Rights of Man* bringing Billy as an impressed man. Simple arpeggio figures usually accompany his songs—e.g. '*Billy Budd, king of the birds!*'; the passage in his duet with Dansker (I, iii) where he sings '*Ay! and the wind and the sails and being aloft and the deck below so small and the sea so wide*'; his duet with the Captain when he thinks he's about to be promoted (II, ii); his Ballad in the Darbies, and his song of farewell (II, iii). All these are obvious instances, so obvious that it would hardly seem worth while mentioning them, were it not for the astonishing transformation of this device at the end of Act II, scene ii.

Billy is in a stateroom at the back of the cabin where the drumhead court

186

has been held, and it is left to Captain Vere to communicate the finding of the court to him. Melville says that 'what took place at this interview was never known'; and Forster and Crozier follow Melville in this, leaving the death sentence to be communicated to him off-stage at the end of the act. Britten could have ignored the problem and allowed the curtain to fall immediately after Vere's exit; but he chose to tackle it musically and keeps the curtain up for over a minute on an empty stage, while slow heavy chords for the whole orchestra, or strings, or woodwind, or brass, or horns, ranging through many gradations of volume, tell of the fatal interview behind the closed door. As can be seen from the few bars quoted, the passage is really a widely spread arpeggio of F major, harmonized with chromatic intensity and so tremendously augmented as to give the effect of a simple signal seen through an extremely powerful telescope—a rainbow of hope:

Ex. 62

As for the chords themselves, smaller groups of them appear in the following scene to accompany Billy's farewell to the world—'*I'm strong, and I know it, and I'll stay strong, and that's all, and that's enough*'—and also at the end of the Epilogue when Vere sings '*But I've sighted a sail in the storm, the far-shining sail, and I'm content.*'

One further example of ambivalence deserves special mention.

Perhaps the most moving moment in the first act is the compassionate trio after the Novice's flogging:

Ex. 63

The saxophone melody reappears in the second act when Claggart threatens the Novice with a further flogging if he refuses to tempt Billy. After this scene the Novice never reappears; but when at the end of Act II, scene ii

Vere sings '*Beauty, handsomeness, goodness, it is for me to destroy you. I, Edward Fairfax Vere, captain of the "Indomitable", lost with all hands on the infinite sea*', the Novice's phrase reappears in a turbulent orchestral interjection and provides a poignant comment on the brutality of the punishment devised by man for man.

Ex. 64

Any analysis of the score of *Billy Budd*, however penetrating, would fail in its purpose if it gave the impression that the opera is abstruse or difficult to listen to. The scoring is light and transparent throughout. As soon as the lights go up on Act I, the music captures the feeling of a ship at sea: the shrill sound of the wind in the rigging, the bustle of life on deck, and the underlying swell of the ocean. The second and third scenes with their two beautifully interlocked serenades, show first the officers and then the men during the quiet interval between supper and sleep when there is time to relax and meditate. The second act starts with an alarum that is none the less exciting for being in vain. After the exhilaration of the chase after the French frigate, the mood of frustration induced by waiting returns with the mist, and in this atmosphere Vere witnesses the drama of Claggart and Budd unroll before his eyes with the inevitability of fate and is powerless to intervene.

The final scenes are brief, but everything in them is placed with an unerring sense of effectiveness. Billy's Ballad in the Darbies is one of Britten's happiest inventions—a slow, sleepy tune to a low-pitched, sluggish accompaniment that changes its chords reluctantly as Billy sings '*Roll me over fair. I'm sleepy and the oozy weeds about me twist*', and reminds one of Hylas' song in *The Trojans*.

Some critics have suggested that it would have been an improvement if Budd, being the hero of the opera, had been cast as a tenor instead of a baritone; but in that case the setting of this Ballad would undoubtedly have suffered, for the low pitch at which it is written and the tone colour of that particular range of a baritone voice are essential to its effect.

The march to which the whole crew assembles at the beginning of Act II, scene iv to witness the hanging is a fugato mainly for percussion, and its characteristic rhythm containing both triplets and quintuplets spills over into the Epilogue, where its pianissimo roll on the timpani accompanies Vere's recital of Billy's burial at sea and forms the wake of the *Indomitable* as she disappears from view in the minds of the audience.

There is one final comment to be made. When it was known that Britten was writing an opera with an all-male cast, there were some who prophesied failure because they claimed that without women's voices the monotony of tone colour would be unendurable. (Even Méhul in his biblical opera *Joseph* (1807), where there are no female characters at all, decided to entrust the part of Benjamin to a soprano.) After the first performance of *Billy Budd* this particular criticism disappeared because in fact it was discovered that Britten's writing for this all-male cast was extraordinarily varied. Apart from the seventeen individual characters, who include five tenors, eight baritones, one bass-baritone and three basses, he calls for a main-deck chorus of thirty-six, a quarter-deck chorus of fourteen, four Midshipmen, whose voices should appear to be breaking, and ten powder-monkeys, who are not required to sing, but whose shrill chattering voices form an important element in the build-up of the big chorus in Act II, scene i when chase is given to the French frigate.

The verdict on *Billy Budd* must be that between them Forster and Crozier wrote one of the best and most faithful librettos to be based on a literary masterpiece, and Britten created a score of outstanding psychological subtlety. The work rouses a wide range of emotions; but perhaps the dominant one is compassion—compassion for the weak and unfortunate, for those who are homeless, the victims of fate; compassion for suffering in all its forms.

VIII
Gloriana

I

In the past there have been few English dynastic operas and few occasions on which a reigning monarch has played a direct part in supporting opera in this country.

Albion and Albanius was a dynastic opera in honour of what Dryden called 'the double Restoration of Charles II' (Albion). Unfortunately Charles's death delayed its production until 1685, when Dryden introduced some appropriate references to James II (Albanius) and it was produced at the Queen's Theatre in Dorset Garden; but its run was brought to an untimely conclusion by the news of the rising of the Duke of Monmouth in the West. In any case, the music provided by Louis Grabu seems to have been so indifferent that no one regretted the opera's demise. The same criticism seems to have been true of Clayton's setting of Addison's *Rosamond* (1707): but there the dynasty celebrated was that of the Duke and Duchess of Marlborough.

When the Hanoverians succeeded to the throne, Handel gained a considerable measure of royal favour; and both George I and II contributed towards the cost of Italian opera in London between 1722 and 1744, the royal subscription of £1,000 being paid annually to The Royal Academy of Music (Undertakers of the Opera), except in 1734 when it was paid personally to Handel himself and during the three years 1739–41 when it lapsed.* Handel's only attempt at a dynastic opera was *Riccardo Primo, Re d'Inghilterra* (1727) written on the occasion of George II's Coronation.

After Handel's death Royal patronage declined for about a century: but, thanks largely to the musical sensibility of the Prince Consort, Queen Victoria displayed an active interest in opera in the early years of her reign. During her husband's lifetime, she frequently attended performances at Covent Garden and Drury Lane, and in particular gave her patronage to the various Pyne-Harrison seasons of English opera that started in 1857. When

* See 'Finance and Patronage in Handel's life' in *Concerning Handel* by William C. Smith Cassell, 1948.

Wallace's *Lurline* was successfully produced at Covent Garden in 1860, she is reported to have advised Louisa Pyne and William Harrison to make money out of it as long as it would run. The Prince Consort's death in 1861 put an end to her public theatre-going; and although there were occasional command performances at Windsor Castle towards the end of the century, her interest in English opera seems to have been confined to urging Sullivan to drop his operetta partnership with Gilbert in order to write a 'grand' opera, a royal command that led to the production of *Ivanhoe* in 1891.

Queen Alexandra took a keen personal interest in opera. She frequently attended performances at the Royal Opera House during Edward VII's reign; so it is not altogether surprising to find that *La Fanciulla del West* (1910) was dedicated *A Sua Maestà La Regina Alessandra d'Inghilterra— rispettoso omaggio di Giacomo Puccini*. But at this period so few English operas were performed in London that it is doubtful if members of the Royal Family could have seen any of them at Covent Garden, even if they had particularly wanted to.

There was accordingly no precedent for the fact that in May 1952 Her Majesty Queen Elizabeth II gave her approval to the suggestion that Britten should write a Coronation opera on the theme of Elizabeth I and Essex, and that later Her Majesty agreed to accept the dedication of the work and attend its first performance at a special gala on 8 June 1953 in honour of her Coronation.

An English historical subject was something of a novelty so far as English opera was concerned. Different historical periods had often been used as settings for romantic operas; but if historical characters and episodes were introduced, they were usually treated with the same freedom as if they were creations of fiction. It is true that during the Commonwealth Sir William Davenant had evolved a new type of operatic representation based on historical subjects that were then about a century old—*The Siege of Rhodes* (1656), *The Cruelty of the Spaniards in Peru* (1658), and *The History of Sir Francis Drake* (1659)—but his example was not followed up. This lack of initiative was all the more surprising since the Elizabethan playwrights, particularly Shakespeare, had given a strong lead in the writing of English historical plays.

Later on, there was the example of Russian musicians to note. The nationalist school of the nineteenth century frequently chose patriotic themes for its operas, e.g. Cavos's *Ivan Susanin* (1815), Glinka's *A Life for the Czar* (1836), Rimsky-Korsakov's *The Maid of Pskov* (1873), Mussorgsky's *Boris Godunov* (1874), and Borodin's *Prince Igor* (1890). A similar tendency might easily have been observable in nineteenth-century Italy, had not local censorship proved so sensitive over the political implications of historical subjects that frequently places, names and dates had to be changed.

THE OPERAS

The twentieth century has produced new standards of historical accuracy and new methods of historical research. Invention and fancy are at a discount: it is selection that counts—and presentation.

Just as Pushkin based his tragedy of *Boris Godunov* on Nicolai Karamsin's *History of Russia*, so William Plomer went to J. E. Neale and Lytton Strachey for his sources. *Elizabeth and Essex* was in fact the starting-point for the libretto of *Gloriana*. Strachey had tried to outline this episode of Elizabeth's old age—'a tragic history' he called it—with something of the detachment and self-sufficiency of a work of art; but the bewildering richness of life in Elizabethan England seems to have defeated him. He himself confessed that 'the more clearly we perceive it, the more remote that singular universe becomes'.* The Elizabethans were more paradoxical, elusive, ambiguous and irrational than the Victorians; and profound social changes in the intervening period had left Strachey little or nothing to debunk. *Elizabeth and Essex* is not a completely successful book; but it contains numerous dramatic cues, and it was this aspect of it that appealed so strongly to Britten and Plomer.

In the process of selection, a number of important persons and episodes had to be discarded. Of the persons, perhaps the most notable was Sir Francis Bacon. His character was too subtle, complex and compelling, and would have demanded too much attention in the opera. But it must be admitted that his total disappearance—and there is no mention of him in the libretto—seriously weakens Essex's character. For a considerable time he loomed like a grey eminence in the background, and many of Essex's squabbles with the Queen seem to have been concerned with her unwillingness to listen to his pleas that Bacon should be granted preferment. Yet when the final crisis came and Essex was arraigned for high treason, Bacon acted without scruple or hesitation as counsel for the prosecution, and by his handling of the case helped to bring his former benefactor to the block on Tower Hill.

A less important person who was omitted was Elizabeth's godson Sir John Harington. In *Portraits in Miniature*, Strachey described one of Sir John's adventures as follows: 'He was summoned by Essex to join his ill-fated expedition to Ireland, in command of a troop of horse. In Ireland, with a stretch of authority which was bitterly resented by the Queen, Harington was knighted by the rash Lord Deputy, and afterwards, when disaster came thick upon disaster, he followed his patron back to London. In fear and trembling, he presented himself before the enraged Elizabeth. "What!" she cried, "did the fool bring you too?" The terrified poet fell upon his knees, while the Queen, as he afterwards described it, "chafed much, walked fastly too and fro, and looked with discomposure in her visage". Then suddenly

---

* Lytton Strachey, *Elizabeth and Essex*.

rushing towards him, she caught hold of his girdle. "By God's Son", she shouted, "I am no Queen, and that man is above me!" His stammering excuses were cut short with a "Go back to your business!" uttered in such a tone that Sir John, not staying to be bidden twice, fled out of the room, and fled down to Kelston, "as if all the Irish rebels had been at his heels".' Plomer ignored this intensely dramatic material.*

There was also the celebrated scene in the Council Chamber, when Essex, obstructed by the Queen in the matter of the Irish appointment, lost his temper and turned his back on her. Crying 'Go to the devil!' she boxed his ears, whereat he clapped his hand to his sword and shouted 'This is an outrage that I will not put up with. I would not have borne it from your father's hands.' Plomer did not use this episode either, though he allowed Essex, to make a brief, though hardly noticeable, reference to it in Act II, scene iii.

Cuffe appears in the libretto as a minor character, a feed to Essex. More extended treatment is given to Cecil, Raleigh and Mountjoy, though the result is still somewhat perfunctory when their operatic characters are compared with historical reality. The two women—the Countess of Essex and Penelope, Lady Rich (Essex's sister)—emerge more successfully, partly because less is actually known about them, and the librettist has had a freer hand accordingly. Many facets of Essex's character are shown: the proud and tetchy nobleman; the headstrong lover; the romantic advocate of the simple life; the ambitious courtier; the sullen conspirator; the unsuccessful general. But he disappears completely after his return from Ireland; and although his final act of treason—his march through the City of London and call to insurrection—is referred to obliquely in Act III, scene ii, he himself is not shown at the head of his followers, nor does he appear in connection with his trial and condemnation.

So far it would seem as if the historical picture presented by the opera is a partial and incomplete one. But one character remains—Elizabeth herself—and as soon as she is considered, the whole perspective of the opera changes and everything seems to fall into focus. It is then clear that the episode of Essex has been used only in so far as it helps to place in relief her character as a Queen and a woman who, though ageing, is still at the height of her intellectual powers; and it is here that Britten and Plomer have scored their most incontrovertible success. Even a scene like the Masque at the Norwich Guildhall (II, i), which ostensibly has nothing to do with the plot, is vital to the opera, since it shows that it was an important part of the Sovereign's functions to accept graciously the ceremonial that inevitably accompanied her public progresses through her kingdom.

* It is true that in the original version of the opera Sir John Harington made a brief unexplained appearance in the epilogue where his participation in the Irish expedition was specifically mentioned, but this was omitted in the 1966 revision.

Elizabeth first appears (I, I) leaving a tilting ground after a tournament. It is one of her public appearances, and she is accompanied by a retinue and preceded by trumpeters. Her pompous entry disturbs a brawl between Essex and Mountjoy, whom she has just honoured for his prowess in the tiltyard with the gift of a golden queen from her set of chessmen. She summons her subjects to hear her judgment and forbids Essex and Mountjoy to continue their private quarrel. A feature of this scene is the loyal chorus:

> *Green leaves are we,*
> *Red rose our golden Queen,*
> *O crownèd rose among the leaves so green!*

The following scene (I, ii) is laid in one of her private apartments at Nonesuch. First, she discusses public and private business with Cecil, whose precepts on government give the impression of wheels moving within wheels. Essex arrives. Dismissing Cecil, she abandons herself to the pleasure of his company, well aware of the weakness of his character as well as the lure of his charm. After singing two lute songs to entertain her, the second of which is a setting of a poem '*Happy were he*' actually written by the original Earl of Essex, he urges his claim to be sent to Ireland to conquer the rebel Tyrone; but she easily evades giving an answer. Left alone, she soliloquizes:

> *If life were love and love were true,*
> *Then could I love thee through and through!*
> *But God gave me a sceptre,*
> *   The burden and the glory—*
> *I must not lay them down:*
> *   I live and reign a virgin,*
> *   Will die in honour,*
> *Leave a refulgent crown!*

In the second act she is shown on two ceremonial occasions: the first at Norwich, where she receives the homage of the citizens and is entertained with a rustic masque; and the second at the Palace of Whitehall, where she enjoys an evening of dancing. These are separated by a short scene in the garden of Essex House where Mountjoy's secret tryst with Lady Rich is interrupted by the appearance of Essex complaining to his wife of his treatment by the Queen. Lady Rich, Essex and Mountjoy discuss the possibility of seizing the reins of State, while Lady Essex counsels caution.

During the dancing at the Palace of Whitehall (II, iii), there is an episode that shows how spiteful Elizabeth could be, particularly to women who were closely connected with men she was interested in. Noticing that Lady Essex is dressed with special magnificence, she gets hold of her dress through a stratagem and puts it on herself. It is a gross misfit, and the effect is grotesque. She parades in it before the Court to the extreme confusion of Lady Essex. In all this the librettist has kept fairly close to historical accuracy, his only major change being to transfer the affair of the dress from Lady Mary Howard, whom the Queen actually suspected of an intrigue with Essex, to Lady Essex herself. The attack on Lady Essex is so savage and wilful that when the Queen sweeps off stage, the feelings of the group of would-be conspirators (Lady Rich, Essex and Mountjoy) are roused to fury. First they seek to comfort Lady Essex—'*Good Frances, do not weep*'—and then Essex bursts out with his bitter taunt, '*Her conditions are as crooked as her carcass!*' Suddenly Elizabeth returns in state with her councillors and announces she has appointed Essex Lord Deputy in Ireland. The moment is brilliantly chosen. At a stroke Essex's heart empties of all rancour. He accepts the honour and charge, and assures his Sovereign '*With God's help I will have victory, and you shall have peace.*'

Vain words! The campaign is a calamity; Tyrone and his Irish kerns are unvanquished. Essex, seeing the extent of his failure, panics and, leaving his troops in the lurch, returns hurriedly, unannounced, to England. He breaks in on the Queen early in the morning at Nonesuch, while she is at her toilet 'in a dressing-gown, unpainted, without her wig, her grey hair hanging in wisps about her face, and her eyes starting from her head'.* Surprised though she is by his sudden irruption, not for one moment does she lose command of the situation. Essex talks of forgiveness, babbles of foes who beset him in England, reminds her of his love—but all to no purpose. Elizabeth knows instinctively he has failed her trust, and after he has left her presence, her view is confirmed by Cecil. She makes up her mind. Essex must be kept under surveillance lest he and his followers prove a danger to the realm. To Cecil she confesses:

> *I have failed to tame my thoroughbred.*
> *He is still too proud,*
> *And I must break his will*
> *And pull down his great heart.*
> *It is I who have to rule.*

Nevertheless, Essex and his followers break out in an abortive attempt to rouse the citizens of London, the rising goes off at half-cock, and Essex is proclaimed a traitor (III, ii).

The final scene (III, iii) begins after Essex's trial for treason, when he has been found guilty and condemned to death. The verdict is communicated to the Queen, who for the moment hesitates to sign her favourite's death warrant. Then she grants an audience to Lady Essex and Lady Rich, who

---

* Lytton Strachey, op. cit.

Gloriana: Lord Mountjoy (Geraint Evans), Lady Essex (Monica Sinclair) and Penelope, Lady Rich (Jennifer Vyvyan) plead for the life of the Earl of Essex with Queen Elizabeth I (Joan Cross) in the original Covent Garden production of 1953

both beg for his life. To Lady Essex she is gracious, promising protection for his children. She is antagonized, however, by Lady Rich with her obstinate, insolent importuning. The Queen sees that further delay would be fatal. Summoning up all her fortitude, she signs the warrant.

At this point in the opera, the stage darkens, the Queen is left alone, and there is an epilogue during which time and place become of less and less importance. This is a passage that is always likely to excite controversy. Whereas Strachey ended *Queen Victoria* with a virtuoso coda in which he drew together many of the varied strands of Victoria's life in a long last flashback, in *Elizabeth and Essex* he was content with a final section depicting the gradual stages of Elizabeth's final dissolution. This has been used by Britten and Plomer as basis for a preview of six brief episodes that were to occur after Essex's beheading and before Elizabeth's death. A fully orchestrated version of Essex's second lute-song '*Happy were he*' accompanies these episodes. Many of the words in this epilogue are based on actually recorded speech. Only three phrases are actually sung by Elizabeth, two of them echoes of Essex's lute-song:

In some unhaunted desert . . .
There might he sleep secure . . .
Mortua, mortua, sed non sepulta . . .

Otherwise the dialogue is melodramatic, the music being interrupted to admit speech over sustained tremolo strings, wind, or percussion. The procedure may be unorthodox, but it is undeniably effective in the theatre. Nearly all the words—particularly Elizabeth's speech to the audience, which is derived from her Golden Speech to Parliament—are given maximum clarity; and the lute-song itself rings out nobly in its full orchestral apotheosis, cracked though it be by the melodramatic inter-ruptions.

The whole opera ends with an off-stage repeat of the chorus '*Green leaves are we*' that dies away into deepest silence.

It should be added that a few revisions were made to the score at the time of its revival at Sadler's Wells Theatre in 1966. These affected mainly the Queen's Soliloquy and Prayer (I, ii) and the Epilogue (III, iii).

II

Gloriana is the only through-composed opera of Britten's in which the acts are not unbroken musical entities. Each of the eight scenes, which make up the opera's three acts, is complete in itself and has attached to it a brief orchestral prelude that is played before the curtain rises. The result is that, while the opera lacks the cumulative musical flow of *Peter Grimes*, or *The Rape of Lucretia*, or *The Turn of the Screw*, it provides a succession of vivid self-contained tableaux.

The prelude to Act I, scene i, which is more extended than any of the others, repeats a device originally used in *Peter Grimes*. It gives a graphic musical description of the jousting in the lists, with each percussive charge introduced by a lively sennet on the brass; and when the curtain rises, the same music is repeated quietly to accompany Cuffe's running description of the tournament that is taking place offstage. After the victor (Mountjoy) has

Ex. 65

197

been presented with a golden prize by the Queen, the crowd turns to her and acclaims her in a solemn hymn.

This is one of Britten's most genial inventions. Its slowly moving intervals, especially the 6ths, 7ths and 9ths, unfold like the petals of a cinquefoil Tudor rose, and the sensation that they overlap each other is enhanced by the movement of the parts within the 5/4 tempo. The chorus is repeated in Act II, scene i and Act III, scene iii; and it is not surprising to find it associated in other ways with the course of the opera. Its theme has already been used as the bass for one of the clanking episodes in the jousting prelude to Act I, scene i:

Ex. 66

it provides an introductory flourish for the different phrases of Elizabeth's Soliloquy (I, ii):

Ex. 67

and there are momentary references to it in Act III, scenes i and iii.

There is a motif associated with the Queen's favour that deserves special attention. It is based on the notes of the triad and appears when Essex is appointed Lord Deputy in Ireland (Ex. 68a), having already occurred in Act I, scene i during the quarrel between Essex and Mountjoy (Ex. 68b):

198

Ex. 68a

Ex. 68b

With an altered coda, it forms the theme of the trio (II, iii) '*Good Frances, do not weep*' (Ex. 69a) and also of Lady Rich's pleading for Essex's life (III, iii), which so exasperates the Queen by its obstinate persistence (Ex. 69b):

Ex. 69a

Ex. 69b

and it is related to the '*Victor of Cadiz*' chorus in the Whitehall Palace scene (II, iii), the theme of which is taken over by the orchestra at the end of the scene, when it is played by the orchestra in the pit, the crescendo of its rising phrases gradually overwhelming the Coranto being played by the Court band on the stage.

Much of the score is sparse and muscular. Sometimes monody is sufficient for Britten's purpose, as is the case with the first part of the unaccompanied song of the Country Girls in the Masque (II, i) '*Sweet flag and cuckoo flower*' with its characteristic Lydian fourth, and the flute solo accompanied by a tabor for the Morris Dance (II, iii) which recalls the earlier *Metamorphoses after Ovid*.

THE OPERAS

At the same time there are finely constructed passages, such as the Ensemble of Reconciliation (I, i), the whole sequence of intimate lyrical ensembles in Act II, scene ii (duet, double duet, and quartet), and the particularly beautiful Dressing-Table Song (III, i) sung by the Lady in Waiting with a chorus of Maids of Honour.

Ex. 70a

Ex. 70b

Equally beautiful are the masquers' *a cappella* choruses in Act II, scene i; and the Courtly Dances in Act II, scene iii, succeed in capturing much of the genuine Elizabethan idiom with their cross-rhythms, false relations, crazy filigree ornamentation, and heavy bounding basses.

There are numerous happy touches of musical characterization. For instance, although the Recorder of Norwich has only eight bars in Act II, scene i, he is nevertheless so firmly drawn that he remains as vividly in one's memory as if he had been a character in Crabbe's *The Borough*; and the statesman's mind of Cecil is depicted in all its curious intricacy and ultimate bathos.

But in the end it is through the imaginative presentation of the Queen, who gave her name to the age and was its chief ornament and glory that the opera triumphs. The fanfares accompanying her entrance and retirement in the opening scene at the tournament (I, i) give her a truly majestic setting, and she leads the Ensemble of Reconciliation with characteristic firmness. In the following scene (I, ii) she is depicted first as a politician with her minister Cecil, then as a woman being wooed by Essex, and lastly as a Queen alone with God. This soliloquy and prayer form perhaps her noblest musical utterance in the whole opera. (The number was slightly revised for the revival of *Gloriana* at Sadler's Wells Theatre in 1966.) In the second act she shines mainly by reflection: first she is seen on royal progress in Norwich (II, i), where she receives the loyal homage of the Norfolk masquers; and later (II, iii) in her palace at Whitehall she participates with her courtiers in a strenuous series of dances. In the last act, the irruption of Essex into her dressing room (III, i) where he discovers her in undress, is the cue for a very moving duet. After this, her hesitation before signing Essex's death warrant (III, iii) leads directly into the epilogue with its recession towards old age, death and dissolution. Her final words sung during the recapitulation of the second lute-song (now scored for full orchestra) '*mortua, mortua, sed non sepulta*', recall Essex's words to her in his earlier song:

> *Where, when he dies, his tomb might be a bush*
> *Where harmless robin dwells with gentle thrush.*

An unfulfilled woman maybe; a devious statesman; but undoubtedly a great Queen.

IX
The Turn of the Screw

I

Henry James's novella, *The Turn of the Screw*, was published in 1898 and immediately established itself as a masterpiece of its *genre*—a melodramatic thriller, a ghost story dealing with innocence and corruption, possession and exorcism.

The fact that the main protagonists are two young orphans, Miles aged about eleven and his sister Flora aged eight, was likely to prove an incentive rather than a discouragement from Britten's point of view; the Essex setting of the story in the country house at Bly (distant cousin perhaps to Borley) brought it into close East Anglian focus; and the deployment of the action on two planes—the natural and the supernatural—was an added attraction. Small wonder that Britten wanted to turn it into an opera if possible. But some critics were doubtful of the chances of success, feeling that James's strength lay in his ability to heighten tension through impression and ambiguity, and in his strategy of avoiding the direct confrontations of vital points at issue by recounting the indirect reactions of the characters concerned, so that visual presentation in stage terms might render it difficult to maintain the special atmosphere he created so inimitably in the telling of the story. They need not have worried. Despite reservations natural to his temperament and times, James was more explicit in *The Turn of the Screw* than many people were prepared to allow.

What, for instance, was the nature of the suspect relationship between Miles and Peter Quint, who before his death served as valet to Miles's uncle and guardian? James says quite clearly: 'there had been matters in his life— strange passages and perils, secret disorders, vices more than suspected', and the fault was 'Quint was much too free'—too free with Miles, too free with everyone. Freedom of this kind might bring sexual licence in its train; and a substratum of sexual implications spreads like a secret poison through the layers of the story. In some ways, Miles in his prepuberty is too young to be corrupted by this relationship whether with Quint the man or Quint the ghost; but the essence of the corruption is in the deliberate cultivation of an element of precocity, and the tragedy lies in the struggle for possession of the child by an adult (or adults) on adult terms.

The parallel relationship between Flora and Miss Jessel, her former governess, is rather shadowy and serves merely as confirmation of the crucial importance of the Miles/Quint axis.

It is the new Governess who (as narrator) provides the central emotional reference point. Apart from her, Bly seems to be a closed community, shut off from the world outside. Not only has she come from Hampshire via London to this lonely spot in Essex, but in London she has met and fallen in love with the children's uncle and guardian. This is the absentee master of Bly, who is the mainspring of the situation, though he stands outside the action and ducks his responsibilities. She admits to being carried away at her original interview with him in Harley Street; and when she reaches Bly and makes friends with the housekeeper, Mrs Grose, there is a significant misunderstanding in the course of their conversation.

'What was the lady who was here before?'

'The last governess? She was also young and pretty—almost as young and almost as pretty, miss, even as you.'

'Ah, then I hope her youth and her beauty helped her! He seems to like us young and pretty!'

'Oh, he *did*. It was the way he liked every one!'

The governess is thinking of the guardian uncle; Mrs Grose of Quint. The ambivalence continues as far as Quint's first materialization on the tower of the house at Bly. The Governess is exploring the grounds, meditating how charming it would be if someone (clearly she has in mind the master of Bly) would appear at the turn of a path, stand before her, smile and approve. 'He did stand there!' she writes, 'but high up, beyond the lawn and at the very top of the tower.' After this shock of surprise, there is a second shock which reveals that the apparition is not the master of Bly at all, but a figure who is later identified as the ghost of the master's body-servant, Quint. This was certainly a traumatic emotional experience for the Governess; but afterwards there are no further examples of transference between the master and the servant. Gradually an alternative outlet for her emotions appears—Miles himself. There is an occasion when, nervous and restless, she listens outside her charge's bedroom door at night. Miles realizes she is there and calls her in. Their poignant midnight talk leads to the point where she is so moved that she throws herself upon the boy in bed, embracing him in what she conceives to be the tenderness of her pity. 'My face was close to his, and he let me kiss him, simply taking it with indulgent good-humour.' The mystic union is clinched a little later when, after Mrs Grose has taken Flora away from Bly, the Governess is left alone with Miles. They take their meals together; and in her story she recalls how they continued silent while the maid was with them—'as silent, it whimsically occurred to me, as some young couple who, on their wedding-journey, at the inn, feel shy in the presence of the waiter. He turned round only when the waiter had left us. "Well—so we're alone!" ' It is the crowning irony of the story that finally it is the Governess who, actuated by the strongest though not necessarily the most innocent motives, wrestles with Quint's ghost for the possession of the boy; and when she thinks she has won, '*I* have you,' she cries, 'but he has lost you for ever!' At that moment Miles has a convulsive reaction and is about to fall, when the Governess catches him; and the final words of her narrative are: 'I caught him, yes, I held him—it may be imagined with what a passion; but at the end of a minute I began to feel what it truly was that I held. We were alone with the quiet day, and his little heart, dispossessed, had stopped.'

So ends the Governess's story,* which Henry James presents as being read to a group of listeners (by a man called Douglas) years after it had been written.

As his librettist, Britten chose Myfanwy Piper. Her husband, John Piper, had been involved in the designs for *The Rape of Lucretia*, *Albert Herring*, *Billy Budd* and *Gloriana*, so she had had numerous opportunities of attending rehearsals of these operas and thereby of learning 'the kind of language that sounded natural and inevitable' in opera, and also 'what to avoid—though not always how to avoid it'.†

At an early stage in the planning, composer and librettist decided to construct the libretto in two acts instead of three, and to have a multiplicity of scenes so as to reinforce the impression (so powerfully conveyed in James's story) that the action covered a considerable period of time and that there were long stretches of normality between the occasional supernatural appearances of the phantoms. John Piper had been chosen as designer, and it was understood that transparencies would play an important part in presenting the ghosts on the stage and in fading out from one scene and into another. The scenario took fifteen scenes from the story—there are twenty-four numbered sections in James's novella—and it became clear that even with the benefit of elaborate lighting effects, these would have to be linked with each other by orchestral interludes (later called Variations). Compression and brevity would have to be the order of the day, since a scene with its orchestral prelude could average only about eight or nine minutes. The eight scenes of the first act provided a complete exposition of the characters and the situation. 'At the end of the act, the Governess, though still not

---

* An interesting suggestion has been made by J. Purdon Martin, MD, FRCP (in 'Neurology in Fiction: *The Turn of the Screw*', *British Medical Journal*, 22 December 1973) that the Governess shows signs of suffering from temporal lobe epilepsy.

† 'Some Thoughts on the Libretto of *The Turn of the Screw*' by Myfanwy Piper, from *Tribute to Benjamin Britten*. Faber and Faber, London, 1963.

The Turn of the Screw: The ghost of Miss Jessel (Arda Mandikian) confronts the Governess (Jennifer Vyvyan) in the original 1954 production by the English Opera Group

active, was in possession of all the facts.'* The first scene of the second act, in which the ghosts of Peter Quint and Miss Jessel come together is an invention of the librettist's for which there is no textual authority in James. The remaining seven scenes of that act keep quite close to James's story and show the build-up of the crisis as precipitated by the Governess's interference.

Two late changes seem to have been made in constructing the libretto. 'After the first three scenes had been completed,' writes Myfanwy Piper,† 'it was decided, for musical reasons, that there must be a prologue, and several quite different prologues were written before we were both satisfied. Much later, although we had decided to leave out the whole episode of the letter to the guardian in Act II, it suddenly seemed essential and had to be written and inserted.' Both scenes are in fact essential—the prologue because without it the motivation of the situation at Bly is unexplained, the relationship between the master and his dead valet difficult to apprehend, and the life-line between Bly and the ordinary world outside not established; the letter scene because it enables the Governess to give free rein to her emotions in writing to someone outside Bly and because the theft of the letter by Miles is crucial to the climax.

There is a considerable amount of dialogue in James's story, and much of this could be usefully adapted for the libretto; but not a single word is spoken by either of the ghosts, and this posed a particularly daunting problem, for Britten decided it was essential for the opera that the ghosts should sing 'and sing words (no nice, anonymous, supernatural humming or groaning)'.‡ Two lucky *trouvailles* came to the librettist's assistance in this wellnigh impossible task: from a comment by James himself on Miss Jessel in one of his prefaces, she culled the phrase 'lost in her labyrinth'; and W. B. Yeats's line, 'The ceremony of innocence is drowned', provided a clinching refrain for the Colloquy between the two ghosts in Act II, scene i.

As for the children, one of James's striking sentences—*The lake was the sea of Azof*—was sufficient cue for some of the school-room tags and mnemonics. Nursery songs also played their part. In order to establish Miles's character musically, Britten needed an important but simple song and selected some doggerel dealing with the different meanings of the Latin word *Malo* which he found in an ancient Latin grammar belonging to Myfanwy Piper's aunt.

Malo, Malo, *I would rather be*
Malo, Malo, *in an apple tree*
Malo, Malo *than a naughty boy*
Malo, Malo *in adversity.*

* Myfanwy Piper, op. cit.
† Myfanwy Piper, op. cit.   ‡ Ibid.

It was the librettist's idea that this song should be sung also by the Governess at the end of the opera when she discovers that Miles is dead.

The last scene of Act I was a crucial spot. In James's story, there occurs at this point the celebrated 'double-take' scene where late at night the Governess creeps into an empty room in a lower storey of the tower at Bly to look out of the window unobserved and sees Miles in the courtyard below gazing up at someone she cannot see on the top of the tower, but whom she instinctively knows to be the ghost of Quint. This was hardly capable of realization in terms of the theatre or of opera, even with the help of transparencies on the stage (though it could undoubtedly provide a magnificent scene in a film). Instead, Myfanwy Piper planned a confrontation between all the six characters—the only sextet in the whole opera, and in fact the only time when the two adults, the two children, and the two ghosts are on the stage together.

The complete scenario of the opera runs as follows:

ACT ONE	ACT TWO
Prologue	Variation VIII
Theme	Scene i: *Colloquy and Soliloquy*
Scene i: *The Journey*	Variation IX
Variation I	Scene ii: *The Bells*
Scene ii: *The Welcome*	Variation X
Variation II	Scene iii: *Miss Jessel*
Scene iii: *The Letter*	Variation XI
Variation III	Scene iv: *The Bedroom*
Scene iv: *The Tower*	Variation XII
Variation IV	Scene v: *Quint*
Scene v: *The Window*	Variation XIII
Variation V	Scene vi: *The Piano*
Scene vi: *The Lesson*	Variation XIV
Variation VI	Scene vii: *Flora*
Scene vii: *The Lake*	Variation XV
Variation VII	Scene viii: *Miles*
Scene viii: *At Night*	

II

There are only six characters in *The Turn of the Screw*—seven if one includes the anonymous part in the prologue, which is written for tenor and can be doubled with Quint—and of these, two are children, the one a girl supposed to be about eight years old, and the other her elder brother whose voice is still unbroken. With the children playing key roles, it seemed inevitable that this should be planned as a chamber opera; and Britten decided to write the score for more or less the same orchestra as in *The Rape of Lucretia* and *Albert Herring*.

It has already been shown that the construction of the opera called for sixteen orchestral interludes. Britten decided that these should consist of a theme and fifteen variations. The theme is based on the following note row:

Ex. 71a

(Curiously enough a similar row was used by Stravinsky a few years later in *The Flood* to build up the chord of Chaos with which the Prelude opens.) Britten uses the row tonally rather than serially; and it should be noted how the theme (Ex. 71b) falls into three assymmetrical phrases and how some of the semiquavers owe allegiance to the previous double dotted crochets, while others anticipate the following notes:

Ex. 71b

The intervals of the note row can be construed as a series of six ascending fourths or six descending fifths, and because of this it might be thought that the first note of each pair would carry a dominant and the second note a tonic implication. This, however, is not the case. The first note in each case is a tonic and the second a subdominant. For instance, the theme as quoted above is in A, and not in D.

The tonal suggestions of a theme like this are obvious; and it might also be thought that they would dictate the key structure—a cycle of key changes each of which would represent, as it were, a further turn of the screw. But this is not so. The key structure of the opera is based on other, and in some ways subtler, considerations. For instance, in Act I the main keys of the interludes and scenes are, in ascending sequence, as follows:

Theme and Scene i	A minor
Variation I and Scene ii	B major
Variation II and Scene iii	C major
Variation III and Scene iv	D major (modulating to G minor)
Variation IV and Scene v	E major (modulating to E minor)
Variation V and Scene vi	F major (modulating to F minor)
Variation VI and Scene vii	G major
Variation VII and Scene viii	A flat major

From this table it will be seen that the first seven scenes are in all the seven white-note keys of the octave, and these cover the *natural* aspects of the exposition. The first black-note key appears in the last scene of the act, when for the first time the two ghosts are heard.

The identification of the *supernatural* world with flat keys and flattened notes is important to the score. On Quint's first apparition (in scene iv), an E flat chord is inserted in the prevailing D major texture:

Ex. 72

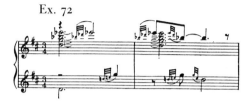

His first sung phrase is in scene viii—a call to Miles on E flat, followed by free-flowing melismata:

Ex. 73a

And Miles answers 'I'm here—O I'm here!' also on a high E flat. In fact, this E flat becomes so surcharged with supernatural overtones that later in the opera its appearance is sufficient to evoke an immediate change of mood.

Act II has four black-note keys as against one in Act I; and with one exception the keys are in descending sequence:

Variation VIII and Scene i	A flat major (modulating to G sharp minor)
Variation IX and Scene ii	F sharp major (modulating to an unestablished F sharp minor)
Variation X and Scene iii	F major (modulating to F minor)
Variation XI and Scene iv	E flat minor
Variation XII and Scene v	E major
Variation XIII and Scene vi	C major
Variation XIV and Scene vii	B flat major
Variation XV and Scene viii	A major (modulating to A flat major)

In the opening Colloquy between the two ghosts, Quint's call (Ex. 73a) is altered to fit the refrain 'The ceremony of innocence is drowned':

Ex. 73b

The other ghost (Miss Jessel) is not so strongly characterized as Quint. Her appearances are frequently underpinned by a sombre, brooding, slowly spread chord:

Ex. 74

Of special interest is the scene (II, iii) where the Governess finds her in the schoolroom sitting at the desk. Part of the sombre, brooding chord has spread into the accompaniment, which pins down the melodic line and prevents it from taking wing:

Ex. 75

This little scene would not be so striking musically were it not for the fact that it provides an exquisite foil for the Governess's letter writing episode that follows. This is treated as a miniature scene within a scene and has a quick instrumental prelude (while the letter to the children's guardian is being written) followed by a quieter section (while she reads aloud what she has written). The musical illustrations show (*a*) the first phrase of the letter as written and (*b*) the same phrase as read aloud:

Ex. 76a

Ex. 76b

The sequel to this scene, where the ghost of Quint tempts Miles to intercept the letter before it is posted, brings a melodramatic passage where Quint has three spoken phrases—'What has she written?' 'What does she know?' 'Easy to take.' Each of these is given different accentuations, which echo a side drum phrase, e.g.:

Ex. 77

As has been mentioned above, some of the children's musical material in the early scenes is based on nursery songs—a straightforward rendering of *Lavender's blue* in the third scene of Act I; and *Tom, Tom, the Piper's Son*, with slightly more extended treatment in scene v. In the schoolroom scene Miles sings his 'Malo' song. This is one of Britten's most touching inventions. Like the Governess's letter song (quoted above), the tune is basically formed out of interlinked thirds. Its apparent simplicity is deceptive, for there is considerable subtlety in the way the melodic line avoids symmetry, and in its harmonization. Quite apart from one or two

Ex. 78

literal repetitions later on, this little tune makes a number of altered appearances. It provides a theme for Variation VI. Played by the horn in a quick and light tempo, it forms a bass to the accompaniment of Quint's short song at the beginning of Act II ('I seek a friend, obedient to follow where I lead'). Played by the cor anglais in its normal rhythm, it is heard at the end of the bedroom scene (II, iv); and immediately afterwards a diminished form is played by the pizzicato strings quickly and urgently as a kind of miniature fugato in Variation XII. It provides an expressive coda to scene v of Act II. And at the end of the opera the theme is sung by the despairing Governess (in the key of A) as she lays Miles's body on the ground.

As is Britten's usual custom, the music is continuous within the acts, and the orchestral interludes and their following scenes are cunningly interwoven. Occasionally the scenes are extremely short—Act II, scene v, for instance, is only twenty-seven bars long. In comparison a scene like The Window (I, v) seems long and moves through a series of different episodes and narrations.

The scenes that have most definite musical unity of their own are perhaps The Bells (II, ii) and The Piano (II, vi). In the former, the sound of the church bells provides a framework for Miles and Flora, as they chant their own Te Deum on their way to Church, diversifying the sacred words with occasional interjections from the schoolroom.

> ... *O ye rivers and seas and lakes:*
> *Bless ye the Lord.*
> O amnis, axis, caulis, collis,
> Clunis, crinis, fascis, follis ...

The bells, with their clear timbre and different changes, emphasize the sinister atmosphere of the scene. The scene called The Piano is constructed on the lines of a miniature piano concerto. A Diabelli-like theme makes its appearance in Variation XIII; and when the lights go up on the schoolroom Miles is seen sitting at the piano, playing. The theme is varied, particularly with repeated notes and scale passages that are played with sufficient virtuosity to justify the comments of the Governess and Mrs Grose 'O what a clever little boy!' Miles starts a second piece and then begins to show off with occasional glissandi. Under cover of his showmanship, Flora slips away to her forbidden tryst with Miss Jessel; and when the Governess and Housekeeper rush off to bring her back and the scene slowly fades, Miles is left playing a triumphant kind of toccata which becomes Variation XIV.

In the final scene (II, viii), the climax between the Governess and Miles and Quint* is played out as a passacaglia over a ground bass built up cumulatively from the basic theme (Ex. 71b). Finally the white-note keys triumph over the black-note keys. Quint's farewell melismata are a semitone higher than in Ex. 73a. The spell is broken; Miles is dead; and all that remains is the altered 'Malo' lament sung by the Governess:

> ... Malo, Malo *than a naughty boy.*
> Malo, Malo *in adversity.*

---

* An interesting note on revisions carried out by Britten on this scene before the vocal score was actually published is to be found in 'Britten's Revisionary Practice: Practical and Creative' by Donald Mitchell. *Tempo*, Autumn/Winter 1963.

X
Noye's Fludde

Sooner or later it seemed inevitable that Britten would want to write a dramatic or semi-dramatic work for church performance. In 1957 he found the subject he required in one of the medieval miracle plays and chose from the Chester Cycle the episode of Noye and his family, the building of the ark, and the flood. Formerly it had been the custom for each separate play in these cycles to be performed by one of the Guilds on a cart known as a pageant, which moved about the town so that several complete performances of the cycle could be given at different spots. As the Introductory Note to Britten's score makes clear: '*Noye's Fludde*, set to music, is intended for the same style of presentation—though not necessarily on a cart. Some big building should be used, preferably a church—but not a theatre—large enough to accommodate actors and orchestra, with the action raised on rostra, but not on a stage removed from the congregation. No attempt should be made to hide the orchestra from sight.'

The action of *Noye's Fludde* made it possible to cast most of the parts for children. Adults are needed only for the Voice of God (spoken), and for Noye (bass-baritone) and his wife (contralto). Sem, Ham and Jaffett and their wives should be played by boys and girls between the ages of eleven and fifteen, though Jaffett (the eldest) may have a broken voice. Mrs Noye's Gossips should be older girls, with strong voices, especially in the lower registers. The animals are played by children. Forty-nine different species are referred to in the Chester Miracle play, of which thirty-five (in pairs) subdivided into seven groups were used in the first production of *Noye's Fludde*, viz.:

I   Lions, Leopards, Horses, Oxen, Swine, Goats, Sheep;
II   Camels, Asses, Buck and Doe;
III Dogs, Otters, Foxes, Polecats, Hares;
IV   Bears, Wolves, Monkeys, Squirrels, Ferrets;
V   Cats, Rats, Mice;
VI ⎫ Herons, Owls, Bittern, Peacocks, Redshanks, Ravens,
and ⎬ Cock and Hen, Kites, Cuckoos, Curlews, Doves, Duck
VII ⎭ and Drake (six pairs in each group).

Noye's Fludde: Noye (Owen Brannigan), Mrs Noye (Sheila Rex) and the animals in the ark

The total cast consists of three adults and ninety children (including four boys who are used as property men).

It now remained to be seen how far the make-up of the orchestra should be influenced by the great preponderance of children in the cast. Britten boldly decided to opt for a large number of young amateur players as well as some professionals. The professional group consists of solo string quintet (Vl I and II, Vla, Vcl, DB), solo treble recorder, piano (four hands), organ, and timpani: the amateur group of strings ripieni (Vl I, II and III, Vle, Vcl I and II, DB), descant recorders in two parts and treble recorders, bugles (in B flat) in four parts, twelve handbells (in E flat), and a specially large collection of percussion, viz.:

Bass Drum	Whip
Tenor Drum	Gong
Side Drum	Chinese blocks
Tambourine	Wind machine
Cymbals	Sandpaper
Triangle	Slung mugs

The score is tempered to the abilities of the young amateur players. For instance, the Introductory Note specifies: 'There are three sorts of amateur *Violins*: the *Firsts* should be capable players, not however going above the 3rd position, and with the simplest double-stops. The *Seconds* do not go out of the first position, while the *Thirds* are very elementary, and have long stretches of just open strings. The *Violas* need to be as accomplished as the *1st Violins*, as do the *1st Cellos*, while the *2nd Cellos* have only the simplest music. The *Double Bass* is very simple.' Similar instructions are given for recorder players and bugles.

As regards the percussion, most of the instruments specified are the stock orchestral ones, with the exception of the Sandpaper and Slung Mugs. The former consists of two pieces of sandpaper attached to wooden blocks and rubbed together: the latter are mugs or cups of varying thickness and size chosen so as to make a rough kind of scale and slung on string by their handles from a wooden stand and hit with a wooden spoon. The total orchestra numbers a minimum of sixty-seven players, of whom ten are professionals and fifty-seven amateurs.

This would seem to imply that no fewer than a hundred and fifty-six actors and instrumentalists are needed to mount a performance of *Noye's Fludde*; but the total forces involved are even larger, for at three points Britten decided to call on the whole congregation to join in singing three hymns that are built into the score—'*Lord Jesus, think on me*'; '*Eternal Father, strong to save*'; and '*The spacious firmament on high*'.

Noye's Fludde begins with the congregation singing four verses of the first of these hymns, at the end of which Noye walks through the congregation on to an empty stage, where he kneels, and God's Voice is heard accusing mankind of sin, prophesying destruction, and promising to save Noye and his family because of Noye's righteousness. After this revelation, Noye summons his family to help in the building of the ark; and the tail-rhyme of the text (stanzas rhyming aaabcccb) is handled so that the triple rhymes (a and c) are set to a simple syncopated tune, while the end rhymes (b) are treated like a refrain from a non-syncopated hymn tune. This refrain varies from verse to verse; but the one that is most characteristic and sticks longest in the mind is Noye's '*At the coming of the flood*':

Ex. 79

The entry of the animals into the ark shows Britten's invention at its most genial. The text's eight half-stanzas (of four lines each) are treated as

216

variations on a basic pattern, consisting of (a) preludial bugle calls, (b) songs by various callers, and (c)—this is Britten's own invention: there is no authority for it in the Chester pageant—a chattering series of repetitions of the phrase '*Kyrie eleison*' sung by the animals themselves as they pass through the congregation to enter the ark:

In the Chester pageant, Noye's wife appears as a perverse and cantankerous character. According to A. C. Cawley,* this comic tradition is at least as old as the picture of Noah's Ark in the Junius manuscript (*c.* A.D. 1000), which shows her standing at the foot of the gangway, while one of her sons tries to persuade her to go on board. Britten takes full advantage of this scene and finds admirably comic variants of the various tunes for Noye's wife and her gossips. The final *coup*, when the three sons bodily carry her, protesting, into the ark and Noye gets his ears boxed for his welcome, leads directly to the storm.

* *Everyman and Medical Miracle Plays.* Ed. A. C. Cawley. J. M. Dent (Everyman Edition) London, 1956.

This is a large-scale instrumental movement composed as a passacaglia. The ground bass consists of pairs of ascending minor thirds rising generally by whole-tone steps. These pass through all twelve notes of the chromatic scale (a reminder of the processes recently used in *The Turn of the Screw*) and are tonally anchored by a strong repeated G. At the outset of the storm, when the first raindrops are heard, the slung mugs impart just the right quality of sound to depict extra-large single raindrops plopping on to hot dry earth:

Ex. 81

The different variations depict rain, more rain, wind, thunder and lightning, more wind, waves, yet more wind, flapping rigging, great waves, ship rocking, panic of the animals; and then at the height of the storm everyone in the ark sings the first verse of '*Eternal Father, Strong to Save*', the congregation joining in the second verse. The storm then gradually subsides, while the orchestra plays fragments of the variations in reverse order, interspersed with fragments of the hymn tune.

The storm is over, but the flood remains.

Noye sends out a Raven, which does not return to the Ark; and the Raven is followed by the Dove. Here graceful use is made of flutter-tonguing for the solo recorder (Ex. 82a). The bird flies off to a lively little waltz tune (Ex. 82b); and when it returns with an olive branch in its beak, the tune is reversed (Ex. 82c), and as it alights on the Ark the opening chords are reversed too (Ex. 82d):

Ex. 82a

Ex. 82b

Ex. 82c

Ex. 82d

Noye blesses God for this sign of peace; and the Voice of God is heard authorizing Noye to allow his family and all the animals to leave the Ark. The animals troop out to a crescendo of Alleluias:

Ex. 83

God's Voice is heard again, promising that the rainbow shall serve as a covenant that in future He will never try to wreak similar vengeance on mankind. This is the cue for the actual appearance of the rainbow accompanied by a peal of handbells (in E flat major); and the final hymn (in G major), '*The spacious firmament on high*', begins. Between the first, second, third, and fourth verses, the Sun, the Moon, and the Stars also appear. The congregation joins in for the fifth verse; and in the sixth Tallis's tune is treated as an eight-part canon at the unison, or octave. At the end, the Voice of God is heard, blessing Noye tenderly. The handbells appear in augmentation; and there is a final dying call from the bugles.

XI
A Midsummer Night's Dream

I

Music was certainly in Shakespeare's mind when he wrote *A Midsummer Night's Dream*, whether the occasion of the first performance was a wedding celebration or a public performance by the Lord Chamberlain's Servants. '*Fairies sing*' is the quarto stage direction before 'You spotted snakes with double tongue'. Bottom sings 'The ousel cock so black of hue' to keep up his spirits; and before his 'exposition of sleep', he calls for 'the tongs and the bones'. The fairies oblige—'*Musicke Tongs, Rurall Musicke*' is the folio direction. A little later, when Oberon wakes Titania, they both call for music—the direction '*Musicke Stille*' appears in the folio—and Oberon invites her to dance:

> *Come, my queen, take hands with me,*
> *And rock the ground whereon these sleepers be.*

When Theseus and Hippolyta enter with their train to hunt, there is a folio direction '*Winde hornes*'.* The second time the horns sound (also marked by a special stage direction), their call awakens the sleeping lovers, who all start up. At the end of the play, just before the Epilogue, Oberon sings a Song (so specified in the text)—'Now until the break of day'.

After the Restoration many of Shakespeare's plays were adapted for the new London patent theatres by producing them with 'Singing, Dancing and Machines interwoven, after the manner of an Opera'. The turn of *A Midsummer Night's Dream* came in 1692, when the text was rearranged, under the title of *The Fairy Queen*, and masques inserted in Acts II, III, IV, and V. The author of this baroque semi-opera is not known, though it is generally thought to have been Elkanah Settle. The music of the masques was written by Henry Purcell. That the production by Thomas Betterton at the Queen's Theatre in Dorset Garden was popular is evident from the fact that it was revived the following year with additional numbers and a second

---

* Singular in the quarto instead of plural.

edition of the libretto called for; that it was extravagant is attested by the stage directions and by the passage in the preface where the author confesses that little material advantage was likely to accrue to the promoters 'considering the mighty Charge in setting it out, and the extraordinary expence that attends it every day 'tis represented'.

(In 1967 a new concert version of *The Fairy Queen* was devised by Peter Pears and edited by Benjamin Britten and Imogen Holst. This was given at the Maltings Concert Hall, Snape, during the 1967 Aldeburgh Festival. The material was arranged in four parts: I. Oberon's Birthday, II. 'Night and Silence', III. 'The Sweet Passion', IV. Epithalamium. In this version, the musical shape of the key-sequences is preserved, and optional wind parts have been added to some of the dances and choruses.)

After Purcell's *The Fairy Queen*, the next operatic version was Richard Leveridge's *The Comick Masque of Pyramus and Thisbe* (1716), where the action was confined to the mechanicals and their play, and the music 'composed in the high stile of Italy', meaning that it was a satire on the prevailing craze for Italian opera in London. Nineteen years later Leveridge's libretto was used for a similarly titled work with music by John Frederick Lampe.

A new operatic version called *The Fairies* was made by John Christopher Smith in 1755. In this case, David Garrick, who presumably wrote the libretto, completely omitted the mechanicals and their 'tedious brief scene' of Pyramus and Thisbe. This was produced at Drury Lane. Sixty years later a further operatic version—this time to a libretto by Frederick Reynolds with music by Henry Bishop, supplemented with songs by Arne and Smith—was produced at Covent Garden.

Other operatic versions were made abroad in the nineteenth century—by G. Manusardi, and F. von Suppé—but all of these pale into insignificance in the light of Felix Mendelssohn and his achievement in writing deliciously appropriate incidental music for the play. As a young boy of seventeen he had seen *A Midsummer Night's Dream* in Berlin and read it in the Schlegel-Tieck translation, and in his resulting enthusiasm he composed a piano duet in classical sonata form for himself and his sister Fanny to play. The following year he orchestrated the piece, and it was first performed as a concert overture at Stettin in 1827. In 1843 he followed up the Overture by composing a full set of incidental numbers, the only difficulty being that this music was written for full orchestra and demanded the utmost care and accuracy in performance, something that few theatre orchestras were capable of achieving.

There were numerous features in Shakespeare's play that were likely to attract Britten as a composer. The contrast between the natural and supernatural elements was something he had already tackled with conspicuous success in *The Turn of the Screw*. The pungent timbre of the

A Midsummer Night's Dream: Oberon (Alfred Deller) and Tytania
(Jennifer Vyvyan) with their attendants in the original 1960 production by
the English Opera Group

boy fairies' unbroken voices made a special appeal to him; and the nocturnal
quality of the action 'following darkness like a dream' called for the exercise
of imaginative powers similar to those already displayed in the *Nocturne*.
Above all, the delicious native freshness of Shakespeare's lyrical poetry will
have put him on his mettle.

It takes a bold man to turn one of Shakespeare's plays into an opera, and it
takes an even bolder man to decide to set Shakespeare's own words. This is
precisely what Britten decided to do. In collaboration with Peter Pears, he
set about shortening and adapting Shakespeare's text. The 2,136 lines of the
original were cut to about half, and the action simplified by starting, not at
the Court of Theseus in Athens, but in the wood with the quarrel between
Oberon and Tytania.* The opera is laid out in three acts, all of which are set
in the wood, except for a transformation scene to Theseus' Palace in Act III.
Shakespeare's material has been rearranged as follows:

* In the opera Britten preserves the quarto spelling, 'Tytania', which affects the
pronunciation of the name.

222

Britten	Shakespeare
Act I	passages from Act II, i; I, i; II, i; I, ii; II, ii
Act II	III, i; IV, i; III, ii
Act III	IV, i; IV, ii; I, i; IV, i; V, i

This means that the first act introduces the various groups of characters who are wandering in the wood, the second act shows the effect on them of Oberon's magic spell, and in the third act, first Oberon undoes the spell, and then the scene is changed to Theseus' Palace so that Theseus and Hippolyta together with the two pairs of lovers can be wedded and the mechanicals can present the play they've been rehearsing as an appropriate nuptial entertainment.

The changes made to Shakespeare's text are minute. One line— 'Compelling thee to marry with Demetrius'—has been invented in order to explain 'the sharp Athenian law' that Lysander and Hermia are trying to evade. And Britten and Pears make one significant addition. Whereas Shakespeare mentions four young fairies by name—Peaseblossom, Cobweb, Moth and Mustardseed—and all four are featured when they first appear (in III, i), Moth gets omitted in the later scene (IV, i). Britten and Pears manage to adjust the position somewhat in his favour by reinstating him in the later scene, though without giving him fresh dialogue. There are naturally a number of minor adjustments. Some of Puck's verse is given to the chorus of fairies. Occasionally lines are re-apportioned among the speakers. The time sequence is adjusted so that the action seems to take place within two consecutive nights, instead of the four happy days mentioned by Theseus at the opening of the play as the waiting period before his nuptials with Hippolyta.

II

The first thing Britten had to decide was how to cast the opera. The quartet of lovers does not predominate, but is central to the action. Helena and Hermia are cast as soprano and mezzo-soprano; Lysander and Demetrius as tenor and baritone. High voices were chosen for the supernatural characters. Cobweb, Peaseblossom, Mustardseed and Moth are boy trebles; and the chorus of fairies, trebles or sopranos. Tytania is a coloratura soprano; and Oberon a counter-tenor (or contralto). Puck posed a special problem; and ultimately Britten, recalling some Swedish child-acrobats gifted with extraordinary agility and powers of mimicry whom he had recently seen in Stockholm, decided to cast Puck as a boy-acrobat with a speaking role. The mechanicals are at the other end of the scale: Flute and Snout are tenors, Starveling baritone, Bottom a bass-baritone, and Quince and Snug basses. Theseus and Hippolyta, who only come upon the scene halfway through the last act, are bass and contralto respectively.

THE OPERAS

The choice of counter-tenor for Oberon posed a problem. There is no doubt that the rarified vocal quality (still a taste to be acquired by some music-lovers) suits the part; but there are tricky problems of balance to be solved, particularly in relation to his duet passages with Tytania.

The score is written for a medium-sized orchestra, larger than the usual English Opera Group chamber ensemble. There are six woodwind (two flutes, oboe, two clarinets, bassoon), four brass (two horns, trumpet, and trombone), two percussion players, two harps, harpsichord and celesta (one player), and strings. There is also a stage band of recorders, small cymbals and wood blocks for the 'tongs and bones' music. It is characteristic of the score that different kinds of texture and orchestral colour are associated with different groups of characters. For instance, the fairies are characterized by harps, harpsichord, celesta and percussion; the lovers, by woodwind and strings; the mechanicals by the lower brass and bassoon.

The first act opens in a wood at night. It is clear that Britten expects the ear like the eye to travel up and down the tall tapering trunks with the shafts of moonlight falling through. As an introduction he has chosen a series of common chords played tremolo by the strings with alternate upward and downward *portamenti* between the chords, giving an impression of the unequal breathing of the slumbering wood:

Ex. 84

It will be seen that these chords are sometimes presented in their root position and sometimes in their first or second inversions, and that their roots (G, F sharp, D, E, A, C sharp, G sharp, E flat, C, B flat, F, B) cover all twelve notes of the chromatic scale. The initial slide from G to F sharp is a step that characterizes much of the fairy music; but the note row itself does not receive such intensive treatment as that of the theme in *The Turn of the Screw* (Ex. 71b). Instead, the episode is used as prelude to Act I and three times as interludes to break up the *a–b–c–b–a* scene construction of the first act (where *a* stands for the fairies, *b* for the lovers, and *c* for the mechanicals). In what should have been the fourth interlude (between scenes iv and v), the chromatic chord sequence becomes the accompaniment to Tytania's air '*Come, now a roundel, and a fairy song*'.

224

These chromatic chords are found in an altered form in the introduction to the initial meeting of Oberon and Tytania in Act I, where the fairies sing

Oberon is passing fell and wrath,
Because that she, as her attendant, hath
A lovely boy stolen from an Indian King,
And jealous Oberon would have the child

above an accompaniment that contains first inversions of the chords of E, G, B, D, C sharp, F, G sharp, C, E flat, F sharp, B flat, the twelfth chord (of A major) being held back for the confrontation of Oberon and Tytania with their simultaneous cry of '*Ill met by moonlight*'.

There is a kind of tart flavour about the music for the fairies. Britten was very conscious that *his* fairies were 'very different from the innocent nothings that often appear in productions of Shakespeare'. He went on to say:* 'I have always been struck by a kind of sharpness in Shakespeare's fairies: besides, they have some odd poetry to speak . . . [They] are, after all, the guards to Tytania: so they have, in places, martial music. Like the actual world, incidentally, the spirit world contains bad as well as good.'

In the fairies' opening chorus there is a curious feature of alternative syllabic accentuation:

Ex. 85

But the 6/4 metre has barely had time to establish itself, in this smoothly flowing line, and in the event the dislocation of the accents is stronger to the eye reading the score than to the ear when listening to a performance. A spikier effect is obtained by more obvious syncopations in a later chorus '*You spotted snakes with double tongue*':

Ex. 86

* From 'A New Britten Opera' by Benjamin Britten. *The Observer,* 5 June 1960.

For Oberon's spell Britten has had recourse to the same sort of tone colour (celesta) that he used to accompany the appearances of the ghosts in *The Turn of the Screw*; but the harmonies are denser and the melody more convoluted. (Note the way the second phrase of the melody is nearly, but not quite, an inversion of the first phrase.)

Ex. 87

This passage provides a lead-in for Oberon's baroque air '*I know a bank where the wild thyme blows*'. The air consists of a number of short contrasting

Ex. 88a

episodes. In the flowers section, the harps trail brief scalic passages like tendrils, which are lengthened for the section where '*the snake throws her enammel'd skin*'. Oberon's reference to Tytania brings a deceptively simple, tartly harmonized tune:

Ex. 88b

The organization and tone quality of this number occasionally call to mind
the Second Lute Song in *Gloriana*.

Puck's spoken lines are generally accompanied by trumpet and drum; and
these trumpet arpeggios are not unlike the sennets that herald some of the
hero's appearances in *Billy Budd*.

For the two pairs of lovers, Britten was faced by the problem of how to
dissolve so much wordy sparring into music. He decided that, speed being
essential, the dialogue should for the most part be set as freely flowing
accompanied recitative. The key to the greater part of the lovers'
declarations and protestations is contained in Lysander's opening questions
to Hermia in Act I:

Ex. 89

The long declamatory line can usually be broken up into motifs where four
adjacent notes in a diatonic scale are arranged in different patterns. This of
course would provide a close fit for the setting of four beat octosyllabic
verse; but as the lovers speak in blank verse or rhyming couplets, the two
extra syllables in each line frequently make it necessary to prolong the
melody by adding two notes to every couple of four-note phrases. This
accompanied recitative is treated very freely; but occasionally there are
nodal points where some sort of ensemble is built up. There is one instance
in Act I where Hermia and Lysander sing a duet built on the phrase '*I swear
to thee* . . .' The most intricate example is the fugato quartet in Act III, where
each lover proclaims that he (she) has found his (her) loved one '*like a jewel,
mine own, and not mine own*':

Ex. 90

The scalic opening of this phrase recalls the hymn-like tune sung by the Male and Female Chorus in *The Rape of Lucretia* and also the final sextet in that opera.

In the middle of Act I the mechanicals meet to discuss the little play they are planning to present at court. When asked what the play treats on, Quince replies, 'Marry, our play is the most lamentable comedy and most cruel death of Pyramus and Thisby.' And this cadence, as echoed by Flute, Snout, Starveling and Snug, becomes a kind of musical epigraph:

Ex. 91

Act II contains only two scenes as against five in Act I: these are (a) the mechanicals rehearsing (which leads to the 'translation' of Bottom) and (b) the lovers wandering through the mazy wood which leads to their final state of enchanted exhaustion. The act starts with a succession of four chords containing all twelve notes of the chromatic scale—the first chord (using three notes) is scored for muted strings, the second (four notes) for muted brass, the third (three notes) for woodwind, and the fourth (two notes) for harps and percussion—typifying the drowsy effects of Oberon's magic spell:

Ex. 92

In the orchestral prelude to the act, some of the notes of the chords are used as starting points for melodic development, but these lines always return to their starting point to reconstitute the original chords. There are three complete variations on the four chords. A fourth is started, but interrupted, as soon as the second chord is reached, by the entry of the mechanicals. At the end of their scene, the four chords return as accompaniment to the doped Tytania's apostrophe to Bottom, '*O how I love thee! how I dote on thee!*' and three further instrumental variations in quickening tempo follow by way of orchestral interlude. In this case the last (fourth) chord heralds the appearance of Puck and Oberon and the lovers straying through the wood, towards the end of which scene each of the four chords is used in turn as Puck leads, first Lysander, then Demetrius, next Helena, and finally Hermia, to the different spots where they lie down and fall asleep. The act

228

ends with the fairies singing a benison '*On the ground sleep sound*' which is
harmonized by the four chords:

Ex. 93

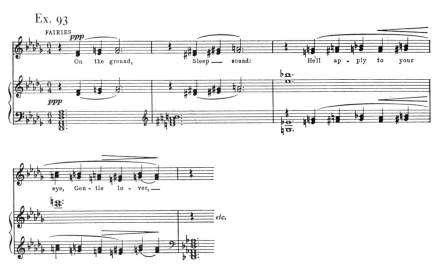

Like the second act, the third is divided into two scenes: but these mark a
complete transition—from the Wood to the Court. A diatonic prelude for
strings establishes the natural mood of day in contrast to the chromaticism
of the supernatural night. This scene brings the reconciliation of Oberon
and Tytania, who dance together to the tune of a beautiful Saraband, where
the melody (cor anglais) is repeated in a mirror-like inversion by the clarinet:

Ex. 94

The reconciliation of the two pairs of lovers follows, in a quartet which has
been referred to above (see Ex. 90). The sound of the Duke's hunting horns
is heard, at first off-stage as the lovers awake, and later as the main substance
of the orchestral interlude before the transformation scene to the Court of
Theseus:

Ex. 95

This final scene starts with a noble introduction for Theseus and Hippolyta, who appear for the first time: a broad, sweeping tune—in fact, one of the most flowing and sustained in the whole opera:

Ex. 96

When the time comes for the mechanicals to present their entertainment in honour of the threefold nuptials, they oblige with what is virtually a little *opera buffa*. (Shade of Leveridge!) The comedy of Pyramus and Thisbe is so amusing in itself that any adequate production of the play is bound to elicit much laughter from the audience. This is the case too when Britten's *opera buffa* is performed; and as at the moment of laughter part of the music is likely to be blacked out, not all listeners may realize how complex and tightly integrated is this little opera within the main opera. Accordingly it is worth while drawing attention to its organization in some detail. There are fourteen numbers in all:

1. Introduction, *Pomposo*. 'If we offend.' (Sextet *a cappella*)
2. *Andante pesante*. 'Gentles, perchance you wonder at this show.' (Prologue)
3. *Lento lamentoso*. 'In this same interlude.' (Wall)
4. *Moderato ma tenebroso*. 'O grim-looked night.' (Pyramus)
5. *Allegretto grazioso*. 'O wall, full often.' (Thisby)
6. Duet, *Allegro brillante*. 'My love thou art.' (Pyramus and Thisby)
7. *Lento lamentoso* (see no. 3). 'Thus have I, Wall?' (Wall)
8. *Allegro giocoso*. 'You ladies.' (Lion)
9. *Andante placido*. 'This lanthorn.' (Moon)
10. *Allegretto grazioso* (see no. 5). 'This is old Ninny's tomb.' (Thisby)
11. *Presto feroce*. 'Oh!' (Lion)
12. (*a*) *Lento*. 'Sweet Moon, I thank thee.' (Pyramus)
    (*b*) *Allegro disperato*. 'Approach, ye Furies fell.' (Pyramus)
13. (*a*) *Allegretto grazioso* (see no. 5). 'Asleep, my Love?' (Thisby)
    (*b*) *Lento*. 'These lily lips.' (Thisby)
14. Bergomask, *Ruvido*

These brief numbers are separated from each other by recitative interjections from Theseus, Hippolyta, and the four lovers, including a particularly involved recitative ensemble for six voices just before the Prologue, laid out on similar lines to the 'chatter' recitatives in *Albert Herring*. As is implicit in the Italian directions,* the music satirizes the style of nineteenth-century romantic opera. The illustrations give an extract from Pyramus' air (no. 4), and the instrumental introduction (note the *flauto obbligato* part!) to Thisby's air (no. 5):

The Bergomask is a jolly dance in two parts—the first moves intermittently between 3/4 and 6/8 (equal quavers); and the second is a very quick rumbustious 2/4, with rushing semiquaver passages and slightly irregular bar-groupings.

It sounds midnight. The epigraph of the Pyramus and Thisby play (see Ex. 91) recurs fortissimo in the orchestra. It subsides; the mortals retire to bed; and the fairies return to a tintinnabulation of instruments simulating the midnight chimes at different pitches and different speeds—thirty-five such chimes in all. The fairies' little tune is deceptively simple, but metrically irregular:

* It is amusing to note that Britten's directions throughout the rest of the score are in English.

The fairies still have a musical trick in reserve; and again it is a metrical one. Britten features a Scotch snap in the slow and solemn setting of the chorus *'Now until the break of day'*. It is a bold stroke to make the weak verse beats

Ex. 99

three times the length of the strong ones, even though the accentuation is not displaced; but it comes off in a fresh and striking fashion. A scherzo passage for Puck brings the opera to a close with gay elusive trumpet calls.

A Midsummer Night's Dream succeeds as an opera, partly because the subject matter is obviously congenial to Britten's temperament, and also because the work is supremely well organized. The differentiation between the natural and supernatural worlds and the characterization of the four groups of persons are so well established musically that the numerous transitions can be made with ease and speed, and no time wasted. It is true that there are moments such as the long stretches of near-recitative dialogue between the two pairs of lovers, where the music seems thin: but this is doubtless deliberate, for whenever the action reaches a climax, the musical tension tightens, drawing in the slack. Throughout the score Britten shows an unerring sense of proportion, and nowhere is this more evident than in the last act where the transformation to the Court of Theseus lifts the opera on to a different level, outside the dream-world of the enchanted wood, and the play within the play turns out to be a miniature opera within the opera proper, with a complex internal organization of its own.

As for the poetry of the original, this has in no way been harmed. Britten himself has confessed* that he did not 'find it daunting to be tackling a masterpiece which already had a strong verbal music of its own', since its music and the music he had written for it were at two quite different levels.

Whereas formerly the only fully satisfactory Shakespeare operas could be said to be Verdi's *Macbeth*, *Otello* and *Falstaff*, now Britten's *Dream* must be added to that short but distinguished list.

* Ibid.

XII
Curlew River

I

Britten's wish to bring some form of operatic entertainment into the church was certainly stimulated by the success of *Noye's Fludde*. He decided, however, that what he next wrote would be for fully professional performers, without an admixture of amateurs, and that the congregation would be an audience in the normal sense, and would not participate in the action. He needed also a new type of convention to obtain a really satisfactory form of presentation.

When casting round for a suitable myth or fable, he recalled one of the Noh plays he had seen on his visit to Japan in February 1956, and which had made a deep impression on him at the time. The occasion was vividly described in the travel diary of Prince Ludwig of Hesse and the Rhine.* 'Before the entrance of an actor through a door at the left a coloured curtain is swept back in an impressive way. The stage itself has a little low door back right, through which stage-hands, chorus etc. enter and depart, bent double; no scenery, except a stylized painting of a pine tree on the back wall of the stage. In the middle, against this back wall, two drummers sit. They let off sudden bursts of clacking and gonging drum-sounds with their hands. To these drummings they recite in strained voices, like people about to vomit. The choir which accompanies the play makes the same kind of sound but sometimes suddenly breaks into chanting song, liturgical and impressive . . . Everything on the stage happens in retarded motion. The actors move with the slowest of steps, artfully lifting the toes of their white-socked feet before setting them down with care and precision. The white feet and the magnificent costumes are reflected clearly by the polished floor of stage and passage. Most of the costumes are obviously historical, made of the most splendid silks and brocades. It seems that the more important a character, the finer his apparel must be, regardless of the appropriate dress

* From 'Ausflug Ost 1956', printed in *Tribute to Benjamin Britten*. Faber, 1963.

for the rôle he is acting. With the finest clothes go the wonderful masks which can change their expression by a tilt of the head.'

To this description of the characteristic style of Noh play production should be added one or two words of explanation. The stylized pine tree at the back of the stage is a symbol that is closely associated with Noh drama in general and in particular with the popular Noh play *Takasago*, which deals with the spirits of the pine tree. As for the musicians, a full complement would be four, consisting of three drummers playing a *taiko* (a flat drum set in a wooden stand on the floor), an *otsuzumi* (an elongated drum held on the knee), and a *kotsuzumi* (an hour-glass-shaped drum held on the shoulder). In addition, there would be a player of the *fue* (flute).

One of the Noh plays that Britten and his party saw in Tokyo was *The Sumida River* (Sumidagawa); and in fact he was lucky enough to attend two performances of the same play in the course of a week or so. A detailed description of the action is given in the Prince of Hesse's travel diary:

'The ferryman is waiting in his boat, a traveller turns up and tells him about a woman who will soon be coming to the river. The woman is mad, she is looking for her lost child. Then she appears and the ferryman does not wish to take a mad person, but in the end he lets her into his boat. On the way across the river the two passengers sit behind each other on the floor as if in a narrow boat, while the ferryman stands behind them, symbolically punting with a light stick. The ferryman tells the story of a little boy who came this way a year ago this very day. The child was very tired for he had escaped from robbers who had held him. He crossed the river in this boat, but he died from exhaustion on the other side. The woman starts crying. It was her son. The ferryman is sorry for her and takes her to the child's grave. The mother is acted by a tall man in woman's clothing with a small woman's mask on his face. Accessories help you to understand what is going on: a bamboo branch in the hand indicates madness, a long stick is the ferryman's punting pole, a very small gong is beaten for the sorrowing at the graveside. As soon as these props are no longer necessary, stage-hands who have brought them to the actors take them away again. The sorrowful declamations of the mother rising and subsiding in that oddly pressed voice, the movement of her hand to the brim of her hat as if to protect her sadness from the outside world, the small "ping" of the little gong which she beats at the child's grave, become as absorbing as does the sudden foot-stamping which emphasizes important passages. The play ends in the chanting of the chorus.'

The memory of this extraordinary theatrical performance had never been far from Britten's mind in subsequent years, for he felt that in some ways it offered a new kind of operatic experience and wondered what could be learned from it. 'The solemn dedication and skill of the performers were a lesson to any singer or actor of any country and any language. Was it not possible to use just such a story—the simple one of a demented mother seeking her lost child—with an English background (for there was no question in any case of a pastiche from the ancient Japanese)? Surely the Medieval Religious Drama in England would have had a comparable setting—an all-male cast of ecclesiastics—a simple austere staging in a church—a very limited instrumental accompaniment—a moral story?'*

Accordingly he approached his friend William Plomer and invited him to adapt *Sumidagawa* as a parable for church performance. Plomer was an excellent choice as librettist, for he had spent several years in Japan and was fully conversant with the Noh drama and its conventions. His main job was to anglicize the action of *Sumidagawa*, which had been written by Motomasa in the first part of the fifteenth century; and in this adaptation the Sumida River became the Curlew River, the scene being changed from the province of Musashi to a church by a Fenland river. The medieval period remained. In the Noh play, the incidental music was the ancient Japanese music that has been jealously preserved by successive generations. For *Curlew River*, this was scrapped; but to preserve the medieval atmosphere Britten decided to start his opera with the plainsong hymn *Te lucis ante terminum*, and from it, as he says,† 'the whole piece may be said to have grown'.

The difficulty of this work of adaptation should not be underrated. It is not that the action is particularly complicated; but the text of *Sumidagawa*, like that of all Noh plays, is impregnated with Buddhistic teaching. 'In this the main theme is the transitoriness of human life, and at the same time is presented a view of all the pain and misery people may endure when they are not rendered superior to it by a recognition of the higher philosophy that teaches that the whole universe is a dream, from whose toils the freed spirit can escape.'‡ To translate this into Christian terms was no easy task; but Plomer showed great discretion in the way he solved the problem. Two examples may be of interest.

Shortly after the arrival of the Madwoman, the Chorus comments—

A thousand leagues may sunder
A mother and her child,
But that would not diminish
Her yearning for her child.

(Curlew River)

In *Sumidagawa*, the Chorus continues in the following Buddhistic vein—

---

* From a note by the composer printed on the jacket of the libretto of *Curlew River*. Faber, 1964.
† Ibid.
‡ From *Plays of Old Japan: The No* by Marie C. Stopes. London, 1912.

The nature of the bond is transient,
The bond is transient in this world, and yet
Parent and child are destined not to live
In loving union even this short while.
But, like the four birds in the fable old,
*Between the cruel separation lies.**

(Sumidagawa)

This passage is missing from *Curlew River.*

The second example comes from the climax of the play, when the spirit of the child is first heard within the tomb, singing with the chorus, and then is heard solo. In *Sumidagawa†* the passage runs as follows:

CHILD: *I adore thee, O Eternal Buddha.*
I adore thee, O Eternal Buddha.

CHORUS: *The voice is heard, and like a shadow too*
Within, can one a little form discern.
[The Spirit of the Child appears.]

MOTHER: *Is it my child?*

CHILD: *Ah! Mother! Is it you?*
[The Spirit disappears.]

Here is the parallel passage from *Curlew River*—

ALL (except MADWOMAN and SPIRIT):
Hear his voice!
See, there is his shape!
[The SPIRIT of the BOY appears in full view above the tomb.]

MADWOMAN: *Is it you, my child?*
[The SPIRIT circles slowly round the MADWOMAN, who appears transformed.]

SPIRIT (off): *Go your way in peace, mother.*
The dead shall rise again
And in that blessèd day
We shall meet in Heaven.
God be with you all.
God be with you, mother.

MOTHER and THE REST:
Amen.

SPIRIT: *Amen.*

Plomer added a double framework. At the beginning of the action, a party of monks and acolytes with their Abbot walk in procession to the acting area,

* Translated by Marie C. Stopes.   † Ibid.

Curlew River: Peter Pears as the Madwoman in the original 1964 production by the English Opera Group

where those monks who are to play the Madwoman, the Traveller and the Ferryman are ceremonially robed. At the end of the action, those who have played the three characters resume their monks' habits, and the Abbot leads the recession away from the acting area. (This device of procession and recession had already been used by Britten in *A Ceremony of Carols*.)

II

Plomer's adaptation of *Sumidagawa* kept sufficiently close to the original to postulate a considerable measure of stylization in the presentation of *Curlew*

River. The cast is all-male, the part of the Madwoman being sung by a tenor. As for the orchestra, Britten abandoned the characteristic layout of the English Opera Group chamber orchestra in favour of a new type of ensemble. From the traditional music associated with the Noh plays, he accepted the need for extended percussion and wrote parts for five small untuned drums, five small bells and one large tuned gong. He likewise accepted the flute as an essential solo instrument and added parts for horn, viola, double-bass, harp and chamber organ.

As is made clear in Colin Graham's notes attached to the rehearsal score, the production of *Curlew River* 'created a convention of movement and presentation of its own'. Graham goes on to suggest that 'movement and production details should be as spare and economical as possible . . . Every movement of the hand or tilt of the head should assume immense meaning and, although formalized, must be designed and executed with the utmost intensity.' The ritualistic nature of the actors' movements was of course intensified by the use of masks for the acting of the story.

The effect of this stylization was to isolate the players within their parts. Seeing this, Britten accepted the consequential implications—that the singers were as much a chamber music ensemble as the instruments and that a conductor was no longer a necessity.

The abolition of the conductor was a momentous step. It meant that each of the vocal parts must be given a measure of freedom that might include an element of rhapsody, and that some way would have to be found in the score of specifying which part (whether vocal or instrumental) had precedence at any particular moment, and, accordingly, who should lead and who should follow in the different episodes of this new musical democracy. The necessary time adjustments could be made with the help of certain rhapsodical effects which could be repeated *ad libitum*; and here Britten found it necessary to invent a new pause mark, which he called the 'curlew'.

According to Imogen Holst's introduction to the rehearsal score, 'the curlew sign over a note or rest shows that the performer must listen and wait till the other performers have reached the next barline, or meeting-point— i.e. the note or rest can be longer or shorter than its written value'. These aids to synchronization (reinforced by continuously cued parts for the players) enabled the singers and instrumentalists to 'conduct' the performance of the opera themselves; and this inevitably meant that, despite careful musical signs and directions, there were more variable factors affecting performances of the work than would be the case in a work relying on the services of a conductor.

Something of the flexibility of the musical style can be gauged from the following passage. The non-alignment of the Madwoman's part, the flute solo, and the harp ostinato shows (in the words of Imogen Holst) that 'their closely linked counterpoint has the freedom of independence':

Ex. 100a

The curlew theme just quoted is later built up into an ensemble by canonic imitation (Ex. 100b); and here the interrelation of the parts is only approximate. It will be seen that the elastic texture of such an ensemble in some ways resembles the 'chatter' recitatives in *Albert Herring* and *A Midsummer Night's Dream.*

Ex. 100b

Another aspect of this non-alignment is provided by the non-synchronized movements of different parts at the unison or the octave. For instance, the opening processional is a plainsong chant:

Ex. 101a

When a little later the monks are ceremonially robed for the play, the chant is taken over by the instruments in an altered guise. The harmonic implications of this asynchronous heterophony are important, for the notes of the melody begin literally to create the notes of the harmonies. Britten

Ex. 101b

accepted this; and if the score of *Curlew River* is compared with that of *Billy Budd* or *A Midsummer Night's Dream*, it will be seen that the new linear supremacy has led to the abandonment of most of the old triadic procedures with their polytonal implications. A simple example of this process is found in the Ferryman's opening music. Note how the horn enunciates the Ferryman's musical theme and how the viola and double-bass provide an accompaniment that usually follows, but occasionally anticipates the theme:

Ex. 102

The effect is sometimes that of underlining, sometimes that of shadowing, but always indicative of subservience. The result is invariably to exalt the musical line at the expense of independent harmony—heterophony in the place of polyphony.

Many of the harmonies in this score are note-clusters fed by prolonging the notes of the melodic line. This can be conveniently observed in the handling of the organ part in various parts of the score. A good example comes when the Spirit of the Boy addresses his mother:

Ex. 103

Beautiful though the curlew theme may be with its rising and falling lilt (Ex. 100), the Madwoman's most memorable moment is her initial cry (offstage) of '*You mock me! You ask me whither I go.*' The Madwoman appears

Ex. 104

on the stage, and the same strange cry becomes '*Let me in! Let me out! Tell me the way!*' The asperities of her agonized search are assuaged only when at the climax of the action her repetition of the curlew theme is ultimately absorbed in the Dorian mode of the plainsong chant *Custodes hominum*, and her thanksgiving of *Sanctae sit Triadi* is echoed by the Spirit of the Boy from inside the tomb—a true resolution to the curlew quest:

Ex. 105

(*N.B.* The remaining vocal & instrumental parts are omitted)

XIII
The Burning Fiery Furnace

Britten's second parable for church performance, *The Burning Fiery Furnace*, followed so closely on the first that it seemed natural for it to be made to more or less the same measure as *Curlew River*. The same convention was used whereby a group of monks led by their abbot, after entering the church in procession, proceed ceremonially to don the necessary costumes for the action, which is then presented in stylized fashion on an open circular-shaped stage, while the monks who are the instrumentalists form a conductorless chamber-music ensemble.

Once again Britten picked William Plomer as his librettist: but this time there was no question of turning a Noh play into a Christian parable. Instead, the two collaborators chose the Old Testament story of the three Jewish exiles (best known by the Babylonian names of Shadrach, Meshach, and Abednego) whom Nebuchadnezzar, at the request of Daniel, appointed as governors of the province of Babylon. Although these three Jews were prepared to accept office under Nebuchadnezzar, they refused to defile themselves with the eating and drinking customs of the country they were living in, or to worship the golden image that Nebuchadnezzar caused to be set up, thereby incurring his stern displeasure. As a punishment he commanded that the strongest soldiers in his army should bind the three of them in their coats, their hose, their hats, and their other garments and cast them into a furnace heated to seven times its usual temperature. This furnace was in fact so fiery that the task force of soldiers was instantly killed by the great heat; but the three Jews survived, praising, glorifying and blessing Jehovah in the furnace, where an Angel in the likeness of the Son of God joined them and protected them from the heat, so that when the king commanded the three of them to be brought out, it was found that they were completely unharmed, not even a hair of their heads having been singed. Nebuchadnezzar was so impressed by this miracle that he made a public proclamation affirming the power of the Jewish deity and restored Shadrach, Meshach, and Abednego to their former position as rulers of the province.

Whereas the action of *Curlew River* had been bisected by the episode of

242

The Burning Fiery Furnace: Nebuchadnezzar (Peter Pears) challenges the three young men in the original 1966 production by the English Opera Group

the actual crossing of the river with its swirling currents depicted by slow glissandi first on the strings and later on the harp, in *The Burning Fiery Furnace* the central point is provided by the sound of 'the cornet, flute, harp, sackbut, psaltery, dulcimer, and all kinds of music' which preludes the dedication of the image of gold set up by Nebuchadnezzar. This passage provided two cues which Britten followed up. In the first place, he added an alto trombone to the group of instruments he had already used in *Curlew River*; secondly, he used these instrumentalists to lead a procession of courtiers through the Church preparatory to the raising of Nebuchadnezzar's golden image—a bold and highly dramatic stroke.

The climax to the work is naturally provided by the miracle of the furnace and its non-destroying flames; but in order to provide a point of contrast, a divertissement is introduced in the early part of the opera, where in the course of Nebuchadnezzar's feast, two boy singers and a tumbler provide a light entertainment. The words of the acolytes are riddling:

The waters of Babylon,
The flowing waters,
All ran dry.
Do you know why?
. . . .
The reason the waters all ran dry
Was that somebody had monkey'd with the water supply;
The reason the gardens grew like mad
Was because of all the water they'd had.

The cocky little tune, with its flute and wooden blocks accompaniment, recalls Bottom's 'tongs and bones' music in *A Midsummer Night's Dream.*

How closely the framework follows the style of *Curlew River* can be seen from comparing the following quotations from the opening plainsong chant and the robing ceremonial with Exs. 101a and 101b above.

Ex. 106a

Ex. 106b

And how closely this plainsong is woven into the music of the three Jewish youths can be seen from the passage where they are left alone after the Astrologer has denounced them to Nebuchadnezzar:

Ex. 107

(*N.B.* The accompaniment is omitted)

After a herald has announced Nebuchadnezzar's decree that an image of gold shall be set up and worshipped, the instrumentalists 'warm up' their instruments for the forthcoming procession, while the three isolated Jewish youths continue to pray. These are the eight key phrases.

Ex. 108a

Ex. 108b

Ex. 108c

Ex. 108d

Ex. 108e

Ex. 108f

Ex. 108g

Ex. 108h

For the actual procession, each tune is lengthened to fill five bars of 4/4 tempo; and the following combinations are used for the different five-bar episodes: (i) horn and Babylonian drum, (ii) horn, trombone, and Babylonian drum, (iii) flute, trombone, viola, and Babylonian drum, (iv) glockenspiel, flute, viola, small cymbals and Babylonian drum, (v) glockenspiel, little harp, and small cymbals, (vi) little harp, horn, and Babylonian drum, (vii) horn, trombone, viola, and Babylonian drum, (viii) glockenspiel, little harp, flute, trombone, and small cymbals, (ix) little harp, flute, horn, viola, and Babylonian drum, (x) glockenspiel, little harp, flute, horn, trombone, viola, small cymbals and Babylonian drum. As can be seen from this scheme, each instrumental combination contains at least one instrument that has appeared in the immediately preceding combination, and towards the end the combinations become cumulative.

This processional march is one of the most original and effective strokes in the whole parable.

The D major/E flat minor polytonality of the march, with its ambivalent F sharp/G flat third, is prolonged by glissandi and tremolandi that greet the raising of the image of gold and the following chorus '*Merodak! Lord of Creation, we bow down before you*'. Melodically it reaches its apogee when the three young Jews are brought before Nebuchadnezzar, who accuses them of refusing to serve the god of gold. (Here the key has modulated to E major/F minor.)

Ex. 109

(*N.B.* The accompaniment is omitted)

The appearance of the Angel in the fiery furnace recalls the miracle of the appearance of the Spirit of the Boy in *Curlew River*, and the Angel's voice is used to prolong some of the phrases of the Song of the Three Children by augmentation and imitation.

Ex. 110

When the performance of the parable is over, the monks help the various characters to disrobe and resume their monks' habits. The Abbot reminds the congregation:

Friends, remember!
Gold is tried in the fire,
And the mettle of man
In the furnace of humiliation.

and then the plainsong chant, *Salus aeterna*, is used as a recessional.

The Burning Fiery Furnace marks a further step in the development of the technique of open-stage opera production.

XIV
The Prodigal Son

I

The libretto for the third of Britten's church operas, like those of the earlier two, was written by William Plomer; and this time the choice of subject fell on a New Testament parable. As Plomer himself wrote, on the jacket of the published libretto:* 'Of all the parables in the New Testament, none has had quite such a universal and ever-renewed appeal as that of the Prodigal Son . . . With its unforgettable climax of reward and rejoicing being lavished not upon virtuous correctness but upon a sinner, this parable celebrates the triumph of forgiveness. The story seems to bring into the clearest possible focus the Christian view of life.'

Once again the action is set within the convention of a group of monks, acolytes and lay brothers, who enter a church in procession and proceed to the acting area where the monks and acolytes are ceremonially robed for their parts, the lay brothers becoming the instrumentalists. The conductor-less orchestra of eight is almost identical with that of *The Burning Fiery Furnace*, the only exceptions being that the flute has been replaced by an alto flute, the trombone by a trumpet in D, and the percussion omits the Babylonian drum that added special colouring to the processional march in *The Burning Fiery Furnace*, but introduces a gourd that underlines the 'walking music' of the Younger Son's journey to and from the great City.

Whereas *Curlew River* and *The Burning Fiery Furnace* are both divided in half—the first by the crossing of the river, and the second by the procession through the church prior to the erection of the image of gold—in *The Prodigal Son* the action requires an A B A change of scene, for the Younger Son is shown leaving his pastoral home to journey to the City and returning home after the Parasites have stripped him of his substance. This is ingeniously accomplished by a kind of panoramic effect, the Younger Son trudging round the circular acting area, while his Father, Elder Brother and their servants recede from view, and later, after the Prodigal has succumbed

---

* Faber, 1968.

248

The Prodigal Son: Peter Pears as the Tempter in the original 1968 production by the English Opera Group

to the temptations of the city, the return journey taking place in reverse.

To those who were familiar with the conventions established by *Curlew River* and *The Burning Fiery Furnace*, *The Prodigal Son* brought an unexpected surprise. Whereas in the two earlier church parables the monks were headed by their abbot in the initial procession, in *The Prodigal Son* the abbot is absent—for the very good reason that he plays the part of the Tempter and makes a surprise entrance through the congregation as soon as the procession of monks has reached the acting area in the church.

This character is an invention of Plomer's and is used to make explicit the motivation of the action:

Ah—you people, listening here today,
Do not think I bid you kneel and pray.
I bring you no sermon,
What I bring you is evil.
. . .
Here you will see them!
Father, servants who obey

His least command,
Two sons, the elder stern,
The younger full of life,
He is the one I'll use
To break this harmony.
What perfect harmony!
See how I break it up.

The scene where the Tempter comes forward and confronts the Younger Son has disquieting undertones. The Younger Son is startled and asks by what right this stranger questions and examines him. The answer comes:

I am no stranger to you,
You know me very well,
I am your inner voice, your very self.

The Tempter proceeds to urge him to act out his secret desires, rather like Nick Shadow addressing Tom Rakewell in *The Rake's Progress*; and the analogy between Plomer's libretto for Britten and the libretto written by W. H. Auden and Chester Kallman for Stravinsky is strengthened when one reaches the Prodigal Son's three temptations in the city. This commentary by the Tempter (spoken recitative) might almost refer specifically to Mother Goose's brothel. '*Your senses have been freed by the pleasures of wine, but you have not yet begun to learn what pleasure means. Now you are offered the delights of the flesh, what you have been praying for. My boy, indulge yourself! Show yourself to be a man.*' But there is no special musical relationship between the two operas. Possibly the device of repetition and fragmentation when the Tempter decides to shatter the idyllic harmony of this Judaean country pastoral:

See how I break it up!
See how I break it up!
See how . . .
See how . . .

recalls the treatment of other similar passages from Britten's operas, such as the scene in *The Turn of the Screw* where Quint tempts Miles to steal the letter (see Ex. 77):

What has she written?
What has she written?
. . .
What does she know?
What does she know?
. . .

Easy to take,
Easy to take,
. . .
Take it!
Take it!

and the scene in *Billy Budd* where Vere suddenly becomes conscious of Claggart's villainy:

John Claggart, John Claggart,
Beware! beware!

The other characters follow closely the lines laid down by the parable in the Bible—the patriarchal Father, the Elder Son with his feeling that he has been unfairly treated, and the servants with their various pastoral pursuits.

II

Like the other two church parables, *The Prodigal Son* opens with a plainsong, *Jam lucis orto sidere*, whose *Amen!* is an ambivalent cue for the

Ex. 111a

Tempter's appearance:

Ex. 111b

It will be seen that the *Amens* of the plainsong as echoed by the Tempter diminish into a lively and insinuating tune with a strange enharmonic twist whereby the penultimate F natural becomes an E sharp. Mention of the country pastoral about to be disclosed pulls the tonality towards B flat; but before that key can be established, the robing ceremony takes place to the same sort of oriental dressing-up of the processional plainsong as was characteristic of *Curlew River* and *The Burning Fiery Furnace* (see Exs. 101b and 106b).

This leads immediately to the discovery of the Father seated under a tree, with his two sons and his servants gathered around him; and the pastoral tonality is established with a warm B flat major chord. The alto flute adds a mellow rhapsodic line to the Father's opening air '*The earth is the Lord's*':

Ex. 112a

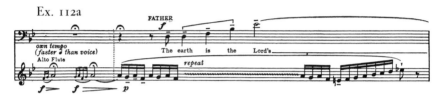

The four notes of the singer's initial phrase are used in different combinations to introduce or close most of his subsequent phrases. Twenty-four permutations of this four-note phrase are possible; and nine of them occur in the Father's air:

Ex. 112b

A little later the phrase is expanded by the Father when he agrees to hand over the Younger Son's portion:

Ex. 112c

And as the Younger Son is robed to indicate his assumption of his share of the inheritance, the four-note theme is harmonized by the harp (later the organ) in a way that recalls similar procedures in *Billy Budd* (see Ex. 62) and elsewhere:

Ex. 112d

When the Prodigal succumbs to the threefold temptations of the City, he is successively stripped of portions of his inheritance. The stripping of the first portion is accompanied by a variant of the chordal version (Ex. 112d above); the stripping of the second portion by the phrase reduced to its basic four notes, but in the remote key of E major;

Ex. 112e

the stripping of the last portion by a bare skeleton outline from the drums. When after repenting his sins the Prodigal decides to return home, his Father greets him with a moving phrase, where only two notes of the original four remain, the other two having been absorbed into the accompanying chord.

Ex. 112f

This is a cadence of true forgiveness.

One of the most exciting features of the score of *The Prodigal Son* is the relationship between the Tempter and the Younger Son, which starts with the scene where the Tempter persuades the Younger Son to act out his desires. At this point the independence of the two vocal parts is complete. (Note how the trumpet takes over the Tempter's earlier exhortation to the people in the audience—see Ex. 111b.)

Ex. 113

(*N.B.* The rest of the instrumental accompaniment is omitted)

On their journey to the City, however, the two voices become more closely related, though the Tempter's still has precedence. (Note how cleverly the Tempter leads the Younger Son into the temptation theme—see Exs. 111b and 113.)

Ex. 114

(*N.B.* The instrumental accompaniment is omitted)

The part of the Elder Son is less important than those of the Father, the Younger Son and the Tempter. Nevertheless he is vividly characterized in the score. His incredulity and lack of sympathy, varied with sudden outbursts of jealousy and impetuosity, are depicted by the use of wide leaping intervals in the vocal line. When he hears that his Father has decided to give his Younger Brother his portion, he breaks out in rising anger:

Ex. 115

(*N.B.* The instrumental accompaniment is omitted)

The members of the chorus have a more varied and more important part to play in *The Prodigal Son* than in either of Britten's earlier church operas.

254

In the first place, they are the servants in the household of the Father and his two sons. (There is no mention in the parable of a wife, or daughter or daughters.) In the City they are the Parasites who urge the Younger Son to excess in drinking, whoring and gambling, and also the Beggars who are starving because of the local famine. In the final scene they become the servants again.

The early choruses in the first part of the opera are composed on the lines of 'chatter' choruses ('*To the fields we go*', etc.) and this makes it easier for them to be faded out, as the servants move off to the fields, or to be faded in later as they return. ('*It is work that keeps*', etc.) In the City section, there are nine different choruses:

I   A quick and rumbustious 9 (2+2+2+3)/8 Parasites' Chorus of welcome. '*Welcome, welcome stranger!*'
II   Siren Voices (off-stage) offering wine.
III   A lively 6/8 Parasites' Drinking Chorus with brilliant whooping trumpet accompaniment, '*Come and try*'.
IV   Siren Voices (off-stage) offering '*nights of ecstasy*'.
V   A slow and heavy 6/8 Parasites' Chorus, '*Nights are days, days are nights.*'
VI   Siren Voices (off-stage) offering gold.
VII   A lively and rhythmic 3/4 Parasites' Gambling Chorus, with two wood blocks featured in the accompaniment, '*Never mind your gold is short*'.
VIII   Siren Voices (off-stage) asking for pity.
IX   Beggars' Chorus in 3/2 slow march time over a ground bass, '*We are starving*'.

This is quite a complex scheme considering that the scene in the City takes little more than a quarter of an hour to perform. When the Prodigal returns home, his Father summons the servants and orders them to bring robe, ring and shoes for his Younger Son to put on, to prepare the fatted calf for banquet, and to take part in a song and dance of welcome. They chant '*O sing unto the Lord a new song*', while dancing a round to the accompaniment of a heavy and sustained ground-bass consisting of an alternately rising and sinking scalic passage, the bars of which fall into an irregular 7+6 grouping:

Ex. 116

(*N.B.* The vocal parts for the Younger Son & Father, and the other instrumental parts, are omitted)

This dance is interrupted by the return of the Elder Son and his expostulation at his Father's extravagance in lavishing more money on his wastrel brother. But his Father reassures him and invites him (and all those present) to share his joy, '*for thy brother was dead, and is alive again, was lost, and is found*'. This is followed by the ceremonial disrobing ceremony, after which the Abbot leads the monks, acolytes and lay brothers in the recessional, *Jam lucis orto sidere* (Ex. 111a).

The Prodigal Son completes, not a trilogy, but a triptych of one-act operas for church performance, each lasting just over an hour and written to a common formula, though using different themes: a Japanese Noh play, an Old Testament story, a New Testament parable. In each case the initial plainsong chant grounds the opera firmly in the traditional context of the Christian Church, and from it grows and proliferates the music that accompanies the acted play—a convention within a convention. The instruments are closely related to the dramatis personae and give each opera its characteristic tone colour: flute in *Curlew River*, alto trombone in *The Burning Fiery Furnace*, trumpet and viola in *The Prodigal Son*. Singers and players form their own self-governing ensemble; and the lead passes from one voice to another, from one instrument to another, as action and music dictate. Despite the economy of means employed, these one-act operas exhibit great variety of musical texture and tension, and the impression they create is indeed full-scale.

XV
Owen Wingrave

For many years Britten had been chary of involving himself in television opera. He particularly objected to the fact that in an orthodox television studio, singers and orchestra had to be in different rooms. But at the beginning of 1969, John Culshaw, head of BBC television music programmes, peruaded him to conduct a BBC2 colour production of *Peter Grimes* by taking singers, supers, orchestra and production staff to the Snape Maltings where the concert hall was adapted to form a temporary television studio. The result was so satisfactory that the BBC decided to try to commission him to write a new opera specially for television. At first Britten hesitated. But in some ways he may have felt that the special restrictions inherent in the medium constituted an almost welcome challenge, and ultimately he consented. By the autumn of 1969 the new opera was well under way.

As subject he chose a short story by Henry James called *Owen Wingrave*. This had originally been published in the Christmas number of *The Graphic* for 1892 and reprinted in an American collection of James's tales entitled *The Wheel of Time* and in a British collection entitled *The Private Life*, both published in 1893. Britten had read it in the early 1950s or even earlier—certainly prior to his setting of *The Turn of the Screw*—and had found that its theme made a special appeal to his pacifist sensibilities, for it showed the viciousness and futility of the wargame as revealed by the history of a family of professional soldiers. 'Wingrave' is the family surname; 'Wingrave' is the story's epigraph—a rare instance in James's writings of a surname being manufactured to reveal a person's dominant characteristics, almost like the naming of humours in the old comedies of masks.

In the preface to the volume in the New York edition of Henry James's works entitled *The Altar of the Dead* which included *Owen Wingrave*, James wrote movingly of the moment of vision which led to the conception of this story 'one summer afternoon many years ago, on a penny chair and under a great tree in Kensington Gardens'. His imaginative process seems to have been awakened by the fact that while he sat there 'in the immense mild summer rustle and the ever so softened London hum a young man should have taken his place on another chair within [his] limit of contemplation, a tall quiet slim studious young man, of admirable type, and have settled to a book with immediate gravity'. James then speculates at what point the young man on the spot became the character in the story, but decides that such questions are answerless:

> My poor point is only that at the beginning of my session in the penny chair the seedless fable hadn't a claim to make or an excuse to give, and that, the very next thing, the pennyworth still partly unconsumed, it was fairly bristling with pretexts. 'Dramatise it, dramatise it!' would seem to have rung with sudden intensity in my ears. But dramatise what? The young man in the chair?

And so the idea of Owen Wingrave was conceived, the young scion of a military family, destined for an army career from birth, but now, just at the moment when he was being coached for Sandhurst, beginning to read for himself, think for himself, and rebel against a choice of career that had been imposed on him by his family, and which he now felt to run counter to his innermost personal convictions.

To get the right compositional focus in his story, James decided to put the professional coach, at whose establishment Wingrave and his friend young Lechmere are being 'crammed', into the centre of the action. Spencer Coyle is a kind and friendly person, through whose eyes and ears the reader follows the story; and in this role he is supported by his sympathetic wife. The characters at Paramore, the Wingraves' ancestral seat—old Sir Philip Wingrave (Owen's grandfather), Miss Jane Wingrave (Owen's aunt), and Mrs and Miss Julian (a widowed mother and her daughter Kate, who are friends of the family)—are seen as if from the outside by the Coyles who are invited to spend part of a weekend at Paramore. Owen takes the visitors round the house and shows them the family portraits. One of these—a double portrait depicting father and son—is the subject of a family legend. Apparently the father killed his son in a fit of anger, because he failed to take up another boy's challenge to fight. Shortly afterwards the father was himself discovered dead on the floor of the same room; and since then Paramore has been haunted by the apparitions of the old man and the boy. This tragedy now repeats itself. Kate Julian, who thinks she's in love with Owen, taunts him with cowardice; and he, to prove his courage, agrees to sleep in the haunted room. He does so and is discovered dead there the following morning.

The cry 'Dramatise it, dramatise it!' continued to ring in James's ears even after *Owen Wingrave* had been completed and published as a story; and in 1907 he decided to turn it into a one-act play, which he called *The Saloon*,

Owen Wingrave: The Paramore set by David Myerscough Jones at Snape Maltings

using the word 'salon' in its old form. When it was finished, he was persuaded to submit it to the Incorporated Stage Society; but the Society rejected it at a meeting on 12 January 1909, when Bernard Shaw agreed to write to James about it. He did so, and his letter sparked off a lively correspondence between the two writers* Shaw deplored James's use of the supernatural, and James put up a characteristically subtle defence. Shaw's argument was summarized in a letter he wrote on 21 January:

> My dear Henry James,
>   You cannot evade me thus. The question whether the man is to get the better of the ghost or the ghost of the man is not an artistic question; you can give victory to one side just as artistically as to the other. And your interest in life is just the very reverse of a good reason for condemning your hero to death. You have given victory to death and obsolescence. I want to give it to life and regeneration. . . .

James's reply was comparatively succinct (for him!):

> There was only one question to me, that is, that of my hero's within my narrow compass, and on the lines of my very difficult scheme of compression and concentration, getting the *best of everything*, simply;

* This correspondence is printed in full in Leon Edel's Foreword to *The Complete Plays of Henry James*. Rupert Hart-Davis, 1949.

Owen Wingrave: Miss Wingrave (Sylvia Fisher), Sir Philip (Peter Pears), Kate (Janet Baker) and Mrs Julian (Jennifer Vyvyan) in the original 1970 BBC television production

which his death makes him do by, in the first place, purging the house of the beastly legend, and in the second place by his creating for us, spectators and admirers, such an intensity of impression and emotion about him as must promote his romantic glory and edifying example for ever. I don't know what you could have more. He wins the victory—that is he clears the air, and he pays with his life. The whole point of the little piece is that he, while protesting against the tradition of his 'race', proceeds and pays exactly like the soldier that he declares he'll never be.

In the event James did not rewrite *The Saloon* as Shaw had urged him to; and it was ultimately produced in London on 17 January 1911 as a curtain-raiser to Cicely Hamilton's *Just to Get Married*. From contemporary notices it appears that the final scene was played (against the author's advice) in far too melodramatic a style, and the play was not a success.

As had been the case with *The Turn of the Screw*, Britten invited Myfanwy Piper to be his librettist. Looking at the available material, she decided to ignore *The Saloon* as being not to her purpose, and concentrated on the story. From the outset she felt herself free to fill out some of James's half-suggestions and to invent new business, particularly when such

260

treatment seemed likely to suit the medium of television. In her hands the action fell into two acts and was presented as follows:

ACT I

Prelude A series of portraits of Owen's military ancestors
Scene 1: The study at Coyle's military establishment (Owen, Coyle, Lechmere)
Interlude I Regimental banners wave brilliantly
Scene 2: Owen in Hyde Park, alone, sitting reading—crosscut with Miss Wingrave and Coyle in her lodgings in Baker Street
Interlude II A sequence of old faded tattered flags
Scene 3: A room at the Coyles' establishment (Mr and Mrs Coyle, Lechmere, Owen)
Interlude III Paramore
Scene 4: Mrs Julian, Kate Julian, Miss Wingrave, and Sir Philip Wingrave receive Owen with deep disfavour at Paramore
Scene 5: A week passes during which Owen is under constant attack from his family
Scene 6: The hall at Paramore. The Coyles and Lechmere have arrived as guests and are received by Owen
Interlude IV The Preparation of Dinner
Scene 7: The serving of dinner (full cast)

ACT II

Prologue Ballad (sung by a Narrator)
Scene 1: The hall at Paramore one evening, and the gallery outside the haunted room
Scene 2: The Coyles' bedroom later that night; and subsequently the gallery outside the haunted room

It looks as if these two acts were planned to serve somewhat different purposes. Act I with its seven different scenes deals primarily with war—the implications of training for war, action in the field, death in battle, glory after death—and offers a background for Owen's violent reaction against his family and its military tradition, whereas Act II is virtually one continuous scene, which presents the legend of the ghosts in ballad form (prologue):

There was a boy, a Wingrave born,
A Wingrave born to kill his foe.
Far away on sea and land,
The Wingraves were a fighting band.
Trumpet blow, trumpet blow,
Paramore shall welcome woe.

This act moves swiftly to its climax where Owen, taunted with cowardice by

Kate Julian, elects to be locked up in the haunted room for the night* and is found dead the next morning.

The difference between the acts is reflected also in a slight shift of focus where Coyle is concerned. At the beginning of Act I, particularly in scenes i, ii and iii, he is shown in his professional capacity as a coach, with Owen and Lechmere as his pupils. From Act I scene vi to the end of Act II he appears in a different capacity as a sympathetic friend of Owen's. The two roles are, of course, closely related: but Coyle's position in the compositional centre of the story means that when he is absent (as is the case in scenes iv and v of Act I) the remaining characters have a tendency to verge on caricature.

The greater part of the libretto is written in a kind of heightened prose, broken into irregular line lengths; the Ballad of the Wingrave Boy makes a good attempt to reproduce true ballad idiom; and some fragments of poetry are specially featured in the second Interlude of Act I when Owen is discovered reading in the Park, not (as James suggested) from a book of Goethe, but (as Myfanwy Piper has decided) from Shelley's *Queen Mab*. It is interesting to find that whereas most of Myfanwy Piper's heightened prose has been set by Britten as air, arioso, or recitative, the Shelley quotations are merely declaimed by Owen at fixed pitches but without exact indication of note lengths or rhythm.

II

Though designed for the intimacy of the television screen, *Owen Wingrave* is not a chamber opera like *The Turn of the Screw*—the score is written for about the same-sized orchestra as Britten's Symphony for Cello and Orchestra. Owen is a baritone; Coyle a bass-baritone; and Lechmere and Sir Philip Wingrave tenors. Three of the women (Miss Wingrave, Mrs Julian, and Mrs Coyle) are sopranos; and Kate Julian is a mezzo-soprano. The ballad is sung by a narrator (tenor). This casting is significant. The choice of middle range takes Owen out of the heroic tenor category with its bravura implications; and it may be recalled that in *Billy Budd* a similar problem was solved in a similar way, Billy being cast as a baritone, with Vere as tenor and Claggart as bass.

The opera opens with an instrumental prelude, which immediately establishes two factors—that the action is concerned with military matters and that for generations past the Wingrave family has been dedicated to the pursuit of war. In the first three bars, the twelve notes of the chromatic scale

* At this point Myfanwy Piper has made an important alteration to James's story. According to James, when Kate taunted Owen and dared him to sleep in the haunted room, he told her he'd already spent the previous night there, to which she retorted 'whether he'd care if he should know she believed him to be trying to deceive them', and so at his express request he was taken by Kate to spend a second night there, and this time she locks him in. This over-subtlety is (rightly) rejected in the opera, where Owen tries only once to sleep in the haunted room—with fatal results.

are split up into three chords forming a martial percussive theme that recalls the sharp sizzling clashes of swords and scimitars:

Ex. 117

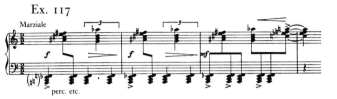

This is immediately followed by a series of eleven musical portraits of Owen's military ancestors as they hang on the gallery and stairs of the family mansion at Paramore. Each portrait is a musical cadenza, usually for a solo instrument; and as each portrait is introduced a fresh note is added to a held chord so that with the eleventh portrait (which is Owen's dead father) the chord contains eleven different notes of the chromatic scale. It is then withdrawn, and the missing twelfth note (D natural) becomes the bass for Owen's personal appearance after his father's portrait has been presented. The structure of this scheme can be gathered from the following example, which shows, first, the eleven-note chord underpinning the portrait of Owen's father (entrusted to woodwind and brass) followed by the martial percussive theme, leading to Owen's actual appearance, where a horn solo is supported by a low D pedal. It should be noted that many of the intervals in Owen's theme are thirds, particularly minor thirds, and where two minor thirds follow each other they imply the existence of a diminished triad. In fact, the solo themes depicting the Wingrave ancestors are riddled with minor thirds, and the military atmosphere seems favourable to the diminished triad. Musically, the diminished triad might be thought to be hesitant, ambivalent, unable to make up its mind: but in this particular context it seems to stand for a constitutional inability to face the ethical and philosophical implications of a system geared to war and aggression as a normal way of life:

Ex. 118

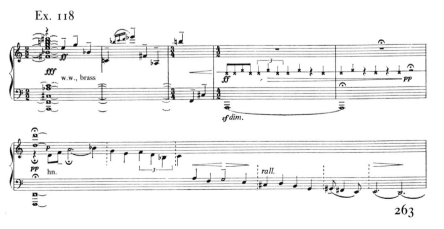

Scene i of the opera follows without pause; and here it will be seen that the notes of the diminished triad occupy an important position in the accompaniment to Coyle's lecture on military tactics:

Ex. 119

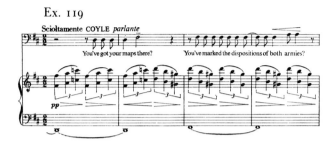

It should be noted that the deployment of the twelve notes of the chromatic scale in Examples 117 and 118 leads to no special serial development. As far as Owen is concerned, the score illustrates his continuing preoccupation to establish himself as a man of peace on a broad diatonic basis; and in this he resembles Billy Budd. To the sensitive listener it is clear that in *Owen Wingrave* Britten has tapped areas of deep feeling that were common also to *Billy Budd*. But whereas in *Budd*, owing to the special limitations of that character, these feelings were not always articulate— indeed, Billy's stammer was an outward indication of his occasional inability to express himself, particularly under stress—in *Owen Wingrave* librettist and composer succeeded in exploring Owen's thoughts and motivation in depth. This is particularly true of Owen's affirmation of belief—'In peace I have found my image, I have found myself'—which comes halfway though Act II. This is close in tonality, feeling and certain aspects of compositional treatment to Billy's moving 'farewell to ye, old Rights o'Man!' which followed immediately on his Ballad in the Darbies. Here is the opening of Owen's air:

Ex. 120

Donald Mitchell has made a most perceptive comparison between the two men's songs:*

> If one compares the actual sound of [these two airs], there is one subtle and significant aspect of the sonority of the 'peace' aria in *Wingrave* that immediately strikes one. Whereas the chords in the *Budd* excerpt are dropped like great anchors into and against the quaver flow of piccolo and harp, in *Wingrave* it is the percussion orchestra which sustains the quaver figuration into which the chords of affirmation are injected. Up to this point in the opera, the percussion has been associated, through the opening pulsation, with the idea of war and violence, with the mailed fists of the Wingraves. Now, however, the percussion orchestra (*plus* vibraphone, *minus* drums) functions in quite a different role, as the very opposite of aggression: as a shimmering radiance, no less, surrounding and decorating the chordal affirmations of Owen's resolve.

The analogy between Owen and Billy can be pursued further still. In *Budd*, after the drumhead court has been held, it is left to Vere to communicate the court's findings to Billy, and he does so behind a closed door in a room off-stage, while the stage itself is left empty, and the unheard communication is accompanied by a slow procession of common chords, instrumented for different sections of the orchestra, or (exceptionally) for full orchestra. In the second act of *Wingrave*, Sir Philip Wingrave summons Owen to an interview. The talk that leads to his disinheritance takes place behind a closed door, while Miss Wingrave, Mrs and Miss Julian, and Mr and Mrs Coyle wait outside in the hall. But this interview is more explicit than the one in *Budd*, since during it Sir Philip's voice is heard from the room off-stage where the interview is taking place, though his words are only half audible, and Owen's part in the discussion is taken over by a horn solo played behind the scenes.

The ballad that forms the Prologue to Act II plays the same sort of role in

* From 'Owen Wingrave and the Sense of the Past', printed in the sleeve note to the recording issued by Decca in 1971.

Ex. 121

the second half of this opera as that of the plainsong chants that frame Britten's three Church Operas. Particularly skilful is the way it is altered to form an eerie pianissimo ostinato that ticks its way through the penultimate scenes of Act II, when Owen has been locked in the haunted room and the Coyles are passing a sleepless night in the guest chamber. And

Ex. 122

even here one is still pursued by the *Billy Budd* analogy when one recalls the prominent part played in that opera by the motif of a rising fifth followed by a rising semitone (see Exs. 55–61). The climax comes when three notes of the theme coalesce to form the chord that accompanies Kate's cry as she finds Owen dead on the floor of the haunted room.

Although the work is fully viable in the opera-house, it must be remembered that it was originally written for television. This has led in places to a compressed style of presentation—bridge passages were shortened, transitions speeded up, scenes mixed. For instance, the scene where Owen is shown sitting alone in the Park, reading from a book that turns out to be a collection of Shelley, is cross-cut with a scene showing Coyle interviewing Miss Wingrave in her London lodging; and at one point all three characters share a common vision of the Horse Guards* trotting by, with what Owen calls 'that rippling, sweet, obedient well-being'. This is a passage that proves to be remarkably successful in television terms. So too in the original BBC 2 production were the colour effects of some of the interludes, particularly the contrast between the clean, brilliant colours of the regimental banners in Interlude I and the old tattered flags with faded colours like smouldering jewels that accompanied Owen's reading from *Queen Mab* (Interlude II). The television production scored a good point too for the way it used what appeared to be an old-fashioned silent film in monochrome as visual accompaniment to the Narrator's Ballad. The intimacy imposed by the small screen was particularly successful in the dinner party scene at the end of Act I. Not only could the cameras give a general view of the dinner party in progress, but they could also focus on each character individually as occasion arose and, if necessary, allow them to reveal their unspoken thoughts.

The television production was seen by millions of viewers when it was broadcast by BBC 2 on 16 May 1971. Two years later the first stage performance was given by the Royal Opera at the Royal Opera House, Covent Garden, on 10 May 1973. There, during its first season, *Owen Wingrave* was seen by audiences numbering several thousands. In future it is to be hoped there will be occasions when the original television production with its admirable cast can be revived. It will always be interesting to compare it with whatever happens to be the most recent 'live' production in an opera-house.

---

* Actually the officers and men of the Royal Military Police Mounted Troop (Aldershot) were filmed for these scenes.

XVI
Death in Venice

I

Britten's close friendship with Peter Pears had led in the course of time to the creation of much music which had been written with the special qualities of his friend's tenor voice in mind. This was true of all the operas after *Paul Bunyan*. The title roles in *Peter Grimes* and *Albert Herring* were specially written for Pears; so too was that of the Earl of Essex in *Gloriana*. In *Billy Budd* Pears was cast as Captain Vere; and this undoubtedly influenced Britten's decision to make the eponymous hero a baritone. Pears gave powerful character interpretations as the Male Chorus in *The Rape of Lucretia*, Quint in *The Turn of the Screw*, the Madwoman in *Curlew River*, Nebuchadnezzar in *The Burning Fiery Furnace*, and the Tempter in *The Prodigal Son*. His only minor character parts had been as Flute in *A Midsummer Night's Dream* and Sir Philip Wingrave in *Owen Wingrave*. Quite an extensive roll-call! When the latter opera was completed, Britten decided to choose for his next opera a subject which would give Pears a major part to play, by way of celebrating a partnership that had lasted for nearly a third of a century.

In the past, when searching for suitable opera subjects, he had occasionally gone to prose texts for his material; and so it was in this case. In Thomas Mann's novella *Der Tod in Venedig*, he found what he was looking for—a particularly rich part for the chief character, Gustav von Aschenbach. In this story Aschenbach appears in a double capacity—as a successful author, or artist, who is presented as if he were a projection of the actual author of *Der Tod in Venedig*, of Mann himself; and also as the chief actor in this story of corruption and death. The elderly man becomes obsessed with Tadzio, a Polish boy of thirteen. Tadzio's part is a silent one in the sense that although he obviously chatters like any normal boy to his boy friends and the members of his family, who are staying at the same hotel on the Lido as Aschenbach, he never says anything to Aschenbach, nor is he reported as saying anything to anyone else who is featured in the story. A dumb part— but not one that was likely to deter Britten, as would be obvious to those who

268

remembered the silent but vital contribution made by Grimes's apprentice boy to the two scenes in the second act of *Peter Grimes*. From the outset it was obvious that if *Der Tod in Venedig* was to be turned into an opera, its success would stand or fall by the composer's ability to provide a convincing part for Aschenbach (who would have to be on the stage for virtually the whole of the action), and by the singer's ability to give an outstanding performance.

Britten had worked before with Myfanwy Piper, who had supplied the librettos for two of his operas based on prose texts by Henry James—*The Turn of the Screw* and *Owen Wingrave*. He now asked her to give Thomas Mann's novella similar treatment. She condensed the action into seventeen scenes; and the composer added an impressionist overture entitled 'Venice' between scenes ii and iii. The opera was to contain about 145 minutes of music: so it was clear that the average length of a scene would be about eight minutes, and the composer would have to arrange his score so that the music moved with swiftness and ease, and the scenes dissolved and flowed into each other without interruption. Indeed, at certain points it might be necessary for the composer to use a kind of musical shorthand in order to maintain the speed necessary to tackle a subject that featured so many complex issues—psychological, philosophical, and mythical.

Aschenbach, one is told, is a successful elderly writer, who since the death of his wife and the marriage of his only daughter has become increasingly insecure and solitary. In the novella Mann described him with so much sympathy and understanding that at times an autobiographical element seemed to creep in. In the opera, as Myfanwy Piper pointed out, there is no narrator, but 'essential information and comment is given by Aschenbach himself in prose (recitative accompanied by the piano), as distinct from the shaped recitative and the songs and lyrical passages.'* Indeed, *Death in Venice* restores recitative—whether dry, accompanied, or near-arioso—to the supreme position it used to hold in the early years of Italian opera, particularly in Venice at the beginning of the seventeenth century. The dry recitative is especially interesting. Here Britten follows a technique he had tried out in the second interlude of *Owen Wingrave* where the pitches are exactly notated, but not the time values of the notes, the implication being that the words are to be given the loose conversational rhythm imposed by the ordinary speaking voice. There are ten such passages in *Death in Venice*. The voice is usually accompanied by the piano, and Aschenbach marks these recitatives by taking from his pocket a small notebook 'the symbol of his novelist's trade'.

In Mann's novella, Aschenbach meets a number of characters—the

---

* From the programme note by Myfanwy Piper for the original English Opera Group performances. Subsequent Myfanwy Piper quotations come from the same source.

Traveller, the Elderly Fop, the old Gondolier, and the Leader of the Strolling Players—who guide him on his journey to what proves to be his death. In her programme note, Myfanwy Piper says:

> Nowhere does Mann suggest that they have supernatural powers, or that they are one and the same person, but he links them by endowing each with the snub nose and grin of death, and the broad-brimmed hat and staff of Hermes, conductor of the dead across the Styx. For dramatic and musical reasons they are all sung by the same performer, and we have extended Mann's list to include the Hotel Manager, the Barber, and the god Dionysus, whose voice is heard by Aschenbach in a dream.

A remarkable feat of doubling!

A major problem was posed by Tadzio, the Polish boy. His beauty catches Aschenbach's eye, and the writer becomes obsessed with this image of perfection; but he fails to establish any sort of contact with the boy, except the occasionally intercepted glance, and remains solely an onlooker and unsuccessful pursuer. Myfanwy Piper comments:

> In the book he has no verbal contact with Tadzio, or his family and friends: nor does he in the opera, and we have emphasized this separateness by formalizing their movements into dance.

Although the seventeen scenes of the opera form a single stream of narrative, for theatrical purposes it was necessary to divide them into two acts. The caesura comes after scene vii, which is entitled 'The Feasts of the Sun'. Watching Tadzio and his boy friends playing on the beach, Aschenbach allows his fancy to stray to ancient Greece:

> He hears the voice of Apollo, turns the children's games into myths and the beach into Socratic Greece, with Tadzio as the olive-crowned victor of the boys' pentathlon. . . . Tadzio smiles at him and Aschenbach realizes that what he feels is love.

This is the climax and end of Act I.

Shortly before the end of the novella, Mann gives a lurid account of a dream Aschenbach suffers one night, an orgy of frenzy associated with the arrival of 'the stranger god', who is presumably Dionysus. In the opera, this nightmare is made more explicit by introducing Apollo in opposition to Dionysus. Mann's exaggerated Bosch-like imagery of frantic females shrieking and clutching snakes, and of horned and hairy males beating on brazen vessels and drums is not followed up in the opera; but his graphic description of the mad rout yelling a cry 'with a long-drawn U-sound at the end' provides the cue for a dionysiac chorus in which the singers vocalize on 'Aa-oo' sounds. The same vowels are echoed in the cries of '*Tadziù!*' called

Death in Venice: The Games of Apollo with Robert Huguenin as Tadzio in the original 1973 production by the English Opera Group

by his friends on the Lido beach, and the gondoliers' cry of '*Aou*'*!*' on the lagoon.

Death in Venice is like a long operatic narration. The following summary of the action is combined with a commentary on the music.

II

Scene 1: A Cemetery in Munich
The opera opens with an accompanied recitative for Aschenbach, which is based on the twelve notes of the chromatic scale. But although the blockage

Ex. 123

of which he sings ('*no words come*') is fundamental to the subsequent action he takes, no attempt is made by the composer to erect this musical series into a compositional system. What is musically important, however, is the emphasis on the minor third—in the above example contoured as F natural to G sharp, G sharp to B natural, B natural to D natural, and C natural to E flat. This recalls Britten's earlier preoccupation with the intervals of the diminished triad in *Owen Wingrave* (see particularly Exs. 118 and 119). An unknown traveller appears on the steps of a mortuary chapel and sings about the marvels of exotic climes, urging Aschenbach to travel to the south. For an instant the two voices combine. Then the traveller disappears; and Aschenbach makes up his mind. A fortissimo major third (F sharp and A sharp) issues from the trombones like a ship's siren—and he is on his way to Venice.

Scene 2: On the Boat to Venice

The boat sets out from Pola; and the beat of the engines is emphasized by a rhythm played on the side drum with brushes. A group of youths on board shout to the girls on shore, and out of this exchange of ribaldries arises the two-part chant of their destination:

Ex. 124

An elderly fop starts a popular song, '*We'll meet in the Piazza*', which is taken up by the youths on board. After the boat has arrived in Venice and the passengers have disembarked, the overture follows.

Overture: Venice

This is the most extended instrumental number in the score. Its initial theme is a kind of barcarolle with a lazily lapping and overlapping motif of ripples (threes against fours). This is based on the '*Serenissima*' phrase and is capable of considerable variation. Out of this barcarolle rises

Ex. 125

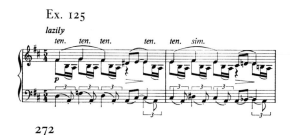

an open-air brass fanfare, the same tune being presented at four different pitches, subject to slight rhythmic changes:

Ex. 126

This Gabrieli-like fanfare alternates with a brazen clangour of bells. And by these simple musical means, the composer's impressionist evocation of Venice with her towers, domes, and palaces rising from the waters of the lagoon is complete.

Scene 3: The Journey to the Lido
Aschenbach takes a gondola from Venice to the Lido, and is rowed by an old Gondolier, who gruffly asserts his independent spirit, somewhat to Aschenbach's discomforture. On the way over, they pass a boatload of boys and girls singing an unaccompanied song '*Bride of the sea . . .*' with a choral refrain of '*Serenissima . . .*' (founded on Ex. 124) which helps to establish a special dimension of distanced sound on the lagoon. On arrival at the Lido quayside, Aschenbach is greeted by the hotel staff, while the old Gondolier is found to have disappeared without being paid. This prompts Aschenbach to a sombre soliloquy—'*Mysterious gondola . . . How black a gondola is— black, coffin black, a vision of death itself and the last silent voyage.*'

Scene 4: The first evening at the hotel
On his arrival at the hotel Aschenbach is welcomed by the Manager and shown to his room—a room with a specially fine view:

Ex. 127

The wide sweep of this panorama recalls the moment in Bartók's *Bluebeard's Castle* when Judith flings open the fifth door to reveal a vast, magnificent landscape. But here it is a lagoonscape instead of a landscape that meets Aschenbach's gaze. The musical motif with its dropping thirds under-pinned by a sequence of rising triads is one of Britten's most characteristic

273

Death in Venice: Peter Pears as Gustav von Aschenbach in the original 1973 production by the English Opera Group

hallmarks. The hotel guests now assemble for dinner. There is a buzz of polyglot conversation, like one of the chatter-choruses in *Albert Herring*; and Aschenbach sees Tadzio for the first time—'*a beautiful young creature, the boy*'—with the other members of his Polish family. This is the cue for a Lydian melody played by the vibraphone, which henceforward is closely associated with Tadzio:

Ex. 128

Scene 5: On the Beach

Aschenbach is on the beach, with various other hotel guests. A strawberry seller passes by. In view of the comparative rarity of female characters in this opera, the high tessitura of her soprano cry is particularly appealing, though the accompaniment is somewhat equivocal:

Ex. 129

Tadzio comes out on to the beach and sits with his family. At first Aschenbach does not know his name, but when his friends call to him to join in their games, they cry '*Adziù!*' and this is echoed '*Ah-oo!*' From

Ex. 130

275

this Aschenbach deduces that the boy's proper name is Tadzio, short for Thaddeus. Tadzio and his friends play games on the beach, which are characterized by the sounds of xylophone, marimba, and glockenspiel. Aschenbach watches and listens, entranced.

Scene 6: The Foiled Departure
Aschenbach crosses to Venice from the Lido (variant of Ex. 125), and the Gondolier's strange cry '*Aou*'!' emphasizes the melancholy of the lagoon. The sirocco is blowing, and Venice is hot and uncomfortably crowded. Aschenbach is depressed and decides to leave. He returns to the Lido by gondola, accompanied as before by a variant of Ex. 125 and the gondoliers' cries of '*Aou*'!' but his departure is foiled by a stupid mistake over the registration of his luggage at the station, and presently he finds himself returning to the Lido, to the hotel he had so recently left. The Manager welcomes him back and shows him into the room he had vacated a few hours previously. The view is virtually unchanged (a slight variant of Ex. 127). Aschenbach sees Tadzio and some of his friends playing on the beach. His spirits revive and he realizes what it was that had made it so hard for him to leave. This leads to a moment of resignation and acceptance that recalls similar moments in other operas by Britten, such as the '*Goodnight!*' scene in *Owen Wingrave*. This particular cadence contains a feeling of impending doom, despite its resolution each time on a major triad:

Ex. 131

Scene 7: The Feasts of the Sun (The Games of Apollo)
This scene allows the framework of the opera to be extended so as to project

the boys' beach games into the antique Olympian world and to provide an important ballet diversion. The greater part of the music consists of a succession of four-part choral hymns, but at several points the solo voice of Apollo (counter-tenor) is heard off-stage, announcing that '*He who loves beauty worships me*'. The chorus elevates Tadzio so that he is equated with Phoebus, and he is given a formal dance solo. This is followed by the feasts of the sun. The boys compete in a variety of sports—running, long jump, discus throwing, javelin throwing, and wrestling—and in each of these Tadzio is the winner. Aschenbach is so excited by this display of skill that he feels he must speak to Tadzio to congratulate him, but as the boy passes him, he fails to do so. Tadzio smiles at him: and when it is too late to respond, Aschenbach bursts out, '*Ah! don't smile like that! No one should be smiled at like that.*' An eight-part upward surge in the orchestra, passing from pianissimo to fortissimo, helps Aschenbach to realize exactly what has happened to him, and he sings the words 'I love you' to an empty stage just before the curtain falls.

The music of the second act starts exactly where the music of the first act left off—a device that Britten had already used in the original version of *Billy Budd*. An extensive passage of dry recitative for Aschenbach, '*So, it has come to this*', leads to—

Scene 8: The Hotel Barber's Shop (i)
Aschenbach is having his hair trimmed. The snipping instrumental accompaniment to the Barber's chatter role paints as vivid a picture as some of the character vignettes to be found in an opera like *Albert Herring*.

Scene 9: The Pursuit
Aschenbach crosses to Venice (variant of Ex. 125). There are posters in the streets advising citizens to take precautions against infection. The foreign newspapers feature rumours of an outbreak of cholera in Venice. The Polish family turns up, and Aschenbach starts to trail them. The pursuit is accompanied by a restless bass *ostinato*. They all reach the Piazza

Ex. 132

and a café orchestra starts to play, led by a violinist who shows off his double-stopping. The Polish family moves on; and Aschenbach follows them into the Cathedral of San Marco, where the choir is singing a *Kyrie Eleison* punctuated by the pealing of bells. This pursuit leads to the point where the Polish family takes to a gondola, followed by Aschenbach also in a

Britten's composition sketch for the end of Act I and the beginning of Act II of *Death in Venice*

gondola, while the two gondoliers' cries of '*Aou*'*!*' attract the cry of a third gondolier in the distance.

Scene 10: The Strolling Players
Back in the hotel after dinner, Aschenbach attends a light entertainment given by a small company of strolling players on the terrace, consisting of:
    I. Waltz (duet for a boy and a girl) '*O mio carino*'
    II. Popular Song (the Leader) '*La mia nonna*'
    III. Laughing Song (Leader and chorus) '*Fiorir rose in mezzo al giasso*'

Scene 11: The Travel Bureau
Aschenbach visits a travel bureau, which is being besieged by a number of tourists anxious to leave Venice because of the threat of plague. When they have gone, the clerk speaks to Aschenbach in an open, uninhibited way. His long prose speech, '*In these last years Asiatic cholera has spread from the delta of the Ganges . . .*', set as accompanied recitative, is one of the most moving numbers in the score. Its climax unequivocally reveals the presence of the plague:

Ex. 133

Scene 12: The Lady of the Pearls (*i.e.* Tadzio's mother)
Aschenbach decides to warn Tadzio's mother of the danger she runs in allowing herself and her family to linger on at the Lido while the plague is tightening its grip, and to urge her to depart. But when an opportunity occurs, he cannot bring himself to speak.

Scene 13: The Dream
In his despair he falls asleep. The voices of Dionysus ('*Receive the stranger*

god') and Apollo ('*No! reject the abyss!*') are heard epitomizing the struggle in his mind. The followers of Dionysus appear, singing their orgiastic cry of '*Ah-oo!*' and this leads to a dionysiac dance, one of the few *allegro molto* passages in the score, based mainly on the Tadzio theme (see Ex. 128).

Scene 14: The empty beach
Aschenbach, shocked by the revelations of his recent nightmare, goes down to the beach and watches Tadzio and his friends, who are playing about in a desultory fashion.

Scene 15: The Hotel Barber's Shop (ii)
To the same sort of accompaniment as before, Aschenbach sits in the Barber's chair to have his greying hair and moustache tinted, and his ageing features made up.

Scene 16: The Last Visit to Venice
In the first flush of his rejuvenation, Aschenbach sings an altered version of the song '*We'll meet in the Piazza*', which was sung by the youths on the boat bound for Venice in scene ii. His pursuit of the Polish family is resumed as in scene ix; and in the excitement of the chase the bass ostinato (Ex. 133) is inverted and treated fugally. At the climax, Tadzio detaches himself from his family and waits for Aschenbach to pass, deliberately looking him full in the face as he does so. But Aschenbach turns away; the Polish family disappears from view; and the strawberry seller reappears with her characteristic cry (Ex. 129). Aschenbach buys some fruit, but finds it musty to his taste. He sits down wearily by a well-head and is appalled by the chaos and sickness that seem to have overtaken him. He apostrophizes himself bitterly; and this leads to the most moving and most extended of all the recitative passages in the opera. In his dilemma he recalls the wisdom of Socrates and the argument in the *Phaedrus* about beauty and its ability to mediate between sensuous man and the world of the spirit—'*Does beauty lead to wisdom, Phaedrus?*' The musical setting recalls the beautiful *Hölderlin Fragments* for tenor and piano that Britten composed in 1958, and which Pears and himself performed so memorably at many of their recitals. Here Aschenbach is accompanied by flourishes from harp and piano, giving

Ex. 134

the singer the simplest cues; and the final notes of each cadence are prolonged by string harmonics. At the end of this deep, tranquil meditation there is a vigorous fugato passage for brass and woodwind based mainly on the landscape theme (Ex. 127)—strings are added at the climax. This is one of the most vigorous musical statements in the score.

Scene 17: The Departure
The Polish family decides to leave and Aschenbach sees their baggage waiting in the hall. He goes out on to the deserted beach, where presently Tadzio appears with some of his friends. The boy has a scrap with one of his friends and a chorus of voices (off-stage) calls '*Adziù!*' (see Ex. 130). Left alone, Tadzio makes a clear gesture towards Aschenbach; but the writer is unable to respond. The boy strolls down the beach towards the distant sea, his vibraphone theme (Ex. 128) soaring higher and higher, becoming increasingly attenuated, and ultimately dying away, while at the moment of death the writer remains slumped in his chair. This orchestral postlude is Britten's most searing Mahlerian *adagio*, bringing to a close his final opera, his farewell to the operatic stage, on a note of pain and passion. It also brings to a close a career in the opera house that spanned more than thirty years.

In a personal tribute to Britten, broadcast on BBC Radio on the day of the composer's death (4 December 1976), Michael Tippett summed up one of Britten's greatest achievements thus:

> [after *Peter Grimes*] he was now willing in himself, and, indeed, determined to be, within the twentieth century, a professional opera composer. That in itself is an extraordinarily difficult thing to do; and one of the achievements for which he will always be remembered in musical history books is that, in fact, he actually *did* it.

Appendix A
Chronological List of Published Compositions*

Key to Publishers:

B. & H. = Boosey & Hawkes Ltd.
F.M.   = Faber Music Ltd.
A.C.B. = A. & C. Black Ltd., The Year Book Press.
N.   = Novello & Co.
O.U.P. = Oxford University Press.
P.P.U. = Peace Pledge Union.

1925

FIVE WALZTES (WALTZES); for piano, composed between 1923 and 1925: edited by the composer 1969   F.M.

1929

THE BIRDS: song for medium voice and piano (words by Hilaire Belloc)—revised in 1934   B. & H.

A WEALDEN TRIO: Song of the Women (words by Ford Madox Ford); for women's voices (SSA) unaccompanied: edited by the composer 1967   F.M.

1930

A HYMN TO THE VIRGIN: anthem for mixed voices unaccompanied (words anon.)—revised in 1934   B. & H.

THE SYCAMORE TREE: for unaccompanied mixed chorus SATB (words traditional): edited by the composer 1967   F.M.

---

* This list does not include folk song arrangements and Purcell realizations. Those withdrawn or unpublished works that were published posthumously are listed separately at the end of Appendix A.

1931

TIT FOR TAT: five settings from boyhood of poems by Walter de la Mare, composed between 1928 and 1931: edited by the composer 1968
F.M.
SWEET WAS THE SONG: for contralto solo and chorus of female voices (from an unpublished Christmas Suite called *Thy King's Birthday*): edited by the composer 1966
F.M.
STRING QUARTET IN D MAJOR: edited by the composer 1975
F.M.

1932

THREE TWO-PART SONGS: for boys' or female voices and piano (words by Walter de la Mare)
O.U.P.
SINFONIETTA: for chamber orchestra, op. 1
B. & H.
PHANTASY QUARTET: for oboe, violin, viola and cello, op. 2
B. & H.

1933

A BOY WAS BORN: choral variations for mixed voices unaccompanied, op. 3 (words selected from *Ancient English Christmas Carols* and the *Oxford Book of Carols*)—revised in 1955
O.U.P.
TWO PART SONGS: for mixed voices and piano—1. I Lov'd a Lass (words by George Wither); 2. Lift Boy (words by Robert Graves)
B. & H.

1934

SIMPLE SYMPHONY: for string orchestra or string quartet, op. 4   O.U.P.
HOLIDAY DIARY: suite for piano, op. 5
B. & H.
FRIDAY AFTERNOONS: twelve songs for children's voices and piano, op. 7 (words selected from *Tom Tiddler's Ground* by Walter de la Mare and from other sources)
B. & H.
MAY: unison song with piano (words anon.)
A.C.B.

1935

SUITE: for violin and piano, op. 6, republished (B. & H. 1976) in a shorter version as *Three Pieces* for violin and piano (I March II Lullaby III Waltz)
B. & H.
TE DEUM IN C MAJOR: for choir and organ
O.U.P.

1936

OUR HUNTING FATHERS: symphonic cycle for high voice and orchestra, op. 8 (text devised by W. H. Auden)　　　　　　　　B. & H.

SOIRÉES MUSICALES: suite of five movements from Rossini for orchestra, op. 9　　　　　　　　B. & H.

1937

VARIATIONS ON A THEME OF FRANK BRIDGE: for string orchestra, op. 10　　　　　　　　B. & H.

ON THIS ISLAND: five songs for high voice and piano, op. 11 (words by W. H. Auden)　　　　　　　　B. & H.

TWO BALLADS: for two sopranos and piano—1. Mother Comfort (words by Montagu Slater); 2. Underneath the Abject Willow (words by W. H. Auden)　　　　　　　　B. & H.

FISH IN THE UNRUFFLED LAKES: song for high voice and piano (words by W. H. Auden)　　　　　　　　B. & H.

MONT JUIC: suite of Catalan Dances for orchestra, op. 12—written with Lennox Berkeley　　　　　　　　B. & H.

PACIFIST MARCH: unison song with accompaniment (words by Ronald Duncan)　　　　　　　　P.P.U.

1938

PIANO CONCERTO in D, op. 13—revised in 1945　　　　　　　　B. & H.

ADVANCE DEMOCRACY: chorus for mixed voices unaccompanied (words by Randall Swingler)　　　　　　　　B. & H.

1939

BALLAD OF HEROES: for high voice, chorus and orchestra, op. 14 (words by W. H. Auden and Randall Swingler)　　　　　　　　B. & H.

VIOLIN CONCERTO op. 15—revised in 1958　　　　　　　　B. & H.

LES ILLUMINATIONS: for high voice and string orchestra, op. 18 (words by Arthur Rimbaud)　　　　　　　　B. & H.

CANADIAN CARNIVAL (KERMESSE CANADIENNE): for orchestra, op. 19　　　　　　　　B. & H.

APPENDIX A

1940

SINFONIA DA REQUIEM: for orchestra, op. 20   B. & H.

DIVERSIONS ON A THEME: for piano (left hand) and orchestra, op. 21—
revised in 1954   B. & H.

SEVEN SONNETS OF MICHELANGELO: for tenor and piano, op. 22
                                                                                                               B. & H.

INTRODUCTION AND RONDO ALLA BURLESCA: for two pianos, op. 23,
No. 1   B. & H.

1941

PAUL BUNYAN: operetta (libretto by W. H. Auden): revised by the composer
in 1975 and given the opus number 17   F.M.

MATINÉES MUSICALES: second suite of five movements from Rossini for
orchestra, op. 24   B. & H.

MAZURKA ELEGIACA: for two pianos, op. 23, No. 2   B. & H.

STRING QUARTET: No. 1 in D, op. 25   B. & H.

SCOTTISH BALLAD: for two pianos and orchestra, op. 26   B. & H.

1942

HYMN TO ST. CECILIA: for mixed voices unaccompanied, op. 27 (words by
W. H. Auden)   B. & H.

A CEREMONY OF CAROLS: for treble voices and harp, op. 28 (words by
James, John and Robert Wedderburn, Robert Southwell, William
Cornish, and from anonymous sources)   B. & H.

1943

PRELUDE AND FUGUE: for eighteen-part string orchestra, op. 29 B. & H.

REJOICE IN THE LAMB: festival cantata for choir and organ, op. 30 (words by
Christopher Smart)   B. & H.

SERENADE: for tenor, horn and strings, op. 31 (words by Cotton, Tennyson,
Blake, anon., Ben Jonson and Keats)   B. & H.

THE BALLAD OF LITTLE MUSGRAVE AND LADY BARNARD: for male voices
and piano (words anon.)   B. & H.

286

1944

A SHEPHERD'S CAROL: for unaccompanied voices (words by W. H. Auden)—written for the BBC programme 'A Poet's Christmas'   N.

CHORALE (AFTER AN OLD FRENCH CAROL): for unaccompanied voices (words by W. H. Auden)—written for the BBC programme 'A Poet's Christmas'*

1945

PETER GRIMES: opera, op. 33 (libretto by Montagu Slater)   B. & H.

FESTIVAL TE DEUM: for choir and organ, op. 32   B. & H.

THE YOUNG PERSON'S GUIDE TO THE ORCHESTRA: variations and fugue on a theme by Purcell, op. 34   B. & H.

THE HOLY SONNETS OF JOHN DONNE: for tenor and piano, op. 35
    B. & H.

STRING QUARTET: No. 2 in C, op. 36   B. & H.

1946

THE RAPE OF LUCRETIA: opera, op. 37 (libretto by Ronald Duncan)—revised in 1947   B. & H.

AN OCCASIONAL OVERTURE IN C: for orchestra, op. 38—[*withdrawn*]

1947

ALBERT HERRING: comic opera, op. 39 (libretto by Eric Crozier)
    B. & H.

PRELUDE AND FUGUE ON A THEME OF VITTORIA: for organ   B. & H.

CANTICLE I: for tenor and piano, op. 40 (words by Francis Quarles)
    B. & H.

A CHARM OF LULLABIES: for mezzo-soprano and piano, op. 41 (words by Blake, Burns, Robert Greene, Thomas Randolph and John Philip)
    B. & H.

1948

SAINT NICOLAS: cantata for tenor, mixed voices, string orchestra, piano, percussion and organ, op. 42 (words by Eric Crozier)   B. & H.

THE BEGGAR'S OPERA: a new realization of John Gay's ballad opera, op. 43
    B. & H.

* Published in *The Score*, No. 28, January 1961.

APPENDIX A

1949

SPRING SYMPHONY: for soprano, alto and tenor soli, mixed chorus, boys'
　　choir and orchestra, op. 44 (words from various sources)　　B. & H.
THE LITTLE SWEEP: opera for young people, op. 45 (libretto by Eric Crozier)
　　　　　　　　　　　　　　　　　　　　　　　　　　　　B. & H.
A WEDDING ANTHEM (Amo Ergo Sum): for soprano and tenor soli, choir and
　　organ, op. 46 (words by Ronald Duncan)　　　　　　　　B. & H.

1950

FIVE FLOWER SONGS: for unaccompanied mixed chorus, op. 47—1. To
　　Daffodils (words by Robert Herrick); 2. The Succession of the Four
　　Sweet Months (Robert Herrick); 3. Marsh Flowers (George Crabbe); 4.
　　The Evening Primrose (John Clare); 5. Ballad of Green Broom (anon.)
　　　　　　　　　　　　　　　　　　　　　　　　　　　　B. & H.
LACHRYMAE—Reflections on a song of Dowland: for viola and piano, op. 48
　　also arranged for viola and strings by the composer in 1975 (op. 48a)
　　　　　　　　　　　　　　　　　　　　　　　　　　　　B. & H.

1951

SIX METAMORPHOSES AFTER OVID: for oboe solo, op. 49—1. Pan;
　　2. Phaeton; 3. Niobe; 4. Bacchus; 5. Narcissus; 6. Arethusa B. & H.
BILLY BUDD: opera, op. 50 (libretto by E. M. Forster and Eric Crozier)
　　　　　　　　　　　　　　　　　　　　　　　　　　　　B. & H.

1952

CANTICLE II—Abraham and Isaac: for alto, tenor and piano, op. 51 (text
　　from the Chester Miracle Play)　　　　　　　　　　　　B. & H.

1953

GLORIANA: opera, op. 53 (libretto by William Plomer)　　　　B. & H.
WINTER WORDS: for tenor and piano, op. 52 (lyrics and ballads by Thomas
　　Hardy)　　　　　　　　　　　　　　　　　　　　　　B. & H.

1954

THE TURN OF THE SCREW: opera, op. 54 (libretto by Myfanwy Piper after
　　Henry James)　　　　　　　　　　　　　　　　　　　B. & H.
CANTICLE III—'Still falls the rain': for tenor, horn and piano, op. 55 (words
　　by Edith Sitwell)　　　　　　　　　　　　　　　　　B. & H.

CHRONOLOGICAL LIST OF PUBLISHED COMPOSITIONS

1955

ALPINE SUITE: for recorder trio   B. & H.

SCHERZO: for recorder quartet   B. & H.

HYMN TO SAINT PETER: for mixed-voice choir with treble solo and organ (words from the Gradual of the Feast of St Peter and St Paul), op. 56a
     B. & H.

1956

ANTIPHON: for mixed-voice choir and organ (words by George Herbert) op. 56b   B. & H.

THE PRINCE OF THE PAGODAS: ballet, op. 57   B. & H.

1957

SONGS FROM THE CHINESE: for high voice and guitar (words by Chinese poets, translated by Arthur Waley), op. 58   B. & H.

NOYE'S FLUDDE: the Chester Miracle Play set to music, op. 59 B. & H.

1958

EINLADUNG ZUR MARTINSGANS: eight-part canon composed for Martin Hürlimann's Sixtieth Birthday.

NOCTURNE: for tenor, seven obbligato instruments and string orchestra (words by Shelley, Tennyson, Coleridge, Middleton, Wordsworth, Owen, Keats, and Shakespeare), op. 60   B. & H.

SIX HÖLDERLIN FRAGMENTS: for tenor and piano, op. 61   B. & H.

1959

CANTATA ACADEMICA — CARMEN BASILIENSE: for soprano, alto, tenor, and bass solos, chorus and orchestra (Latin text from the charter of the University of Basle and from older orations in praise of Basle), op. 62
     B. & H.

FANFARE FOR ST EDMUNDSBURY: for three trumpets   B. & H.

MISSA BREVIS IN D: for boys' voices and organ, op. 63   B. & H.

1960

A MIDSUMMER NIGHT'S DREAM: opera, op. 64 (libretto by Benjamin Britten and Peter Pears after William Shakespeare)   B. & H.

APPENDIX A

1961

SONATA IN C: for cello and piano, op. 65   B. & H.
JUBILATE DEO: for mixed-voice choir and organ   O.U.P.
FANCIE: for unison voices and piano (words by Shakespeare)   B. & H.
WAR REQUIEM: for soprano, tenor and baritone solos, chorus, orchestra, chamber orchestra, boys' choir and organ (text: *Missa pro Defunctis* and poems by Wilfred Owen), op. 66   B. & H.

1962

PSALM 150: for two-part children's voices and instruments, op. 67
       B. & H.
A HYMN OF SAINT COLUMBA—Regis regum rectissimi: for mixed voice choir and organ   B. & H.

1963

SYMPHONY FOR CELLO AND ORCHESTRA, op. 68   B. & H.
CANTATA MISERICORDIUM: for tenor and baritone solos, small chorus and string orchestra, piano, harp, and timpani (Latin text by Patrick Wilkinson), op. 69   B. & H.
NIGHT PIECE (NOTTURNO): for piano solo   B. & H.

1964

CADENZAS TO HAYDN'S CELLO CONCERTO IN C   B. & H.
NOCTURNAL (AFTER JOHN DOWLAND): for guitar solo, op. 70   F.M.
CURLEW RIVER: a parable for church performance (text by William Plomer after the Japanese Noh Play, *Sumidagawa*), op. 71   F.M.

1965

SUITE NO. 1 IN G MAJOR: for cello (edited by Mstislav Rostropovich), op. 72
       F.M.
GEMINI VARIATIONS: for flute, violin, and piano (four hands), op. 73
       F.M.
SONGS AND PROVERBS OF WILLIAM BLAKE: for baritone and piano, op. 74
       F.M.
VOICES FOR TODAY: anthem for full choir and boys' chorus, with optional organ accompaniment, op. 75   F.M.
THE POET'S ECHO: six poems of Pushkin for high voice and piano, op. 76
       F.M.

CHRONOLOGICAL LIST OF PUBLISHED COMPOSITIONS

1966

THE BURNING FIERY FURNACE: second parable for church performance (text by William Plomer), op. 77 F.M.

THE GOLDEN VANITY: vaudeville for boys' voices and piano, words by Colin Graham after the old English ballad, op. 78 F.M.

CADENZAS FOR MOZART'S PIANO CONCERTO IN E FLAT MAJOR (K. 482) F.M.

HANKIN BOOBY: folk dance for wind and drums F.M.

1967

OVERTURE 'THE BUILDING OF THE HOUSE': for orchestra and chorus, op. 79 F.M.

THE OXEN: carol for women's voices and piano (text by Thomas Hardy) F.M.

SUITE NO. 2 IN D MAJOR: for cello (edited by Mstislav Rostropovich) F.M.

1968

THE PRODIGAL SON: third parable for church performance (text by William Plomer) op. 81 F.M.

1969

CHILDREN'S CRUSADE: ballad for children's choir and orchestra (text by Bertolt Brecht), op. 82 F.M.

SUITE IN C: for harp solo, op. 83 F.M.

1970

WHO ARE THESE CHILDREN? for tenor and piano (Lyrics, Rhymes and Riddles by William Soutar), op. 84 F.M.

1971

OWEN WINGRAVE: opera (text by Myfanwy Piper after the story by Henry James), op. 85 F.M.

CANTICLE IV—*Journey of the Magi*: for counter-tenor, tenor, baritone and piano (text by T. S. Eliot), op. 86 F.M.

SUITE NO. 3 for cello (edited by Mstislav Rostropovich), op. 87 F.M.

1973

DEATH IN VENICE: opera (text by Myfanwy Piper after Thomas Mann), op. 88   F.M.

1974

CANTICLE V—*The Death of Saint Narcissus*: for tenor and harp (text by T. S. Eliot), op. 89   F.M.
SUITE ON ENGLISH FOLK TUNES '*A time there was . . .*' for orchestra, op. 90
                                                                                                             F.M.

1975

SACRED AND PROFANE: eight medieval lyrics for unaccompanied voices (SSATB), op. 91   F.M.
A BIRTHDAY HANSEL for voice and harp (texts by Robert Burns), op. 92
                                                                                                             F.M.
PHAEDRA: dramatic cantata for mezzo-soprano and small orchestra (text from Robert Lowell's translation of Racine's *Phedre*), op. 93   F.M.
STRING QUARTET NO. 3, op. 94   F.M.

1976

EIGHT FOLKSONG ARRANGEMENTS with harp or piano   F.M.
WELCOME ODE: for young people's chorus and orchestra, op. 95   F.M.

Works Published Posthumously by Faber Music

1928

QUATRE CHANSONS FRANÇAISES: for soprano and orchestra (texts by Paul Verlaine and Victor Hugo) published in 1982

1932

PHANTASY: for string quintet (edited by John Evans) published in 1983

CHRONOLOGICAL LIST OF PUBLISHED COMPOSITIONS

1936

THREE DIVERTIMENTI: for string quartet (I March II Waltz III Burlesque) revised by the composer from *Alla Quartetto Serioso: 'Go play, boy, play'* (1933) and published in 1983

RUSSIAN FUNERAL: march for brass and percussion, published in 1981

1937

REVEILLE: fanfare for violin and piano, published in 1983

1937–9

CABARET SONGS: for voice and piano (words by W. H. Auden): 1. Tell me the truth about love 2. Funeral Blues 3. Johnny 4. Calypso. Published in 1980

1939

YOUNG APOLLO: fanfare for pianoforte solo, string quartet and string orchestra, op. 16: withdrawn by the composer but published in 1982

1947

MEN OF GOODWILL: variations on 'God rest ye merry, Gentlemen', for orchestra: composed for a BBC radio feature broadcast on Christmas Day 1947: published in 1982

1962

THE TWELVE APOSTLES: for solo voice, unison chorus and piano: published in 1981

Appendix B
First Performances of the Operas

PAUL BUNYAN

5 May 1941	New York, Columbia University Brander Matthew Hall

Operetta in a prologue and two acts. Text by W. H. Auden

In the Prologue

OLD TREES	Chorus
YOUNG TREES	Ellen Huffmaster, Jane Weaver, Marlowe Jones, Ben Carpens
THREE WILD GEESE	Harriet Greene, Augusta Dorn, Pauline Kleinhesselink

In the Interludes

NARRATOR	Mordecai Bauman

In the Play

THE VOICE OF PAUL BUNYAN	Milton Warchoff
CROSS CROSSHAULSON	Walter Graf
JOHN SHEARS	Leonard Stocker
SAM SHARKEY	Clifford Jackson
BEN BENNY	Eugene Bonham
JEN JENSON	Ernest Holcombe
PETE PETERSON	Lewis Pierce
ANDY ANDERSON	Ben Carpens
OTHER LUMBERJACKS	Alan Adair, Elmer Barber, Arnold Jaffe, Marlowe Jones, Charles Snitow, Robert Zeller, W. Fredric Plette, Thomas Flynn, Joseph Harrow
WESTERN UNION BOY	Henry Bauman
HEL HELSON	Bliss Woodward

294

JOHNNY INKSLINGER	William Hess
FIDO	Pauline Kleinhesselink
MOPPET	Harriet Greene
POPPET	Augusta Dorn
THE DEFEATED	Ben Carpens, Eugene Bonham, Adelaide Van Wey, Ernest Holcombe
SLIM	Charles Cammock
TINY	Helen Marshall
THE FILM STARS AND MODELS	Eleanor Hutchings, Ellen Huffmaster, Ben Carpens, Lewis Pierce
FRONTIER WOMEN	Marie Bellejean, Eloise Calinger, Irma Commanday, Alice Gerstz Duschak, Marian Edwards, Elizabeth Flynn, Rose Harris, Ethel Madsen, Jean Phillips, Evelyn Ray, Irma Schocken, Adelaide Van Wey, Jane Weaver, Ida Weirich, Marjorie Williamson

Conductor: Hugh Ross
Producer: Milton Smith

PETER GRIMES

7 June 1945   London, Sadler's Wells Theatre

Opera in three acts and a prologue derived from the poem of George Crabbe. Text by Montagu Slater

PETER GRIMES	Peter Pears
ELLEN ORFORD	Joan Cross
AUNTIE	Edith Coates
NIECE I	Blanche Turner
NIECE II	Minnia Bower
BALSTRODE	Roderick Jones
MRS. SEDLEY	Valetta Iacopi
SWALLOW	Owen Brannigan
NED KEENE	Edmund Donlevy
BOB BOLES	Morgan Jones
THE RECTOR	Tom Culbert
HOBSON	Frank Vaughan
DOCTOR THORPE	Sasa Machov
A BOY (GRIMES'S APPRENTICE)	Leonard Thompson

Conductor: Reginald Goodall
Producer: Eric Crozier
Scenery and Costumes: Kenneth Green

THE RAPE OF LUCRETIA

12 July 1946   Glyndebourne Opera House, Sussex
Opera in two acts. Text by Ronald Duncan, after *Le Viol de Lucrèce* by André Obey.

MALE CHORUS	Peter Pears
FEMALE CHORUS	Joan Cross
COLLATINUS	Owen Brannigan
JUNIUS	Edmund Donlevy
TARQUINIUS	Otakar Kraus
LUCRETIA	Kathleen Ferrier
BIANCA	Anna Pollak
LUCIA	Margaret Ritchie

Conductor: Ernest Ansermet
Producer: Eric Crozier
Designer: John Piper

ALBERT HERRING

20 June 1947   Glyndebourne Opera House, Sussex
                               (E.O.G. production)
Comic opera in three acts. Text by Eric Crozier, freely adapted from a short story by Guy de Maupassant.

LADY BILLOWS	Joan Cross
FLORENCE PIKE	Gladys Parr
MISS WORDSWORTH	Margaret Ritchie
THE VICAR	William Parsons
THE SUPERINTENDENT OF POLICE	Norman Lumsden
THE MAYOR	Roy Ashton
SID	Frederick Sharp
ALBERT HERRING	Peter Pears
NANCY	Nancy Evans
MRS HERRING	Betsy de la Porte

EMMIE	Lesley Duff
CIS	Anne Sharp
HARRY	David Spenser

Conductor: Benjamin Britten
Producer: Frederick Ashton
Scenery and costumes: John Piper

THE BEGGAR'S OPERA

24 May 1948 — Cambridge, Arts Theatre
(E.O.G. production)
A new musical version of John Gay's ballad opera (1728), realized from the original airs, in three acts.

BEGGAR	Gladys Parr
MR. PEACHUM	George James
MRS. PEACHUM	Flora Nielsen
POLLY	Nancy Evans
CAPTAIN MACHEATH	Peter Pears
FILCH	Norman Platt
LOCKIT	Otakar Kraus
LUCY LOCKIT	Rose Hill
JENNY DIVER	Jennifer Vyvyan
MRS. VIXEN	Lesley Duff
SUKY TAWDRY	Lily Kettlewell
MRS. COAXER	Catherine Lawson
DOLLY TRULL	Gladys Parr
MRS. SLAMMEKIN	Elisabeth Parry
MOLLY BRAZEN	Anne Sharp
BETTY DOXY	Mildred Watson
HARRY PADDINGTON	Roy Ashton
BEN BUDGE	Denis Dowling
WAT DREARY	John Highcock
MAT OF THE MINT	Norman Lumsden
JEMMY TWITCHER	Norman Platt
NIMMING NED	Max Worthley

Conductor: Benjamin Britten
Producer: Tyrone Guthrie
Assistant Producer: Basil Coleman
Scenery and costumes: Tanya Moiseiwitsch

APPENDIX B

THE LITTLE SWEEP

14 June 1949   Aldeburgh, Jubilee Hall
                              (E.O.G. production)

The opera from *Let's Make an Opera!*, an entertainment for young people. Text by Eric Crozier. One act.

BLACK BOB	Norman Lumsden
CLEM	Max Worthley
SAM	John Moules
MISS BAGGOTT	Gladys Parr
JULIET BROOK	Anne Sharp
GAY BROOK	Bruce Hines
SOPHIE BROOK	Monica Garrod
ROWAN	Elisabeth Parry
JONNY CROME	Peter Cousins
HUGH CROME	Ralph Canham
TINA CROME	Mavis Gardiner

Conductor: Norman Del Mar
Producer: Basil Coleman
Scenery and costumes: John Lewis

BILLY BUDD

1 December 1951   London, Covent Garden

Opera in a prologue, four acts,* and an epilogue. Test by E. M. Forster and Eric Crozier, after the story by Herman Melville.

CAPTAIN VERE	Peter Pears
BILLY BUDD	Theodor Uppman
CLAGGART	Frederick Dalberg
MR. REDBURN	Hervey Alan
MR. FLINT	Geraint Evans
LIEUTENANT RATCLIFFE	Michael Langdon
RED WHISKERS	Anthony Marlowe
DONALD	Bryan Drake
DANSKER	Inia Te Wiata
NOVICE	William McAlpine
SQUEAK	David Tree
BOSUN	Ronald Lewis

* In the 1960 revision, these four acts were reduced to two.

FIRST MATE	Rhydderch Davies
SECOND MATE	Hubert Littlewood
MAINTOP	Emlyn Jones
NOVICE'S FRIEND	John Cameron
ARTHUR JONES	Alan Hobson
FOUR MIDSHIPMEN	Brian Ethridge, Kenneth Nash, Peter Spencer, Colin Waller
CABIN BOY	Peter Flynn

Conductor: Benjamin Britten
Producer: Basil Coleman
Designer: John Piper

GLORIANA

8 June 1953   London, Covent Garden
Opera in three acts. Text by William Plomer. Composed in honour of the Coronation of HM Queen Elizabeth II, and first given at a gala performance in the presence of HM The Queen.

QUEEN ELIZABETH I	Joan Cross
EARL OF ESSEX	Peter Pears
LADY ESSEX	Monica Sinclair
LORD MOUNTJOY	Geraint Evans
PENELOPE LADY RICH	Jennifer Vyvyan
SIR ROBERT CECIL	Arnold Matters
SIR WALTER RALEIGH	Frederick Dalberg
HENRY CUFFE	Ronald Lewis
LADY-IN-WAITING	Adele Leigh
A BLIND BALLAD-SINGER	Inia Te Wiata
THE RECORDER OF NORWICH	Michael Langdon
A HOUSEWIFE	Edith Coates
SPIRIT OF THE MASQUE	William McAlpine
MASTER OF CEREMONIES	David Tree
CITY CRIER	Rhydderch Davies
TIME	Desmond Doyle
CONCORD	Svetlana Beriosova

Conductor: John Pritchard
Producer: Basil Coleman
Designer: John Piper
Choreographer: John Cranko

APPENDIX B

THE TURN OF THE SCREW

14 September 1954   Venice, Teatro la Fenice
                                    (E.O.G. production)

Opera in a prologue and two acts. Text by Myfanwy Piper, after the story by Henry James.

PROLOGUE	Peter Pears
THE GOVERNESS	Jennifer Vyvyan
MRS. GROSE	Joan Cross
QUINT	Peter Pears
MISS JESSEL	Arda Mandikian
FLORA	Olive Dyer
MILES	David Hemmings

Conductor: Benjamin Britten
Producer: Basil Coleman
Designer: John Piper

NOYE'S FLUDDE

18 June 1958   Orford Church, Suffolk
                                  (E.O.G. production)

The Chester Miracle Play set to music. One act.

THE VOICE OF GOD	Trevor Anthony
NOYE	Owen Brannigan
MRS. NOYE	Gladys Parr
SEM	Thomas Bevan
HAM	Marcus Norman
JAFFETT	Michael Crawford
MRS. SEM	Janette Miller
MRS. HAM	Katherine Dyson
MRS. JAFFETT	Marilyn Baker
THE RAVEN	David Bedwell
THE DOVE	Maria Spall
MRS. NOYE'S GOSSIPS	Penelope Allen, Doreen Metcalfe, Dawn Mendham, Beverley Newman
PROPERTY MEN	Andrew Birt, William Collard, John Day, Gerald Turner

Conductor: Charles Mackerras
Production and setting: Colin Graham
Costumes: Ceri Richards

A MIDSUMMER NIGHT'S DREAM

11 June 1960   Aldeburgh, Jubilee Hall
                               (E.O.G. production)

Opera in three acts. Text adapted from William Shakespeare's play by Benjamin Britten and Peter Pears.

OBERON	Alfred Deller
TYTANIA	Jennifer Vyvyan
PUCK	Leonide Massine II
PEASEBLOSSOM	Michael Bauer
COBWEB	Kevin Platts
MUSTARDSEED	Robert McCutcheon
MOTH	Barry Ferguson
FAIRIES	Thomas Bevan, Thomas Smyth
THESEUS	Forbes Robinson
HIPPOLYTA	Johanna Peters
LYSANDER	George Maran
DEMETRIUS	Thomas Hemsley
HERMIA	Marjorie Thomas
HELENA	April Cantelo
BOTTOM	Owen Brannigan
QUINCE	Norman Lumsden
FLUTE	Peter Pears
SNUG	David Kelly
SNOUT	Edward Byles
STARVELING	Joseph Ward
MASTER OF CEREMONIES	John Perry
ATTENDANT	Jeremy Cullum
PAGES	Robert Hodgson, Nicholas Cooper

Conductor: Benjamin Britten
Producer: John Cranko
Scenery and costumes: John Piper, assisted by Carl Toms

APPENDIX B
CURLEW RIVER

13 June 1964	Orford Church, Suffolk
	(E.O.G. production)

Parable for church performance in one act. Text by William Plomer, after the Japanese Noh-play *Sumidagawa* by Motomasa.

THE ABBOT	Don Garrard
THE FERRYMAN	John Shirley-Quirk
THE TRAVELLER	Bryan Drake
THE MADWOMAN	Peter Pears
THE SPIRIT OF THE BOY	Robert Carr
HIS VOICE	Bruce Webb
THE PILGRIMS	John Barrow, Bernard Dickerson, Brian Etheridge, Edward Evanko, John Kitchener, Peter Leeming, Philip May, Nigel Rogers

Production and setting: Colin Graham
Costumes: Annena Stubbs

THE BURNING FIERY FURNACE

9 June 1966	Orford Church, Suffolk
	(E.O.G. production)

Parable for church performance in one act. Text by William Plomer.

THE ASTROLOGER (THE ABBOT)	Bryan Drake
NEBUCHADNEZZAR	Peter Pears
ANANIAS	John Shirley-Quirk
MISAEL	Robert Tear
AZARIAS	Victor Godfrey
THE ANGEL	Philip Wait
ENTERTAINERS AND PAGES	Stephen Borton, Paul Copcutt, Paul Davies, Richard Jones, Philip Wait
THE HERALD AND CHORUS LEADER	Peter Leeming
CHORUS OF COURTIERS	Graham Allum, Peter Lehmann Bedford, Carl Duggan, John Harrod, William McKinney, Malcolm Rivers, Jacob Witkin

Production and setting: Colin Graham
Costumes and properties: Annena Stubbs

THE PRODIGAL SON

10 June 1968   Orford Church, Suffolk
                                (E.O.G. production)
Parable for church performance in one act. Text by William Plomer.

THE TEMPTER	
(THE ABBOT)	Peter Pears
THE FATHER	John Shirley-Quirk
THE ELDER SON	Bryan Drake
THE YOUNGER SON	Robert Tear
CHORUS OF SERVANTS,	Paschal Allen, Peter Bedford,
PARASITES AND BEGGARS	Carl Duggan, David Hartley,
	Peter Leeming, John McKenzie,
	Clive Molloy, Paul Wade
YOUNG SERVANTS AND	Robert Alder, John Harriman,
DISTANT VOICES	Peter Heriot, Richard Kahn,
	David Morgan

Production and setting: Colin Graham
Costumes and properties: Annena Stubbs

OWEN WINGRAVE

18 May 1971   BBC 2
Opera for Television. Text by Myfanwy Piper based on the short story
by Henry James.

OWEN WINGRAVE	Benjamin Luxon
SPENCER COYLE	John Shirley-Quirk
LECHMERE	Nigel Douglas
MISS WINGRAVE	Sylvia Fisher
MRS. COYLE	Heather Harper
MRS. JULIAN	Jennifer Vyvyan
KATE JULIAN	Janet Baker
GENERAL SIR PHILIP	
WINGRAVE	Peter Pears

COLONEL WINGRAVE AND GHOST	Peter Pears
YOUNG WINGRAVE AND GHOST	Stephen Hattersley
YOUNG WINGRAVE'S FRIEND	Geoffrey West
NARRATOR	Peter Pears

Conductor: Benjamin Britten
Producer: Brian Large and Colin Graham
Designer: David Myerscough Jones
Costumes: Charles Knode

DEATH IN VENICE

16 June 1973   The Maltings, Snape
(E.O.G. production)
Opera in two acts. Text by Myfanwy Piper, based on the short story by Thomas Mann.

GUSTAV VON ASCHENBACH	Peter Pears
THE TRAVELLER	
THE ELDERLY FOP	
THE OLD GONDOLIER	
THE HOTEL MANAGER	John Shirley-Quirk
THE HOTEL BARBER	
THE LEADER OF THE PLAYERS	
THE VOICE OF DIONYSUS	
THE VOICE OF APOLLO	James Bowman
THE POLISH MOTHER	Deanna Bergsma
TADZIO	Robert Huguenin

and members of the English Opera Group, the Royal Ballet, and children of the Royal Ballet School

Conductor: Steuart Bedford
Producer: Colin Graham
Designer: John Piper
Costumes: Charles Knode
Choreographer: Sir Frederick Ashton

Appendix C
Short Bibliography

This short bibliography is confined to books (and booklets) by Britten, and on him and his music. A useful list of articles in periodicals and books, and of special Britten issues of magazines, up to 1952, is given in the Mitchell/Keller symposium listed below.

ABBIATI, FRANCO. *Peter Grimes.* (A volume in the series 'Guide Musicali dell'Istituto d'Alta Cultura'.) Milan, n.d. [1949].

BRITTEN, BENJAMIN. *On Receiving the First Aspen Award.* London, Faber and Faber, 1964.

BRITTEN, BENJAMIN, AND IMOGEN HOLST. *The Story of Music.* London, Rathbone Books, 1958. Reissued as *The Wonderful World of Music,* London, Macdonald, 1968.

CROZIER, ERIC (ed.) *Peter Grimes.* (Sadler's Wells Opera Books, no. 3) London, The Bodley Head, 1945.

(ed.) *La Création de l'Opéra anglais et 'Peter Grimes'.* French translations of Sadler's Wells Opera Books nos. 1 and 3 by C. Ormore and Annie Brierre. Paris, Richard-Masse, 1947.

(ed.) *The Rape of Lucretia: a symposium.* London, The Bodley Head, 1948.

GISHFORD, ANTHONY (ed.) *Tribute to Benjamin Britten on his Fiftieth Birthday.* London, Faber and Faber, 1963.

HOLST, IMOGEN. *Britten.* (A volume in the 'Great Composers' series.) London, Faber and Faber, 1966. 2nd edition 1970, 3rd edition 1980. [See also under BENJAMIN BRITTEN]

HOWARD, PATRICIA. *The Operas of Benjamin Britten.* London, Barrie and Rockliff, 1969.

HURD, MICHAEL. *Benjamin Britten.* (A pamphlet in the 'Biographies of Great Composers' series.) London, Novello, 1966.

KELLER, HANS. *The Rape of Lucretia: Albert Herring.* (A booklet in the 'Covent Garden Operas' series.) London, Boosey and Hawkes, 1947. [See also under DONALD MITCHELL]

MITCHELL, DONALD, AND HANS KELLER (eds.) *Benjamin Britten: a Commentary on his works from a group of specialists.* London, Rockliff, 1952. (Also: Greenwood Press reprint, U.S.A., 1972.)

[MITCHELL, DONALD (ed.) and JOHN ANDREWES] *Benjamin Britten: A Complete Catalogue of his Works.* London, Boosey & Hawkes, 1963. Revised edition 1973 with supplement to 1976, Boosey & Hawkes and Faber Music Ltd.

[PEARS, PETER]. *Armenian Holiday: August 1965.* Privately printed. n.d. [1965].

STUART, CHARLES. *Peter Grimes.* (A booklet in the 'Covent Garden Operas' series.) London, Boosey & Hawkes, 1947.

WHITE, ERIC WALTER. *Benjamin Britten: a Sketch of his Life and Works.* London, Boosey & Hawkes, 1948.
Benjamin Britten: eine Skizze von Leben und Werk. German translation by Bettina and Martin Hürlimann. Zurich, Atlantis Verlag, 1948.
Benjamin Britten: a Sketch of his Life and Work. New edition, revised and enlarged. London, Boosey and Hawkes, 1954.
Benjamin Britten: His Life and Operas. London, Faber and Faber in association with Boosey & Hawkes, 1970.

YOUNG, PERCY M. *Benjamin Britten.* ('Masters of Music' series.) London, Benn, 1966.

Since this Bibliography was compiled in 1970, and particularly since Britten's death in 1976, a great number of books have been written about the composer. The Editor has included these for reference purposes, though they obviously have no bearing on Eric Walter White's text.

BLYTH, ALAN. *Remembering Britten* (with notable contributions from Britten's family, friends and colleagues). London, Hutchinson, 1981.

EVANS, PETER. *The Music of Benjamin Britten.* London, Dent, 1979.

HEADINGTON, CHRISTOPHER. *Britten.* (A volume in the series entitled 'The Composer as Contemporary'—General Editor: John Lade.) London, Eyre Methuen, 1981.

HERBERT, DAVID. *The Operas of Benjamin Britten.* London, Hamish Hamilton, 1979.

KENNEDY, MICHAEL. *Britten.* (A volume in the 'Master Musicians' series.) London, Dent, 1981.

MITCHELL, DONALD and JOHN EVANS. *Benjamin Britten: Pictures from a Life 1913–1976.* London, Faber and Faber, 1978.

MITCHELL, DONALD. *Britten and Auden in the Thirties: The Year 1936.* (The T. S. Eliot Memorial Lectures delivered at the University of Kent at Canterbury in November 1979.) London, Faber and Faber, 1981.

WHITTALL, ARNOLD. *The Music of Britten and Tippett: Studies in Themes and Techniques.* Cambridge University Press, 1982.

Appendix D
From **The Britten Estate Executors' Press Statement** (1977)

A.1   The Aldeburgh Festival has already celebrated its thirtieth year, and with the continuing artistic guidance of Peter Pears and the other Artistic Directors who will be joined next year as an Artistic Director by Mstislav Rostropovich, the world-famous cellist and long standing friend of Aldeburgh, the Executors are confident that the Festival's future is assured.

A.2   As for the development of the Britten–Pears Library at The Red House, where Britten lived and worked for twenty-five years, and which was already established by him and Peter Pears during the last years of Britten's life, the Executors are acutely aware of the unique opportunity which exists to keep together *in an integral collection* the contents of Britten's working library, his personal archives and papers and—most important of all—a major collection of his musical manuscripts and sketches which includes some of his most celebrated compositions, such as WAR REQUIEM, THE TURN OF THE SCREW, and DEATH IN VENICE. That the integrity of the collection should be preserved and that the manuscripts should remain at Aldeburgh are aims to which the Executors are vigorously committed, and these form the basis of their current negotiations in both the national and international interest—so they believe—with HM Treasury. The importance of Aldeburgh is central to the understanding of the composer, who himself said in 1964: 'I belong at home—there—in Aldeburgh. I have tried to bring music *to* it in the shape of our local Festival; and all the music I write comes *from* it. I believe in roots, associations, in backgrounds, in personal relationships. . . . I do not write for posterity. . . . I write music, now, in Aldeburgh, for people living there, and further afield, indeed for anyone who cares to play it or listen to it. But my music now has its roots in where I live and work.'

A.3   Britten's concern for the development of musicianship and musical understanding among young people scarcely needs stressing in view of such famous works as LET'S MAKE AN OPERA! and NOYE'S FLUDDE—which beguilingly 'educate' those participating in their performance—and the

YOUNG PERSON'S GUIDE which has educated successive generations of listeners. It is those primarily creative ideals, matched with the standards of excellence and musical insight embodied in the performances given by Peter Pears and Benjamin Britten throughout the long years they worked together as partners—not to speak of Britten's activities as a performer and conductor of many other composers' works beside his own—which the Britten–Pears School for Advanced Musical Studies will nourish, sustain and expand. The School, to which Britten devoted much thought before his death, is fortunate in having Peter Pears as its Director of Singing, and Cecil Aronowitz as Director of Strings. It is also planned to develop the area of Academic Studies, which will introduce specialist study and research in the Britten field as well as in those other musical territories especially associated with Aldeburgh. Imogen Holst has been closely associated with the planning of the School and its curriculum. The School will gradually evolve from its present part-time status into a full-time institution, and one which will not only fill a unique national need for a Music School post-graduate in character, but also constitute a permanent living memorial to Benjamin Britten and to his work as a performer with Peter Pears. His Executors intend to support the development of this project, which will further the challenging musical standards on which Britten insisted throughout his life. Public financial support on a substantial scale is needed now, and extensive preparations are being made for launching the Britten Memorial Appeal in the autumn.

A number of Britten's manuscripts, including the full score of the *War Requiem*, were presented to the Nation and received by HM Treasury in lieu of capital transfer tax. They are at present on loan from the British Library, together with the composer's other manuscripts, at the Britten–Pears Library in Aldeburgh.

The Benjamin Britten Memorial Appeal was launched in order to convert the south block of the Maltings complex for the Britten–Pears School for Advanced Musical Studies, as the national memorial to the composer. The School was officially opened by HM Queen Elizabeth, The Queen Mother (Patron of the Aldeburgh Festival) on 28 April 1979, and it continues to flourish under its Founder-Director, Sir Peter Pears, C.B.E.

J.E. 1982

Index

Figures in **bold** indicate major entries and figures in *italics* refer to illustrations.
Works by Britten are indexed only under Britten, Edward Benjamin.

INDEX

INDEX

INDEX